AF659103

Perspectives in Animal Ecology and Reproduction

Perspectives in Animal Ecology and Reproduction

— Volume 4 —

Editors
Dr. V.K. Gupta
Dr. Anil K. Verma

2007
DAYA PUBLISHING HOUSE
Delhi - 110 035

© 2007 V.K. GUPTA (b. 1953–)
ANIL K. VERMA (b. 1963–)
ISBN 81–7035–459–5

All rights reserved. Including the right to translate or to reproduce this book or parts thereof except for brief quotations in critical reviews.

Published by : **Daya Publishing House**
1123/74, Deva Ram Park
Tri Nagar, Delhi - 110 035
Phone: 27383999
Fax: (011) 23244987
e-mail : dayabooks@vsnl.com
website : www.dayabooks.com

Showroom : 4760-61/23, Ansari Road, Darya Ganj,
New Delhi - 110 002
Phone: 23245578, 23244987

Laser Typesetting : **Classic Computer Services**
Delhi - 110 035

Printed at : **Chawla Offset Printers**
Delhi - 110 052

PRINTED IN INDIA

Editorial Board

– Editors –

Dr. V.K.Gupta
Senior Scientist,
Regional Research Laboratory (CSIR),
Canal Road, Jammu – 180 001
India
E-mail: vgupta_rrl@yahoo.com

Dr. A.K. Verma
Head,
Department of Zoology,
Government P.G. College,
Udhampur (J&K), India
E-mail: anilvermaverma@lycos.com

– Members –

Prof. Clement A. Tisdell
Professor Emeritus,
School of Economics,
The University of Queensland, Brisbane,
Australia
E-mail: c.tisdell@economics.uq.edu.au

Prof. Vanitha Kumari
Head,
Department of Zoology,
Bharathiar University,
Coimbatore, India
E-mail: regr@bharathiaruni.org

Prof. Tej Kumar Sherstha
Head,
Central Department of Zoology,
Tribhuvan University,
Kathmandu, Nepal
E-mail: drtks@ccsl.com.np

Dr. Sunil Kumar
Deputy Director (Sr. Grade),
Division of Reproductive Toxicology & Histochemistry,
National Institute of Occupational Health,
Meghani Nagar, Ahmedabad – 380 016, India
E-mail: sunilnioh@yahoo.com

Dr. Justin Gerlach
Chief Scientist,
The Nature Protection of Seychelles,
133 Cherry Hinton Road,
Cambridge CBI 7BX,UK
E-mail: justgerlach@aol.com

Foreword

Last few decades have seen unprecedented growth in knowledge concerning ecology and reproductive biology of the animals. In order to keep pace with this explosion of knowledge in these areas of biology the readers and researchers have to look for these scattered information from various sources, a deed which demand a lot of time for the purpose. In such a situation through a series of volumes of the book entitled, "*Perspectives in Animal Ecology and Reproduction*" edited by Dr. V.K. Gupta and Dr. Anil K. Verma, authors have tried to bridge this gap to a certain extent.

A lot many articles exclusively written by numerous experts of these areas of animal research have been used in this volume (Volume IV) and the editors have painstakingly polished and incorporated them in this volume. The editors have really done a commendable job in producing these excellent volumes, which, I hope would be able to serve the purpose of the students, teachers and researchers interested in the fields of animal ecology and reproductive biology.

I wish to congratulate both of them for the big task they have undertaken in consolidating the knowledge of ecology and reproduction in the animal world.

Prof. Baldev Sharma
Dean, Faculty of Life Sciences,
Head,
Department of Zoology,
University of Jammu, Jammu – 180 001

Preface

Every living creature has an overwhelming desire to breed/ reproduce under conducive environmental conditions but fascinatingly, the genders are profoundly at odds concerning with whom they mate and the amount of efforts they put into raising their progeny/offspring's. The sexual activities are either restricted to a particular season (seasonal breeders) or spread over a considerable period of time in an year and qualify to be called continuous breeders. This book is an attempt to review and describe in details the ecological aspects and reproductive behavioural patterns displayed by diverse organism ranging from the pseudocoelomic through true coelomic forms.

Sex in the natural world is aggressive, competitive and sometimes lethal. The opportunities created by sex differ for males and females. This volume although, is mainly devoted to ecological and reproductive traits in various animals but also includes their general introductions and descriptive accounts, taxonomic richness and population density in details. The attention of researchers is also drawn towards the prevailing anthropogenic, pollutional and parasitic threats to the fishes, crocodiles and avifaunal elements confined to various/diverse habitats in India and abroad.

This volume is prepared by keeping in view the requirement of researches of various institutions and the undergraduate and post graduate students of several universities. Topics are addressed with terminology that is as simple as possible so that readers will not be impeded by arcane words but will be able to see what the phenomenon is without difficulty.

The present book (Volume IV) not only covers all the important topics that is required in a book, but also introduces the reader with latest developments and information in the field and inclusion of the topic: "*Computer in Wildlife: A Tool to Create a Communication with Birds*" is an attempt in this regard. There are 27 papers in two sections, 16 included under the head animal ecology and remaining 11 devoted to the section of animal reproduction, written by national and the international experts of the field and the treatment bears the seal of authority.

We hope that this volume of the book like the first three volumes, will be of great use to all the groups of individuals and organizations a few mentioned above. We welcome suggestions for further improvement of the future volumes of the book.

Jammu, India

V.K. Gupta

Anil K. Verma

Contents

Section II: Animal Reproduction

Section I
Animal Ecology

Chapter 1

Nematology in India: Problems and Perspectives

A.K. Chaubey, Nidhi Tyagi and Satyandra Kumar

Nematology Laboratory, Department of Zoology,
C.C.S. University, Meerut – 250 004
E-mail: kumar_ac2001@rediffmail.com

The green revolution in India has led to a quantum jump in agricultural production, thereby placing the country in a comfortable position as regard to domestic food demand. However, yield losses (50–80 per cent) in agricultural crops from the parasitic phytonematodes are not uncommon. In India, there is hardly any area where the crops are not infested by plant parasitic nematodes. The annual yield loss ranges from 10–100 per cent in cereals, whereas in vegetables the yield loss was 27.3–90 per cent. In one of the State (Rajasthan), it has been estimated that 60 million rupees are lost in the wheat production due to a single nematode species *Heterodera avenae*. Although the horticulture and fruit plants are also severely infected with the plant parasitic nematodes yet the yield loss is not estimated. On an average, plant parasitic nematodes cause 15–35 per cent losses in crop yield. Sasser (1979) estimated 15 per cent average crop loss in tropical countries only due to *Meloidogyne*. Both yield loss and population dynamics are greatly influenced by initial population density (Pi), host factors such as susceptibility, tolerance and the environmental factors like soil type, planting time, fertilizer and water availability together with too many biotic and abiotic factors.

Nematodes are taxonomically, ecologically and geographically diverse group, with an estimated 5×10^5 species in the world. They usually comprise more than 90 per cent of the metazoan fauna (McIntyre, 1971). Almost all the plants are susceptible to attack by parasitic phytonematodes. One of the orders Tylenchida is the largest and economically important group of plant parasitic nematodes. Some families of order Tylenchida (Paratylenchidae, Meloidogynidae and Heteroderidae) are of the great economic importance as parasites of agricultural crops, ornamental, horticulture and forest trees.

Although probably best known as parasites, most nematodes are free-living and occur in enormous numbers in almost every habitat suitable for life. India is a large country having almost all types of climates, ranging from the extreme hot of its peninsular tip to the extreme cold of the trans Himalayan region, with abundant forest, large coastline from the extremely wet North-East to the extremely dry Western regions both hot and cold deserts constitute different biomes and a lot of ecological niches to thrive on. Nematode dispersal is mainly by water, wind, transportation by man and animals. Irrigation waters can provide an annual inoculum of 0.144×10^6 to 15.362×10^6 nematodes/hectare (Faulkner and Bolander, 1970). They have a sensitive relationship with their environment. Nematodes respond to the spatial and temporal dynamics of resources (sources of nutrition and characteristic of the habitat) and stressors (fundamental attributes of ecosystems) in soil habitats. These attributes individually and collectively influence the structure and function of nematode biodiversity.

All plant parasitic nematodes are equipped with a hollow, protrusible structure called stylet that is used to inject secretions into host tissues and pierce plant cell wall to withdraw nutrients (Hussey and Grundler, 1998). Some phytonematodes feed upon and quickly destroy plant cells as migratory parasites whereas other nematode species become sedentary in later life stages and modify plant cells to provide a sustained local source of nutrition for parasitic success. The feeding activities of the nematodes usually resulted in stunted growth; reduced crop yields and occasionally plant death. Plant parasitic nematodes are therefore, considered worst hidden enemies of a wide host range of plants.

The vertical and horizontal distribution of nematodes must also be taken into account because some species appear to prefer certain depths. Richter (1969) found that *Trichodours* occurred in greater numbers in the deeper soil layer compared with species of *Tylenchorhynchus* and *Pratylenchus*. Flegg (1968a) found that *Xiphinema diversicaudatum* and *Xiphinema vittenezi* decreased in numbers with increasing depth whereas *Longidours macrosoma* increased with depth up to 70 cm. Koen (1966) found that the seasonal variations from summer to winter and the consequent change of soil temperature and soil moisture influence the pattern of vertical distribution of *Meloidogyne javanica* in the soil. The composition of the nematode community depends on the vegetation, soil type, season, soil moisture level, amount of soil organic matter and too many other environmental factors biotic and abiotic both. Temperature and moisture are two most important factors affecting the distribution (vertical and horizontal), population dynamics (spatial and temporal), activity and survival of nematodes. Nemic communities are much more diverse than those of other soil mesofauna (Wasilewska, 1971). Not all the phytonematodes are harmful to farmers but some are beneficial and can be used as biocontrol agents (Enrique and Edwin, 2002; Chaubey *et al.*, 2004) and some may be used as bio-indicators of the condition of the soil environment (Chaubey and Malhotra, 1988).

Among the edaphic factors like texture, soil temperature, moisture, pH, organic carbon and too many other factors (organic and inorganic constituents), micro- and macro-nutrients of the soil play a pivotal role in influencing the population of plant parasitic nematodes (Norton, 1978; Anita and Chaubey, 2004; Anita, 2004).

Nemic Diversity

Out of estimated 5×10^5 species of nematodes only 10 per cent are known to parasitic to plants. Yeates *et al.* (1993) made a comprehensive review of the nematode feeding habits and categorized soil nematodes into eight feeding types:

1. Plant Feeding
2. Fungal Feeding
3. Bacterial feeding
4. Substrate ingestion

5. Predation on protozoa and soil animals
6. Unicellular eucaryote feeding (diatoms and other algae)
7. Dispersal stages of animal parasites of both invertebrates and vertebrates.
8. Omnivores (mainly Dorylaimid nematodes).

Above all, on the basis of habit and habitat, the plant parasitic nematodes are broadly categorized into two major groups *i.e.* ecto- and endoparasites. The representative of semi- or endoparasites (migratory or sedentary) of aerial parts are known to occur in India (Table 1.1). However, their distribution is not uniform and depends upon several ecological particularly host and environmental factors. Among migratory endoparasitic species (*Pratylenchus*) are most common and the majority of them are polyphagous. Some of the species like *P. zeae*, *P. coffae* are widely distributed and have wide host range as compared to others, which are restricted in their distribution, and narrow host range. Distribution of *Hirchimanniella* is co-related with rice in India. *H. oryzae*, *H. mucronata*, *H gracilus* are most common species of this genus. The genus *Radopholus* is represented by *R. similis* and *R. colbrani*. *R. similis* is reported in several southern and northern states but most common in coastal region of Kerala and Tamil Nadu where it attacks on banana, coconut and other palm varieties. Majority of several species of *Scutellonena* reported from India are ectoparasites of root and only *S. grande* and *S. bradys* are endoparasite. *S. grande* was reported to feed on wheat and on grasses from Haryana state whereas *S. bradys* is parasite of tuber crop specially yam in south India.

Rotylenchulus reniformis (sedentary-semiendoparasite) is the most widely distributed species feeding upon number of cultivated and weed plants. *R. parvas* is reported from Haryana as the parasite of Morni hills where as *Tylenchus semipanetrans* associated with citrus, grapevine, mango, loquat and jasmine (*Jasmine samba*) in India. *Trophotylenchulus* represented by three species *viz.*, *T. andhraensis T. piperis*, *T. waliai* are reported from Andhra Pradesh, Kerala and Haryana respectively. Few non-cyst (heteroderids) *Bilobodera flexa* and *B. mesoangustus* (Sharma and Siddiqui, 1992) are reported from Andhra Pradesh and Haryana. *Cryphodera kalesari* (Bajaj *et al.*, 1989), only non-cyst forming endoparasites heteroderids, reported from Kelesar forest of Haryana.

Table 1.1: Some Economically Important Phytonematodes of India

Mode of Parasitism	*Genus and Species*
Aerial Parasite	*Subanguina chrysopogoni, S. minuta*
Migratory Endoparasite	*Apratylenchoides homoglans, Hirschmanniella abnormis, H. exacta, H gracilis, H. kaverii, H. magna, H. mucronata, H. orycrena, H. oryzae, H. shamini, Pratylenchus barkati, P. bhattii, P. coffeae, P. crassi, P. crenatus, P. cruciferus, P. dasi, P. delatteri, P. ekrami, P. exilis, P. fallax, P. flakensis, P. gibbicaudatus, P. goodeyi, P. impar, P. loosi, P. manaliensis, P. manohari, P. microstylus, P. mulchandi, P. neglectus, P. nizamabadensis, P. penetrans, P. ranjani, P. scribneri, P. similes, P. teres, P. thornei, P. typicus, P. vulnus, P. zeae, Pratylenchoides crenacauda, Radopholus colbrani, R. similes*
Sedentary Semi-endoparasite	*Belobodera flexa, B. mesoangustus, Cryphodera kalesari, Rotylenchulus reniformis, R. parvus, Scutellonema bradys, S. grande, Tylenchulus semipenetrans*
Sedentary Endoparasite Non gall-farming	*Aphelenchoides bessyi, A. ritzemabosi, A. fragariae*
Gall-farming	*Meloidogyne incognita, M. incognita acrita, M. javanica, M. arenaria, M. graminicola, M. lucknowica, M. bravicauda, M. africana, M. exigua, M. thamesi, M. graminis, M. triticoryzae, M. piperi, M. indica*

The aerial nematode *Anguina tritici* although wiped out from many states of India but still is a problem of small and marginal farmers. *A. cecidoplastes* was reported from Tamil Nadu, forming galls on the leaves, stem and flower of *Bothriochloa pertusa*. Some other nematodes *Subanguina chrysopogoni* (Bajaj *et al.*, 1989) and *Ditylenchus phyllobia* are reported from Haryana and Tamil Nadu respectively. *Aphlenchoides besseyi* is restricted to eastern and southern states whereas *A. ritze mabosi* and *A. fragariae* are reported from Himachal Pradesh and Jammu-Kashmir respectively. *Ditylenchus angustus* thrives only in deepwater rice of Assam, some part of eastern U.P., Bihar and West Bengal.

Milestones

Anguina tritici, perhaps the first plant parasitic nematode was seen and reported by John Turbevill Needham (1743, published in

1744). During middle of the 19th century, Tylenchids (a group of nematodes) were beginning to receive attention. In England, Berkeley (1855) observed 'Vibrios' (root knot nematodes) in galls of a glasshouse cucumber. *Meloidogyne* species were the first plant parasitic forms observed in the United States by May (1888). DeMan (1876, 1884) produced excellent monographs on soil, plant and freshwater nematodes. *An Introduction to Nematology* (joint contribution of B.G. Chitwood and his wife M.B. Chitwood, 1950) is a milestone in the history of nematology. T. Goodey's (1933) book *Plant Parasitic Nematodes and the Diseases they Cause* aroused great interest in the study of plant nematology.

In India Barber (1901) reported root knot nematode infesting tea for the first time and for that he is known as the father of the *Indian Nematology*. Later Butler (1919) reported *Ditylenchus angustus* causing 'urfa disease' of rice. Ayyar (1926, 1933a,b) was also among early nematologist as he reported root knot nematode infesting vegetable and other crops. Dastur (1936) reported 'white tip' disease of rice caused by *Aphelenchoides besseyi*. All these reports constitute important landmark in the history of nematology in India. Sitaramiah *et al.* (1971) and others studied the ecology of the plant parasitic nematodes. Scant information, on the role of environmental regulatory processes and spatial distribution, is available in India (Khan *et al.*, 1971; Malhotra and Chaubey, 1990; Chaubey, 1994; Chaubey and Anita, 2001, 2002).

Ecological and Parasitic adaptations have the key to phylogenetic relationships while abiotic factors influence pathogenecity (Siddiqui, 1986). Khan (1981) has stirred interest in some reviews. Seasonal fluctuations of nematodes have been studied by Chaubey (1992, 1994); Mittal and Prasad (1998); Talwana *et al.* (2000) and others. Oostenbrink (1966) discussed the seasonal fluctuations in *Tylenchorhynchus dubius* and correlated with plant growth factors. Trevathan *et al.* (1985); Chaubey (1992); Yeates *et al.* (1998) and Anita and Chaubey (2003) applied statistical tools to find out the relationships between plant parasitic nematodes and soil environmental factors. The densities of plant nematodes have been correlated with a visual index of soil texture in an *Alfa alfa* field (Carter, 1975; Prasad and Rao, 1980; Goodel and Ferris, 1980; Agu and Ogbuji, 2000). Some workers (Elsen *et al.*, 1998, Chaubey *et al.*,

2003, Anita and Chaubey, 2004; Chaubey and Sharma, 2005) have studied the population density and biochemical constituents of the plant parasitic nematodes. The impact of climatic changes on the population distribution of nematodes has been studied (Yeates *et al.*, 1998). The response and function of the micro-organisms (including nematode) in the soil food web, organic matter amendment, nutrient status, soil physical conditions and biological activity of soil greatly influence the nemic population (Bossio and Scow, 1998; Bongers and Ferris 1999; Akhtar and Malek, 2000; Chen and Ferris, 2000; Ferris and Matute, 2003).

Spatial and temporal distribution of plant parasitic nematodes have been studied by many workers (Malhotra and Chaubey, 1990; Sharma and Nene, 1992); differences in pH (Chaubey and Dwivedi, 1991; Chaubey and Anita, 2000, 2001; Singh and Garg, 2003); temperature (Trudgill, 1995; Chen and Dickson, 1997; Muzne *et al.*, 1998; Patel and Patel, 1998; Choudhury and Sivakumar, 1999; Kamra and Sharma, 2000; Castillo and Vovlas, 2002); soil moisture and other meterological conditions (Chaubey and Dwivedi, 1991; Das and Rahman, 1996; Walia and Mathur, 1997; Verma *et al.*, 2001; Verma and Jain, 2001; Anita and Chaubey, 2003) have been correlated the relative abundance of a species in studies comparing soil samples from different places. The natural soil environment harbours the multitude of microorganisms. Many studies have been conducted on the uptake of nutrients like organic carbon/organic matter (Zaki and Bhatti, 1989; Chaubey and Dwivedi, 1993; Barman and Das, 1996; Pandey, 2000; Chaubey and Anita, 2001; Verma and Jain, 2001); N.P.K. (Balaji and Vaitheeswaran, 1989; Shanthi and Sundarababu, 1995; Das and Mishra, 1999; Hussani *et al.*, 2001; Sundersingh, 2001; Anita and Chaubey, 2002); sodium (Malhotra and Chaubey, 1993; Dunkan and Abou-Setta, 1995); sulphur (Khan and Khan, 1996; Bahadur and Tamilselvan, 2001; Anita and Chaubey, 2002).

The impact of tracer elements (micro-nutrients) on the size and activity of natural soil microbial communities have been well documented. Agricultural field studies revealed that the elevated loading of tracer element resulted in decreased microbial community size (Chander and Brookes, 1991; Forstgard *et al.*, 1996; Chahal *et al.*, 1997; Konopaka *et al.*, 1999; Pannanen, 2001; Anita and Chaubey, 2002; Kelly *et al.*, 2003; Anita, 2004).

Nematodes Infesting Cereals

Wheat, rice, barley, sorghum and maize are the important cereal crops, of which wheat and rice has been the most important economic crop in India. The productivity of these crops is adversely affected by insect pests and plant diseases, among which nematode problems pose a serious threat (Table 1.2). *Anguina tritici*, reported from India is the most important seed-gall nematode of wheat whereas cyst/ white females of *Heterodera avenae* are visible on the roots of the same host crop. *Hirschmanniella oryzae, Ditylenchus angustus, Meloidogyne graminocola, Aphelenchus avenae, A. besseyi* and *Pratylenchus zeae* were found to be widespread and abundant in rice. *Helicotylenchus* sp., *Hoplolaimus* sp., *Pratylenchus zeae, P. brachyurus, Punctidera chalcoensis* and *Heterodera zeae* were the predominant plant-parasitic nematodes on maize. Worldwide, the loss in yield of maize due to plant parasitic nematodes is estimated at over 10 per cent (Sasser and Freckman, 1987). More than 15 species of *Pratylenchus* are associated with maize.

Table 1.2: Reported Yield Losses in Rice in India due to Some Common Nematode Infesting Cereals Pests*

Nematode	*% Yield Loss*	*Reference*
Aphelenchoides besseyi	20	Muthukrinshnan *et al.*, 1974
Ditylenchus angustus	50	Singh and Garg, 1949; Singh, 1953
Ditylenchus angustus	10–30	Rao *et al.*, 1968
Hirschmanniella orvzae	30–87	Mathur and Prasad, 1972
Hirschmanniella oryzae	45–95	Muthukrinshnan *et al.*, 1977
Hirschmanniella oryzae	14–19	Venkitesan and Charles, 1979
Hirschmanniella oryzae	8–23	Kuriyan and Sheela, 1981
Hirschmanniella oryzae	40	Rao *et al.*, 1985
H. mucronata	25	Panda and Rao, 1971
H. mucronata	21	Prasad and Rao, 1979a
H. mucronata	43	Prasad and Rao, 1979b
H. mucronata	6–50	Prasad *et al.*, 1986
Hoplolaimus indicus	5–19	Ranana and Rao, 1977
Pratylenchus indicus	18–55	Prasad and Rao, 1978b
Meloidogyne graminicola	21–64	Rao and Biswas, 1973
Heterodera oryzicola	21–42	Kumari and Kuriyan, 1981
Heterodera oryzicola	5–56	Jayprakash and Rao, 1982
Heterodera oryzicola	38	Rao, 1984

* Transcript from *Tropical Pest Management* 1987, 33(2): 127–136.

Barley (*Hordeum vulgare*) is one of the world's major grain crops. This crop is infested with *H. avenae* and the national programms of both India and Pakistan has conducted research on the control of this parasite (Handa *et al.*, 1985; Pankaj and Dhawan, 1991). However not much work has been conducted on pest status of the major nematode parasite of this crop. Sorghum (*Sorghum vulgare*) is infested with *Tylenchorhynchus martini, T nudus, P. zeae, H. gambiensis* and *H. sorghi, M.incognita, P. zeae* and *T. vulgaris* infesting the pearl millet (*Pennisetum glaucum*). Information on nematodes associated with finger millet (*Setaria italica*) is limited but three species of cyst forming nematodes (*Heterodera gambiensis, H. sorghi* and *H. delvii*) and the reniform nematode, *Rotylenchulus reniformis* have been reported as parasites (Seshadri, 1970; Krishna Prasad and Krishnappa, 1982). Pigeon pea is reported to be infested with *H. cajani, M. incognita* and *M. javanica* whereas chickpea is infested with root knot nematode. Cyst/white females of *H. cajani* are reported to be visible on roots of cowpea. Several other species of plant-parasitic nematodes have been found associated with cereal crops. Prasad *et al.* (1959) reported cyst nematode (*H. avenae*) of wheat and barley from some places of Rajasthan which later established in some other several areas in Haryana, Punjab, Himachal Pradesh, Jammu and Kashmir, Uttaranchal and Delhi states of north India (Table 1.3). Some areas are identified as severely infested with *Heterodera* spp. in the country (Table 1.4).

Table 1.3: Distribution of Some Important Cereal Cyst Nematode in India

Species	*State*
Heterodera. Avenae	Delhi, Haryana, Himachal Pradesh, Jammu and Kashmir, Leh, Punjab, Rajsthan, Uttar Pradesh
H. zeae	Andhra Pradesh, Bihar, Delhi, Gujarat, Haryana, Himachal Pradesh, J ammu and Kashmir, Maharashtra, Madhya Pradesh, Punjab, Rajasthan, Uttar Pradesh
H. cajani	Andhra Pradesh, Bihar, Delhi, Gujarat, Haryana, Himachal Pradesh, Karnataka, Madhya Pradesh, Maharastra, Punjab, Rajasthan, Tamil Nadu, Uttar Pradesh, West Bengal
H. graminis	Delhi, Himachal Pradesh, Rajasthan, Uttar Pradesh
H. oryzicola	Haryana, Kerala, Madhya Pradesh, Orissa, West Bengal

Contd...

Table 1.3–Contd...

Species	State
H. gambiensis	Karnataka (Bangalore)
H. sorghi	Haryana, Madhya Pradesh, Maharashtra, Uttar Pradesh
H. swarupi	Rajasthan, Uttar Pradesh
H. filipjevi	Haryana
Globodera rostochiensis	Karnataka, Kerala, Tamil Nadu

Table 1.4: Hotspot and Pest Free Areas Identified for Cyst Nematodes (*Heterodera* spp.) in India*

Nematode	Major Crop	Hotspot Identified	Pest Free Areas
H. avenae	Wheat, Barley	–	Solan, Una, Kangra (H.P.)
H. cajani	Pigeon pea, mung, soyabean, sesamum	Bharuch (Gujarat) (more than 50 cyst/plant)	Hamirpur (U.P.), Murshidabad (W.B.), Hisar and Ambala (Haryana)
H. zeae	Maize	Nil	Nil
H. oryzicola	Banana	Pampadunpara, Puttadi, Konni, Pathanamthitta (Kerala)	Alpuzha and Kannur (Kerala)
H. oryzicola	Paddy	Malipada and Sakigopal (Orissa)	24 Paraganas (North), Hoogly (W.B.)

**Source*: Lal and Sehgal, 2001.

In India the nematode problem is more serious and complex than in the other developing countries. The reasons for this are many and varied. The country is a subtropical region where the climate is suitable for the activity and reproduction of nematodes throughout the year. Sandy and warm soils are very favorable for nematode development especially in irrigated areas that are used continuously for crop production. Perennial crops and crops grown in the same field year after year are often heavily attacked by nematodes.

Nematodes Infesting Vegetable

Many vegetable crops, in several botanical families such as Solanaceae, Cucurbitaceae, Leguminosae, Liliaceae,

Chenopodiaceae, Compositae, Umbelliferae, Cruciferae and Malvaceae are reported as heavily infested and greatly damaged by the nematode pests. Root knot nematodes belonging to the *Meloidogyne* genus are one of the most economically important plant pathogen (Siddiqi, 2000) especially in high value intensive vegetable crops. So far, over 80 species have been described, but more than 90 per cent of the estimated damage worldwide is considered to be cause by four root knot nematodes: *M incognita* (Chitwood), *M. javanica* (Chitwood), *M arenaria* (Chitwood), and *M. hapla* (Chitwood) (Siddiqi, 2000). Since several species of *Meloidogyne* are commonly found together, unambigous identification to species level is essential for the successful management of these nematodes (Table 1.5).

Table 1.5: Important Root Knot Nematode (*Meloidogyne* species) Distributed in India

Species	*State*
M. areneria	Bihar, Delhi, Gujarat, Haryana, Himachal Pradesh, Maharashtra, Punjab, Tamil Nadu, Uttar Pradesh
M. graminicola	Assam, Haryana, Himachal Pradesh, Karnataka, Kerala, Madhya Pradesh, Orissa, Punjab, Uttar Pradesh, West Bengal
M. hapla	Himachal Pradesh, Tamil Nadu
M. incognita	Assam, Bihar, Delhi, Gujarat, Haryana, Himachal Pradesh, Karnataka, Kerala, Madhya Pradesh, Orissa, Rajasthan, Tamil Nadu, Uttar Pradesh, West Bengal
M. indica	Gujarat
M. javanica	Delhi, Gujarat, Haryana, Himachal Pradesh, Karnataka, Kerala, Madhya Pradesh, Rajasthan, Uttar Pradesh, West Bengal
M. piperi	Kerala
M. triticoryzae	Delhi, Uttar Pradesh

Annual yield losses to potato (*Solanum tuberosum*) due to damage by plant parasitic nematodes are estimated to exceed 12 per cent worldwide (Sasser and Freckman, 1987). The major nematode pests of potato are the root-knot nematodes (*M. incognita, M. javanica* and *M. hapla*), the cyst nematodes (*Globodera rostochiensis* and *G. pallida*), and *Ditylenchus destructor*. Lesion nematodes (*Pratylenchus* spp.) may also cause yield loss. Sweet potato (*Ipomoea batatas*) is severely infested with *M. incognita*. Several other species of plant-parasitic nematodes such as *Pratylenchus* spp. and *Rotylenchulus reniformis* have been found damaging the production of vegetables. Common beans (*Phaseolus vulgaris*) are infested with *M. incognita, M. javanica, H. glycines* and *R. reniformis.* Besides some hotspots have been identified from different parts of the country (Table 1.6).

Table 1.6: Hotspot Identified for Root Knot Nematode (*Meloidogyne* spp.) in India

Nematode	Major Crops	Hotspot Identified
M. incognita	Brinjal	Morkhal (Gujarat) (13,100 J_2/100 g soil)
	Cotton	Sabarkantha (Gujarat) 576 J_2/200 cm^3 soil, Siwani, Bolyan, Dabawali, Sirsa, Durba (Haryana)
	Jute	Palaghachha, Goraghachha, Ayespur, Bansbara, Narayanpur, Paharpur, Kumarsahatpur, Bidyapur, Mundarpukur (W.B) (~ 500 J_2/200 cm^3 soil)
M. graminicola	Paddy	Malipada, Sakhigopal (Orissa)
M. indica	Citrus	Samsherpura, Banaskantha (Orissa)

**Source*: Lal and Sehgal, 2001.

Nematodes Infesting Ornamental/Horticulture Plants

India has rich heritage of ornamental/horticulture plants that maintain ecological balance and healthy environment of town and cities. Apple, prunus, banana, citrus, mango, guava, papaya and pear constitute the major fruit-tree crops in India. Although fruit-trees have received very little attention from nematologists, several species of plant-parasitic nematodes have been recorded associated with the ornamental and horticulture plants. Siddiqi (1964) reported *Tylenchorhynchus semipenetrans* from citrus plant and perhaps it is the first report from India from any horticulture plant. Monocroping of horticulture crops increased the nematode populations to alarming dimensions and hence plant parasitic nematodes have become one of the major limiting factors in horticultural crops. Little information is available on parasitic phytonematodes associated with ornamental plants and the net damaged by them. Indubala (2000) has reported some dominant ectoparasites *viz., Xiphinema setariae, Longidorus nirulai, Hoplolaimus indicus, Aphelenchus avenae, Hemicrico-nemoides mangiferae,* and *Meloidogyne incognita* infesting rose, chrysanthemum and croton. Sunflower, a worldwide grown oilseed and ornamental plant is reported to be susceptible to nematode attack *viz., M. incognita, H. avenae, Helicotylenchun multicinctus, Tylenchorhynchus* sp., *Hoplolaimus* sp., *Xiphinema* (Amarnatha and Krishnappa, 1990). *Xiphinema, Longidorus, Hoplolaimus* are common to rose.

The root knot nematode, *M. incognita* is a serious problem on grapevine, papaya, banana, pineapple and passionfruit. *M. indica* and *M. javanica* is a major problem of citrus in some states (Andra Pradesh, Delhi). *Rotylenchulus reniformis* is a major problem on grapevine, papaya and pineapple. *Radopholus similis* causes rhizome rot of banana and becomes a serious problem in Tamil Nadu, Kerala, Karnataka and Andhra Pradesh (Parvatharedi and Singh, 1980). The distribution of most commonly occurring nematodes in all types of vegetation is an evidence of polyphagus. The citrus nematode *Tylenchulus semipenetrans* and *Xiphinema* spp. are widespread and cause considerable damage in Punjab. Banana plantations in West Bengal have been severely attacked by nematodes, especially *M. incognita, R. reniformis, Hel. multicinctus* and *Radopholus similis,* although a total of 52 nematode species have so far been reported from this crop in India.

Scant information is available on ecological and population density of phytonematodes infesting mango and guava of western U.P. (Anita, 2005; Anita and Chaubey, 2004). *Paratrichodorus minor, Hemicricinemoides mangiferae, Helicotylenchus indicus, Tylenchorhynchus nudud, Dorilaimus* sp., *Rotylenchus reniformis, Tylenchorhynchus brassicae* are reported from the root zones of mango plantations whereas *T. nudus, Hel. lobus, X. diversicadatum, Pratylenchus* sp., *Trichodorus* sp., *Longidorus attenuadus* infesting guava plants in western U.P. Out of which *H. mangiferae* (96.43 per cent) while *T. nudus* (93 per cent) was dominant around mango and guava plants respectively. The highest population of *H. mangiferae* and *T. nudus* was present at upper strata followed by middle and deeper strata of the soil.

Nematodes Infesting Other Crops

Tobacco, cotton, jute and sugarcane are the most important cash crops in our country. In India *Pratylenchus zeae* is the predominate genus in sugarcane followed by *Hoplolaimus, Tylenchorhynchus, Helicotylenchus* (Mehta, 1989; Sundaraja and Mehta, 1994).

Tylenchorhynchus spp. was found to occur in large number in Gujarat, Madhya Pradesh and Uttar Pradesh (Mehta, 1989). Some important members of dorylaimids associated with this crop are *Xiphinema, Longidorus, Paratrichodorus* and *Trichodorus.* Cotton is infected with some important nematodes *viz., Helicotylenchus* sp., *Tylenchorhynchus* sp., *Rotylenchulus reniformis* and *M. incognita* that act as a limiting factor for crop production in certain areas of the country. Heavy infestation of *M. incognita* and *M. javanica* has been found in tobacco plantations throughout the country.

Nematode Management/Control

Chemical

Nematode disease detection and species identification are basic to any management/control strategy. The use of chemicals to control plant-parasitic nematodes has always been an expensive remedy that could be economically justified only on high-value crops such as pineapple, banana, tobacco, potato and some ornamental plants. The first products to be used as nematicide were volatile compounds with fumigant action that required specialized application equipment. Few of these compounds are now in use because of the problems that were discovered concerning toxicity and in some locations unwelcome persistence in the environment that resulted in contamination of groundwater extracted for drinking. The development of organophosphate and oximecarbamate (non-volatile) has enabled the use of chemical treatments for a wider range of situations without the need for specialized application equipment (Table 1.7). Unlike most of the fumigants, these products are not phytotoxic at effective doses and multiple applications to growing crops can be made. Timing, soil moisture, degree of incorporation and rainfall/irrigation can influence nematicide efficacy. With non-volatile compounds economies can be made by treating plants in rows or by spot applications around individual plants rather than as overall broadcast treatments. Liquid formulations are easy to apply in drip irrigation systems where dosages can be carefully controlled; however, concentrated formulations are highly toxic and should be used only by qualified personnel. An integrated approach to nematode control, nematicides can have an important function in decreasing numbers of nematodes to levels at which other remedies may be used to their best advantage.

Table 1.7: Some Important Chemical Compounds Used as Nematicide to Control the Plant Parasitic Nematodes

Compounds	*Chemical Name*	*Trade Name*	*Dose*	*Remark*
Halogenated hydrocarbon				
Chloropicrin	Trichloronitromethane	Larvacide, Picfume	12 oz/100 sq ft	–
Methylbromide	Bromomethane	Ebafume, Bedfume, Pestmaster	1–2 lb/100 sq ft	–
DD	1,3-dichlopropene-1, 2-dichloropropane	DD, Vidden D, Telone	250–500 l/ha	–
EDB	1,2-dibromoethane	Domfume W-40, W-85; Nemax 42, 85; Soilfume-85	60 l/ha	Highly effective to all nematodes
DBCP	1,2-dibromo-3-chloropropane	Nemagon, Fumazone	25–50 l/ha	Not highly phytotoxic
Organophosphate				
Parathion	0,0-diethyl 0-(p nitrophenyl) phosphorothioate	Ekatex, Folidol, Thiophos	–	–
Dichlophenthion	0-(2,4dichlorophenyl) 0,0-diethyl phosphorothioate	VC-13, Hexanema	8–12 kg a i/ha	Less phytotoxic
Thionazin	0,0-diethyl-0-2-pyrazinyl phosphorothioate	Zinophos, Nemaphos	4-8 kg a i/ha	Effective against root knot and sting nematode
Fensulphothion	0-0-diethyl-0-p (methyl sulphonyl) phenyl phosphorothioate	Terracur P, Dasanit	4–8 kg a i/ha	Against root knot and turf nematode
Phenamiphos	Ethyl-4 (methylthio)-m-tolyl isopropyl phosphoramidate	Nemcur	4–8 kg a i/ha	Against root knot nematode

Contd...

Table 1.7–Contd...

Compounds	Chemical Name	Trade Name	Dose	Remark
Ethoprop	0-ethyl S,S-dipropyl phosphorodithiate	Mocap	4–8 kg a i/ha	
Phorate	0-0,diethyl S-(ethylthimethyl) phosphorodithiate	Thimet	4–8 kg a i/ha	Against stunt, lesion and pin nematodes
Dithiocarbamates				
Metham Sodium	Sodium N-methyldithiocarbamate	Vapam, Sistan, Vitafuma Unifume	40 fl oz/100 sq ft	Highly phytotoxic
Aldicarb	2,methyl-2-(methylthio)propionaldehyde 0(methylcarbamoyl) oxime	Temic	2–4 kg a i/ha	Against root knot nematode
Carbofuran	2,3-dihydro-2,2dimethyl-7-benzofuranyl methyl carbamate	Furadon	4–8 kg a i/ha	Against root knot nematode
Methomyl	S-methyl N-[(methylcarbamoyl)oxy] thioacetamidate	Lannate	4–8 kg a i/ha	
Ozanyl	Methyl N,N-dimethyl-N-[(methyl carbamoyl) oxy-1-thiooxamidate	Vydate, D-1410	4–8 kg a i/ha	Against foliar nematode

Biocontrol Agents

Pesticides have been used in the past to increase agricultural output but the use of pesticide was a short-term solution to the problem of nematodes. The nematode population increases several months after the nematicides use or a new nematode race developed due to environmental selection pressure caused by the use of resistant varieties. Most of the nematicides are expensive and are being banned due to their harmful effect on man and animals, their persistence in the soil, contamination of the foul water and the other population of the environment. Diverse groups of microorganisms occur in some soils in sufficient numbers to provide effective control of several nematode pests (Table 1.8). This natural control tends to develop when nematode-susceptible crops are grown in monocultures and under perennial crops. Detailed studies have been made on the biological control provided by a limited number of fungi and bacteria appear to be predictable and effective. However, in several cases the level of control achieved is inadequate and other control measures are required to reduce nematode populations below damaging levels. Research on soils in which nematode multiplication is suppressed by the activities of nematophagous microorganisms has helped identify potential biological control agents and key factors, which affect their activity.

Biocontrol agents have got some inherent advantages of being safe, host specific, efficacious, eco-friendly and long lasting. Despite the significant advantages of biological pesticides, there are some factors that reduced their practical effectiveness and commercial exploitation. These include lack of suitable technique for mass production of many potential bio-control agents (*Pasteuria penetrans* and *Paecilomyces lilacinus*) requirement of high dose, slow action, host specificity need for synchronizing the time of application with susceptible stage of pest, lack of knowledge and influence of environmental factors etc.

The genetic engineering offers the possibilities of developing new generation of biological nematicides based on the construction of genetically modified organism that can retain the advantage of classical biological agents but suffer fewer drawbacks. The two management techniques are in use now a day- the manipulation of organism that are natural pathogens of pests in order to increase their effectiveness as biological control agents. Another one is the genetic manipulation of obligate parasites of nematodes like *P. penetrans* may give rise to mutants that can be cultured *in vivo* and still retains parasitic abilities.

Table 1.8: List of Some Important Biocontrol Agents Harmful to Plant Parasitic Nematodes

Organisms	*Target Species*	*Reference*
Virus		
	Meloidogyne incognita	Loewenberg *et al.*, 1959
	Tylenchorhynchus martini	Ibrahim *et al.*, 1973
Rickettsia		
	Heterodera goettingiana	Shepherd *et al.*, 1973
	H. glycine	Endo, 1979
	Xiphinema index	Xiphinema index
Bacteria		
Bacillus thuriengiensis	*M. graminicola*	Parsed *et al.*, 1972
B. penetrans	*Meloidogyne* sp., *Pratylenchus*	Adam and Eichenmuller, 1963
Pasteuria penetrans	*Heterodera* sp.	Bhattacharya and Swaroop, 1988
Pseudomonas denitrificans	*X. americanum*	Adam and Eichenmuller, 1963
Duboscquia penetrans	*Pratylenchus pratensis*	Thorne, 1940
Fungi		
Arthrobotrys robusta	*Ditylenchus myceliophagus*	Kayrol, 1978
Meria coniospora	*Meloidogyne* spp.	
Catenaria auxiliaria	*H. schachtii*	Duddington *et al.*, 1956
Dactylaria candida	*H. schachtii*	Hams and Wilkin, 1961
D. endermata	*Globodera rostechiensis*	Hutchinson and Mai, 1954
D. thoumasia	*M. incognita*	Mankau, 1961
Phialophora malorum	*H. schachtii*	Rozypal, 1934
P. heteroderae	*G. rostochiensis*	
Cylindrocarpon readicicola	*H. avenae*	Goffart, 1932

Contd...

Table 1.8–Contd...

Organisms	*Target Species*	*Reference*
Townshend *et al.*, 1989C, *destructant*	*G. rostochiensis*	Margan-Jones *et al.*, 1986
Paecilomyces lilacinus	*M. incognita, G. pallida, Tylenchulus* sp., *Radopholu* sp., *Rotylenchulus* sp.	Reddy and Khan, 1988; Gaspard, 1990
Streptomyces avermitilis	*M. incognita*	Garabediurn and vanGundy, 1983
Verticilium chlamydosporium	*H. avenae, H. glycines, M. arenaria*	Kerry and Crump, 1980, Gintis *et al.*, 1983, Morgan and Jones, 1983
VAM	*G. solanocearum, M. hapla, M. incognita*	Fox and Spasoff, 1972
Protozoans, Collembola and Mites		
Caliculosporidum penetrans	*Pratylenchus, M. javanica, Xiphinema, Helicotylenchus microlobus*	Williams, 1960; Prasad and Mankau, 1969; Esser and Sobers, 1964
Theratomyxa weberi	*G. rostochiensis, H. schachtii*	Weber *et al.*, 1952
Onychiurus armatus	*H. cruciferae*	Murphy and Doncaster, 1957
Mites	*T. dubius, Meloidogyne* sp.	Sharma, 1971, Linford and Oliveira, 1938
Enchytraeids sp.	*H. schachtii*	Schaerffenberg and Tendle, 1951
Predaceous Nematodes		
Clarkus papillatus	*Meloidogyne* spp., *H. schachtii*	Steiner and Heinly, 1922; Thorney, 1927
Iotonchus brachylaimus	*H. schachtii*	Cassiy, 1931
Mylonchulus sigmaturus	*T. semipenetrans*	Cohn and Mordechai, 1974
Prionchulus panetalus	*M. incognita*	Small, 1979
Thornia sp.	*T. semipenetrans*	Boosalis and Mankau, 1965

Biological agents for control of plant parasitic nematodes have largely concerned the use of facultative and obligate parasites, but the impacts of rhizosphere bacteria and root-colonizing fungi on the multiplication of nematodes have also been investigated. It is essential to understand the biology and ecology of these organisms for the selection and application of virulent strains. The type of interaction greatly affects the level of population control, damage to the crop and the method of application. Biological agents have generally provided insufficient control and their integration with other control methods is essential. The host plant greatly affects the efficacy of several agents but this interaction has been little studied. Similarly, the mode of action and the epidemiology of most agents are poorly understood. Such organisms are essential for the rational development and application of control strategies using biological agents. Now a day much work is going on in various institutes on this aspect.

Alternate Strategies

The management of plant-parasitic nematodes continues to present problems. For some intensively managed systems, control by the use of nematicides may be justified economically but there is increasing pressure to substitute chemical methods with safer non-chemical alternatives. For subsistence and marginal producers who crop their land continuously, control of nematodes with chemicals has never been feasible and alternative methods have to be developed. The feature of crop intensification is the increased pressure of pests and diseases. The traditional practices of long rotations and fallows are no longer possible in these situations.

The greatest abundance of plant-parasitic nematodes is to be found in the soil zones where there is a high density of plant roots. Eggs and free-living stages occur in spaces between the soil particles. The problems of exerting an influence on nematodes and their eggs in this environment are considerable. Such influences are based on nematode vulnerability to starvation, anaerobic conditions, heat (above 40°C), desiccation and also to certain biotic influences such as natural enemies and antagonists and repellents or poisons issuing from plant roots or decaying organic matter. The selection of one or more of these potential controlling influences to sustain crop productivity is a challenge that is being taken up with great vigor in many countries.

The integrated pest management (IPM) systems for plant-parasitic nematodes must be a key component of this strategy. With the continuing removal of the most efficacious nematicides in most countries, nematode management systems should include a number of compatible strategies and tactics. The available strategies and respective tactics for combating nematodes are:

1. Quarantine that should be strictly followed.
2. Reduction of initial population densities (elimination of established nematodes, cultural practices, including crop rotation, inter- and intra-cropping, soil amendments, trap crops, fallow, time of planting and harvesting, farm hygiene and general culture; vertical resistance; chemical nematicides, biological agents; and physical treatments such as heat-solarization, flooding, and tillage for destruction of residual, infected roots).
3. Suppression of nematode reproduction through enriching the soil with organic matter.
4. Restriction of current crop damage (use of tolerant cultivars or development of other tactics under the first three strategies).
5. Establishment of nematodes assay centers at District headquarter at least.

The integration of strategies and tactics must be based on their relative efficacy, their compatibility/antagonism, and assessments of regional and site-specific nematode problems. Where nematode assay services are unavailable, assessment of some of the major infestations can be made through on-farm, critically timed bioassays, including root galling for *Meloidogyne* spp., or cyst counts for *Heterodera*/*Globodera* spp. Although considerable progress has been made, much research on damage functions and low-input management tactics/options for nematodes and associated disease complexes remains to be accomplished. Where damage thresholds and related information are unavailable, nematode-hazard indices (relative probability of moderate to severe damage) might be considered. An expert systems approach can facilitate the integration and development of these management tactics.

Future Prospect

Root-knot nematodes (*Meloidogyne* spp.) are major pests of various crops in developing countries including India. The plant parasitic nematodes can be controlled by various methods but nematicides are generally not applied because of being too expensive and harmful due to its residual effects. Crop rotations are time-consuming and not applicable to polyphagous species on cash crops. Biological control is not yet practical at large scale. Soil solarization is a feasible technique in hot summers that was generally done in our country by the farmers before three decades.

The main constraint to the development of applied nematology in our country is the lack of an interaction between researchers, district agriculture officers and farmers. Extension service is poor and very little information is known on the real problems because they do not have the facilities and/or the habit of going into the field. Even if some good research is done or practical results are obtained these are not transferred to users and remain confined to laboratories.

The perspectives for the future is of an increasing population resulting in further intensification of agriculture. In the longer term continued unchecked population increase will completely outstrip the capacity of even sophisticated agriculture to satisfy world needs and mass starvation is inevitable. In the shorter term we know intensification of agriculture will increase the damage and problems caused by nematodes. Nematicides will remain expensive and the synthetic ones are likely to be unwise to rely on it to solve most of the problems. Biological control will probably be made to work only in a limited range of special situations.

Applied nematologists need a broad knowledge of crop production systems, plant physiology, etc. relevant to their interests. In practice, they need to collaborate with other specialists, including plant breeders, to develop nematode resistant varieties of various economically important crops. In such collaboration, it is vital that nematologists are expert in their science and work from basic principles. Scientific progress is interactive and based on a series of small steps. For the future, nematicides will be less available and genetic engineering and biological control agents will probably only be effective against nematode problem. Therefore, most marginal farmers will need to rely on more traditional methods including rotation, resistant varieties, improved hygiene, etc. Integration of these approaches with other methods such as solarization may be productive and the role of leading farmers in helping to pioneer such advance should not be underestimated. Overall, training at all levels will be vital and will need to include nematologists, extension workers and farmers.

Acknowledgement

Authors are thankful to the Head, Department of Zoology for providing lab facilities. Senior author (AKC) is thankful to Department of Science and Technology, New Delhi for financial assistance through its research scheme (No. SR/SO/AS–32/2003) to carry out the work on nematode problems of Western U.P.

References

Agu, C.M. and Ogubuji, R.O., 2000. Effects of *Meloidogyne javanica* and *Rhizoctonia solani* alone and in combination on soybean in ferrallitic soil of Eastern Nigeria. *Tropenlandwirt*, 101: 91–97.

Akhtar and Malek, A., 2000. Roles of organic soil amendments and soil organisms in the biological control of plant parasitic nematodes (a review). *Biore. Technol.*, 74: 35–47.

Amarnatha, B.S. and Krishnappa, K., 1990. Plant parasitic nematodes associated with sunflower in Karnataka. *Indian J. Nematol.*, 20(2): 193–196.

Anita, 2004. Eco-biology of dominant parasitic phytonemas infesting mango and guava of western U.P. *Ph.D. Thesis*, Ch. C.S. University, Meerut, 285 pp.

Anita and Chaubey, A.K., 2002. Effect of soil edaphic factors on major parasitic phytonematodes around roots of *Mangiferae indica*. In: *Proc. 4th Indian Agric., Scientists and Farmers Cong.*, 71 p. (Abstr.).

Anita and Chaubey, A.K., 2003. Influence of zinc and sulphur on infra population of *Hoplolaimus indicus* infesting *Mangiferae indica* and *Psidium guajava*. In: *Proc. 5th Nat. Symp. of Soc. Pl. Protec. Sci.*, 188–189 pp. (Abstr.).

Anita and Chaubey, A.K., 2004. Influence of major edaphic factors on *Hemicriconemoides mangiferae* infesting *M. indica* L. *J. Exp. Zool.*, India, 7(2): 285–293.

Anita and Chaubey, A.K., 2004. Impacts of organic matter on the population dynamics of *Hemicriconemoides mangiferae* infesting *Mangifera indica*. *Biol. Mem.*, 30(2): 108–111.

Ayyar, P.N.K., 1926. On root-knot nematodes infesting vegetables and other crops in south India. *Madras Agric. J.*, 49: 113–118.

Ayyar, P.N.K., 1933a. Some experiments on the root gall nematode *Heterodera marioni* in south India. *Madras Agric. J.*, 21:97–107.

Ayyar, P.N.K., 1933b. Further experiments on root gall nematode *Heterodera marioni* (cornu) Goodey in south India. *Madras Agric. J.*, 3: 1064–1071.

Bajaj, H.K., Walia, R.K., Dabur, K.R. and Bhatti, D.S., 1989. *Cryphodera kalesari*, a new heteroderid nematode species from Haryana, India. *Syst. Parasitol.*, 14: 113–116.

Balaji, A. and Vaitheeswaran, M., 1988. Effect of Potassium stress on *Hibiscus cannabinus* infected by the root-knot nematode, *Meloidogyne incognita. Indian J. Nematol.*, 18(2): 216–218.

Barber, C.A., 1901. A tea eel worm diseases in south India. *Madras Agric. Branch J. Bull.*, 45: 227–234.

Barman, M. and Das, P., 1986. Effect of chemical seed dressing and organic amendment alone and in combination for the management of root knot nematode, *Meloidogyne incognita* on green gram. *Indian J. Nematol.*, 26 (1): 72–76.

Berkeley, M.J., 1855. Vibrio forming excrescences on the roots of cucumber plants. *Gard. Chron.*, 14: 220.

Bongers, T. and Ferris, H., 1999. Nematode community structure as a bio–indicator in environmental monitoring. *Trends in Ecology Evolution*, 14: 224–228.

Boogert-vandem, P.H.J.F., Vevlis, H., Etterma, C.H. and Bouwman, L.A., 1994. The role of organic matter in the population dynamics of the endoparasitic nematophagus fungus *Drechmeria coniospora* in microorganisms. *Nematologica*, 40: 249–257.

Boosalis, M.G. and Mankau, R., 1965. Parasitism and predation of soil microorganisms. In: *Ecology Soil-borne Plant Pathogens*, (Eds.) K.P. Baker and W.C. Snyder, pp. 374–379. Univ. Calif. Press, Berkeley, 571 pp.

Bossio, D.A. and Scow, K.M., 1998. Impacts of carbon and flooding on soil microbial communities, phospholipids, fatty acids profiles and substrate utilization patterns. *Micro. Ecol.*, 35: 265–278.

Budhar, I.N. and Tamilselven, N., 2001. Effect of sulphur on yield attributes and yield of rained green *Vigna radiatus* L. *Madras Agric. J.*, 88(7–9): 504–505.

Butler, E.J., 1919. The rice worm (*Ditylenchus angutus*) and its control. *Mem. Dept. Agr.*, India, 10(1): 1–37.

Carter, W.W., 1975. Effect of soil texture on the interaction between *Rhizocolonia solani* and *Meloidogyne incognita* on cotton seedlings. *J. Nematol.*, 7: 234–236.

Castillo, P. and Vovlas, N., 2002. Factors effecting egg hatch of *Heterodera meditterranea* and differential responses of olive cultivars to infestation. *J. Nematol.*, 34(2): 146–150.

Chahal, P.P.K., Chahal, V.P.S., Kaur, N.P. and Arora, C.L., 1997. Effect of iron interaction with *Rhizobium* sp. and *Meloidogyne incognita* in relation to uptake of nitrogen, iron, zinc in mungbean. *Nematol. Medit.*, 25: 37–40.

Chander, K. and Brookes, P.C., 1991. Effects of heavy metals from past application of sewage sludge on microbial biomass and organic matter accumulation in a sandy loam soil and silty loam U.K. soil. *Soil Biochem.*, 23:927–932.

Chaubey, A.K. and Anita, 2000. Influence of pH and organic carbon on population densities of major ectoparasitic phytonemas infesting mango and guava in Western U.P. In: *Proc. 3rd Indian Agric. Scits. Farmer's Cong.*, 56 p.

Chaubey, A.K. and Anita, 2001. Influence of pH and organic carbon on population densities of major ectoparasitic phytonemas infesting mango and guava in western U.P. In: *Proc. 3rd Indian Agric. Scientist and Farmers Congress*, 56 pp.

Chaubey, A.K. and Anita, 2002. Studies on amino acid in altitudinally variant *Hoplolaimus indicus* infesting fruit plants. *Biochem. Cell Arch.*, 2(1&2): 93–99.

Chaubey, A.K., Anita and C. Sharma (2004). E P N–An effective bio-tool: Its biology culture and application. *J. Exp. Zool.*, India., 7(1): 1–12.

Chaubey, A.K. and Dwivedi, B.K., 1991. Influence of potassium in high altitude pine forest soil on *Hoplolaimus indicus* in Gharwal Himalayas. *Bioved.*, 2(1): 29–32.

Chaubey, A.K. and Dwivedi, B.K., 1993. Population dynamics of *Hoplolaimus indicus* around *Pinus roxburgii* under influence of soil temperature and organic matter. *Curr. Nematol.*, 5(2): 6–12.

Chaubey, A.K. and Malhotra, S.K., 1988. An account of possible role of phytonematodes as indicator of drought condition. In: *Proc. Nat. Symp. Thr. Hab.*, 9 p.

Chaubey, A.K., 1992. Ecological and biochemical studies on some phytonematodes associated with certain economically important plant at high altitude and Allahabad. *D. Phil. Thesis*, Allahabad University, pp. 445.

Chaubey, A.K., 1994. Behavioral response of *Tylenchorhynchus bravilineatus* to thermal and moisture stress around root zones of *Citrus aurentium* at high altitude. *Curr. Nematol.*, 5(2): 149–158.

Chaubey, A.K. and Sharma, C., 2005. Estimates of amino acid in phytonematodes of various altitudes. *J. Appl. Biosci.*, 31(1): 25–29.

Chaudhary, B.N. and Shivakumar, C.V., 1999. Seasonal fluctuations of a commonly occurring predatory mononch, *Mylonchulus minor* (Nematoda: Mononchida). *Indian J. Nematol.*, 29(1): 69–73.

Chen, Z.X. and Dickson, D.W., 1997. A case study of nematode of biocontrol using *pasteuria penetrans. J. Nematol.*, 29: 573.

Chen, J. and Ferris, M., 2000. Growth and nitrogen mineralization of selected fungi and fungal feeding nematodes on sand amended with organic matter. *Pl. Soil*, 218: 91–101.

Chitwood, B.G. and Chitwood, M.B., 1950. *An Introduction to Nematology*. Monumental Printing, Baltimor.

Das, D. and Mishra, S.D., 1999. Effect of *Heterodera cajani, Rotylenchulus reniformis and Meloidogyne incognita* on the uptake of N, P, and K in Pigeon pea. *Curr. Nematol.*, 10(1–2): 13–15.

Das, D. and Rahman, F., 1996. Effect of soil temperature and soil moisture on population fluctuation of certain plant parasitic nematodes on rice at Assam. *Curr. Nematol.*, 7(1): 39–43.

Dastur, J.R., 1936. A nematode disease of rice in the central provinces. *Indian Acad. Sci.*, 4(2): 108–122.

De Man, J.G., 1876. Onderzoesingen over vrij in de aarde Lavende nematodes Tijds. Chr. Netherland. *Dierk. Veren.* 2: 78–196.

De Man, J.G., 1884. "Die eintheimschen frei in der reinen Erde und im Sussen–Wasser Lebenden Nematoden der Niederlandischen Fauna", pp. 206.

Duncan, L.W. and Abou-Setta, M.M., 1995. Use of sodium acetate to increase nematicidal efficacy in vitro. *Nematropica*, 25(2): 175–178. Eds. CAB International Press, Oxford, pp. 213–243.

Elsen, A., Speijer, P.R., Swennen, R. and DeWaele, D., 1998. Effect of cultivar and altitude composition, root damage and yield of Musa in Uganda. *Nematologica*, 44: 490–491.

Enrique, E. Perez and Edwin, E.L., 2002. Use of Entomopathogenic Nematodes to Suppress *Meloidogyne incognita* on Greenhouse Tomatoes. *J. Nematol.*, 34(2): 171–174.

Faulkner, L.R. and Bolander, W.J., 1970. Acquisition and distribution of nematodes in irrigation waterways of the Columbia Basin in Eastern Washington. *J. Nematology*, 2: 362.

Ferris, H. and Matute, M.M., 2003. Structural and functional succession in the nematode fauna of a soil food web. *Applied Soil Ecology*, 23: 93–110.

Flegg, J.J.M., 1968a. The occurrence and depth distribution of *Xiphinema* and *Longidorus* sp. in South Eastern England. *Nematologica*, 14: 189–264.

Frostegard, A., Tunlidand, A. and Baath, E., 1996. Changes in microbial community structure in two soils experimentally contaminated with metals. *Soil Biol. Biochem.*, 28: 55–63.

Goodey, T., 1933. *Plant Parasitic Nematodes and the Diseases they Cause*. E.P. Dutton and Co. Inc., London, pp. 306.

Goodel, P. and Ferris, H., 1980. Plant parasitic nematodes distribution in an *Alfa alfa* field. *J. Nematol.*, 12: 136–141.

Handa, D.K., Mathur, R.L., Mathur, B.N., Yadav, B.D., 1985. Estimation of losses in barley due to cereal cyst nematode in sandy and loam soils. *Indian J. Nematol.*, 15: 163–166.

Hussani, M.M., Mohanty, B.K. and Nanda, S.K., 2001. Effect of nitrogen, phophorus and potash on growth and yield of chilli cv. Pusa Jwala. *Hort. J.*, 14(1): 69–74.

Hussey, R.S. and Grundler, F.M., 1998. Nematode parasitism of plants. In: *Physiology and Biochemistry of Free Living and Plant Parasitic Nematodes*, (Eds.) R.N. Perry and J. Wrought.

Indubala, M., 2001. Morphological studies of certain parasitic phytonemas associated with some ornamental plants. *M. Phil. Thesis*, C.C.S. University, Meerut, 50 pp.

Kamra, A. and Sharma, S.B., 2000. Soil temperature regimes and nematode distribution in India. *Indian J. Nematol.*, 30(2): 219–224.

Kelly, J.J., Haggblom, M.M. and Tate, R.L., 2003. Effects of heavy contamination and remediation on soil-microbial communities in the vicinity of a zinc smelter as indicated by analysis of microbial community, phospholipid, and fatty acid profiles. *Bio. Fertil. Soils*, 38: 65–71.

Khan, A.M., Adhami, A. and Saxena, S.K., 1971. Population changes of some stylet bearing nematodes associated with Mango (*Mangiferae indica* L.). *Indian J. Nematol.*, 1: 99–105.

Khan, E., 1981. One new genus and four new species in the super family Longidoroidea (Nematoda). *Indian J. Nematol.*, 16: 185–193.

Khan, M.R. and Khan, M.W., 1996. Development of root knot and root nodules on Cowpea as influenced by sulphur dioxide. *Nematol. Medit.*, 24: 33–35.

Koen, H., 1966. The influence of seasonal variation on the vertical distribution of *Meloidogyne javanica* in sandy soil. *Nematologica*, 12: 297–301.

Konopka, A., Zakharava, T., Bischoff, M., Oliver, L., Nakatsu, C. and Turco, R.F. 1999. Microbial biomass and activity in lead contaminated soil. *Appl. Environ. Microbiol.*, 65: 2256–2259.

Krishnaprasad, K.S. and Krishnappa, K., 1982. Occurrence of *Heterodera gambiensis* from India. *Indian J. Nematol.*, 12: 405.

Malhotra, S.K. and Chaubey, A.K., 1990. Spatial nemic distribution patterns and (environmental regulatory processes in nematology in India. In: *Ecosegregation in Parasitocoenosis*, (Eds.) S.K. Malhotra, pp. 34–42.

Malhotra, S.K. and Chaubey, A.K., 1993. High salinity tolerance of Eucalyptus and interaction with soil and plant nematodes of pathogenic significance. In: *Towards the Rational Use of High Salinity Tolerant Plants, Vol. 2,* (Eds.) Leigth and Massoum. Kluwer Acad. Publ., The Netherlands, 239–245 pp.

May, J.N., 1888. Club roots. *Amer. Florist.,* 3: 649.

McItyre, A.D., 1971. Observations on the status of subtaidal meiofauna research. In: *Proc. 1st Int. Conj Meiofauna. Smithsonian Contribution to Zoology,* 46: 149–154.

Mehta, U.K., 1989. Nematode pests of sugarcane. In: *Nematode Pests of Crops,* (Eds.) Bhatti, D.S. and R.K. Walia. CBS Pub. and Dist., p. 159–176.

Mittal, A. and Prasad, D., 1998. Seasonal population fluctuations of plant-parasitic nematodes in soybean-wheat rotation. *Pak. J. Nematol.,* 16(1): 51–53.

Muzne, S., Hinai, A. and Mani, A., 1998. Seasonal population changes and management of *Tylenchulus semipenetrans* using amendments and Phinamiphos. *Nematol. Medit.,* 26: 179–181.

Needham, J.T., 1743. A letter concerning certain chalky tubulous Concretions malm with some microscopial observations on the Farina of the red lily and worms discovered in smutty corn. *Philos. Trans. Roy. Soc.,* 42: 173–174, 634–641.

Norton, D.C., 1978. *Ecology of Plant Parasitic Nematodes.* John Wiley and Sons, N.Y., 268 pp.

Ooestenbrink, M., 1966. Major characteristics of the relation between nematodes and plants. *Medit. Landb. Wagen.,* 66(4): 46.

Pandey, R., 2000. Additive effect of free organic materials and nematodes on the reproduction of *Meloidogyne incognita* and yield of *Mentha arvensis. Nematropica,* 30(2): 155–160.

Pankaj and Dhawan, S.C., 1991. Penetration and multiplication of cereal cyst nematode *Heterodera avenae* (pathotype 1) in resistant and susceptible cultivars of barley. *India J. Nematol.,* 21: 172–173.

Parvathareddi, P. and Singh, D.B., 1980. Plant parasitic nematodes associated with banana in South India. *Proc. Nat. Sem. Banana Prod. Tech., Tamil Nadu Agric. Univ.,* Coimbatore, pp. 161–163.

Patel, B.K. and Patel, H.R., 1998. Effect of soil solarization, rabbing, nematicides and green manuring on growth and development of bidi tobacco seedlings, root knot disease, weeds and phytonematodes in nursery. *Indian J. Nematol.*, 28(1): 15–21.

Pannanen, T., 2001. Microbial communities in boreal coniferous forest humus exposed to heavy metals and change in soil pH: A summary of the use of phospolipids, fatty acids, Biology and H-3-thymidine incorporation methods in field studies. *Geoderma.*, 100: 91–126.

Prasad, K.S.K. and Rao, Y.S., 1980. Relative toxicity of chemicals to infective larvae of rice root nematode, *Meloidogyne graminicola. Indian J. Nematol.*, 10: 216–224.

Prasad, N., Mathur, R.L. and Seghal, S.P., 1959. Molya disease of wheat and barley in Rajasthan. *Curr. Sci.*, 28: 453.

Richter, E., 1969. Zur Vertikelen verteilung von nematoden in einem Sandboden. *Nematologica*, 15: 44–54.

Sasser, J.N., 1979. Economic importance of *Meloidogyne* in tropical countries. In: *Root Knot Nematodes* (*Meloidogyne* spp.). *Systematics, Biology and Control*, (Eds.) F. Lamberti and C.E Taylar. Academic Press, London, New York, Francisco, pp. 359–374.

Sasser, J.N. and Freckman, D.W., 1987. A World Perspective on Nematology: The Role of the Society. In: *Vistas on Nematology*, (Eds.) Veech, J.A. and Dickson, D.W. Society of Nematologists, Hyattsville, Maryland, pp. 7–14.

Seshadri, A.R. (1970). Nematology. In: *Agricultural Year Book–vistas in crop yields.* Pp. 370–411. Indian Council of Agricultural Research, New Delhi.

Shanti, A. and Sundrababu, R., 1995. Effect of phosphorus on the interaction of vesicular arbuscular mycorrhizal fungi with *Meloidogyne incognita* on Cowpea. *Nematol. Medit.*, 23: 263–265.

Sharma, S.B. and Nene, Y.L., 1992. Spatial and temporal distribution of plant parasitic nematodes Pigeon pea in alfisols and vertisols. *Nematropica*, 22(1): 13–19.

Sharma, S. B. and Siddiqi, M. R. 1992. *Bilobodera flexa* gen. n.sp. (Nematoda: Heteroderidae) from Andhra Pradesh), India. *Afro-Asian J. Nematol.*, 2: 59–63.

Siddiqi, M.R., 1964. Studies on nematode root rot of citrus in Uttar Pradesh, India. *Proc. Zool. Soc. India,* 17: 67–75.

Siddiqi, M.R., 2000. *Tylenchida: Parasite of Plants and Insects, 2nd Edition.* CAB International, Wallingford, U.K., xvii+833 pp.

Siddiqi, M.R., 1986. *Tylenchida: Parasite of Plant and Insect.* Farnyam Royal U.K. Commonwealth Agriculture, Bureau, 645 pp.

Singh, P.K. and Garg, Y.K., 2003. Distribution of available micronutrients in sodic soil profile. *Biol. Memoirs,* 29(1): 30–35.

Sitaramaih, K., Singh, R.S., Singh, K.P. and Sikora, R.A., 1971. Plant parasitic and soil nematodes of India. U.P. Agriculture University, Pantnagar and United States Department of Agriculture, *Exp. Stn. Bull.,* 3: 70 pp.

Sundararaj, P. and Mehta, U.K., 1994. Distribution pattern of important species of plant parasitic nematodes in the sugarcane ecosystem of India. *Indian J. Nematol.,* 24: 176–180.

Sundersingh, S.D., 2001. Effect of irrigation regimes and nitrogen levels on growth, yield and quality of baby corn. *Madras Agric. J.,* 88(7–9): 367–370.

Talwana, H.A.L., Speijer, P.R. and Waele, D.D.E., 2000. Spatial distribution of nematode population densities and nematode damage in roots of three banana cultivars in Uganda. *Nematropica,* 30: 19–31.

Trevathen, L.E., Caurezma, Teren, J.A. and Gourley, L.M., 1985. Relationship of plant nematodes and edaphic factors in Colombian grains Sorghum production. *Nematropica,* 15(2): 145–153.

Trudgill, D.L., 1995. Host and plant temperature effects on nematode development rates and nematode ecology. *Nematologica,* 41: 398–404.

Verma, K.K. and Jain, R.K., 2001. Influence of soil moisture levels on root knot nematode (*Meloidgyne incognita*) infesting cotton (*Gossypium hirsutum*). *Indian J. Nematol.,* 31(1): 8–11.

Verma, M.L., Bhandari and Thakur, B.C., 2001. Effect of varying soil moisture and irrigation water requirement of almond CV Nanpareil. *Horti. J.,* 14(1): 1–6.

Walia, K. and Mathur, S., 1997. Influence of edaphic factors on soil acarines and plant parasitic nematodes and their interaction. *Indian J. Nematol.*, 27(1): 46–52.

Wasilewska, I., 1971. Nematodes of the dunes of the Kampinos Forest. II. Community structure based on numbers of individuals, state of biomass and respiratory metabolism. *Ecol. Pol.*, 19: 651–668.

Yeates, G.W., Bongers, T., DeGoede, R.G.M., Freckman, D.W. and Georgieva, S.S., 1993. Feeding habits in nematode families: An outline for soil ecologists. *J. Nematol.*, 25: 315–331.

Yeates, G.W., Hoag, Evans, K.A. and Neilson, R., 1998. Impacts of climatic changes on the distribution of the *Paratrichodorous minor* (Nematoda: Trichodoridae) as estimated using climax. *Nematologica*, 44: 293–301.

Zaki, F.A. and Bhatti, D.S., 1989. Effects of castor (*Ricinus communis*) leaves in combination with different fertilizer doses on *Meloidogyne javanica* infecting tomato. *Indian J. Nematol.*, 19(2): 171–176.

Chapter 2

Freshwater Cladocera (Crustacea : Branchiopoda) from Muriyad Wetlands, Trichur, Kerala

K.K. Subhashbabu & John Thomas K.***

**Kerala Limnological Research Institute, Christ College Campus, Irinjalakuda – 680 125, Kerala*
E-mail: kallikadavil@yahoo.com

***Animal Behaviour and Wetland Research Laboratory, Department of Zoology, Christ College, Irinjalakuda – 680 125, Kerala*
E-mail: jkurishinkal@rediffmail.com

ABSTRACT

Cladoceran fauna from Muriyad wetlands is very rich and diversified. Total 30 species belonging to 7 families and 21 genera are recorded from this wetland. Of this, one species *Disparalona adiyodii* is new to science and male of the *Alonella clathratula* Sars is reported first time from India. Further more, 13 species are new records from Kerala-State. Detailed descriptions of selected species are given.

Keywords: *Cladocera, Disparalona adiyodii, Muriyad wetlands, Kerala.*

Introduction

Information regarding the freshwater Cladocera from Kerala-State particularly from the wetland area is meagre. Only a few studies are so far reported from this area. Studies on Cladocera were initiated in Kerala by Brehm (1953) who reported two marine species from Ernakulam, during the period of his Yale-North India expedition. As a part of the comprehensive study on freshwater Cladocera from India, Michael and Sharma (1988) reported 20 species from this State, on the basis of collections provided by Dr. C.K.G. Nayar and Prof. D.G. Frey. Ragunathan (1989) reported 11 species from Wyanadu District, Kerala. Recently, Subhashbabu and Nayar (2004) recorded 23 species from Periyar Lake and its adjacent areas. Other related studies on freshwater Cladocera from this region are, Thressiamma *et al.* (1991) who studied the population density of *Moina micrura* Kurz cultured in different media, Subhashbabu and Nayar (1993, 1997) on biology and life-history of *Ceriodaphnia cornuta* Sars and *Simocephalus serrulatus* Koch respectively.

Muriyad wetlands is a part of Vembanad-kole Ramsar site, located between 10°25′32″ and 10°18′47″ north latitudes and 76°12′48″ and 76°17′19″ east longitudes. The Karuvannur and Keecheri rivers drain the kol and finally discharge in to the Arabian Sea. The kol wetlands runs parallel to the sea and are low-lying tracts located 0.5 to 1 meter below the mean sea level.

Material and Methods

During the period of 2 years (2000–2002) of investigation in the wetlands of Muriyad, total 148 samples were collected by using plankton net with 50 µm and 70 µm mesh size. The limnetic samples were collected by using drag net while the littoral samples were collected by scoop net. The samples were immediately brought to the laboratory and fixed in 4 per cent formalin solution. Dissections were made by using low power dissection microscope and drawings were made by camera-lucida. Measurements were taken by using calibrated ocular micrometer. Identification was done with the help of relevant monographs and literature.

List of Cladocerans Recorded from Muriyad Wetlands

Family: Sididae

1. *Pseudosida szalayi* Daday*
2. *Latonopsis australis* Sars*
3. *Diaphanosoma excisum* Sars
4. *Diaphanosoma sarsi* Richard

Family: Daphiinidae

1. *Ceriodaphnia cornuta* Sars
2. *Simocephalus serrulatus* (Koch)*
3. *Simocephalus acutirostratus* King*
4. *Scapholeberis kingi* Sars*

Family: Moinidae

1. *Moina micrura* Kurz
2. *Moinodaphnia macleayi* (King)

Family: Bosminidae

1. *Bosminopsis deitersi* Richard

Family: Macrothricidae

1. *Macrothrix odiosa* (Gumey)
2. *Macrothix triserialis* (Brady)
3. *Macrothrix Spinosa* King

Family: Ilyocryptidae

1. *Ilyocryptus spinifer* Herrick

Family: Chydoridae

1. *Disparalona adiyodii* sp nov
2. *Alonella clathratula* Sars#
3. *Alona davidi punctata* (Daday)*
4. *Alona cheni* Sinev*
5. *Alona pulchella* King*
6. *Alona verrucosa* (Sars)
7. *Karua alona* (King)*
8. *Chydorus sphaericus* (O.F. Muller)*
9. *Ephemeroporus barroisi* (Richard)
10. *Dadaya macrops* (Daday)
11. *Dunhevedia serrata* Daday*

12. *Graptoleberis testudinaria* (Fischer)*
13. *Leydigia australis ceylonica* (Daday)*
14. *Oxyurella singalensis* (Daday)
15. *Indialona globulosa* (Daday)

*: First report from Kerala State

#: First report from India.

Taxonomic Descriptions

Pseudosida szalayi Daday [Plate 2.1, (A–E)]

Size

Length: 1.03 mm, Height: 0.466 mm.

Parthenogenetic Female

Body elongated and fleshy [Plate 2.1(A and B)]. Valve very soft with characteristic shell gland. Head rounded and small. Cervical sinus present but no cervical gland. Antennule whip like and attached to the ventral side of the rostrum [Plate 2.1(C)]. Postabdomen comparatively small and rounded with two median process on dorsal side of the tip [Plate 2.1(E)]. No true anal denticles but 9–10 clusters of spinules were present on the lateral side of the postabdomen [Plate 2.1(D)]. Claw large and setulated entire length with 3 basal spines, third one was small while the first and second was large.

Remarks

This is a very common species dwelling among the weeds. Thomas (1961) while re-describing *P. bidentata* has synonymized *P. szalayi* with *P. bidentata*. He suggested that, the median process noticed on the postabdomen could be an artifact due to preservation. Venkataraman (1993) reported the same species from Tamil Nadu and noticed the presence of dorsal median process. Although, Michael and Sharma (1988) did not given any taxonomic importance to the structure, Smirnov (1992) treated these two species as distinct based on the presence of this process. In the present study, all the specimens obtained have this process well developed indicating the taxonomic significance.

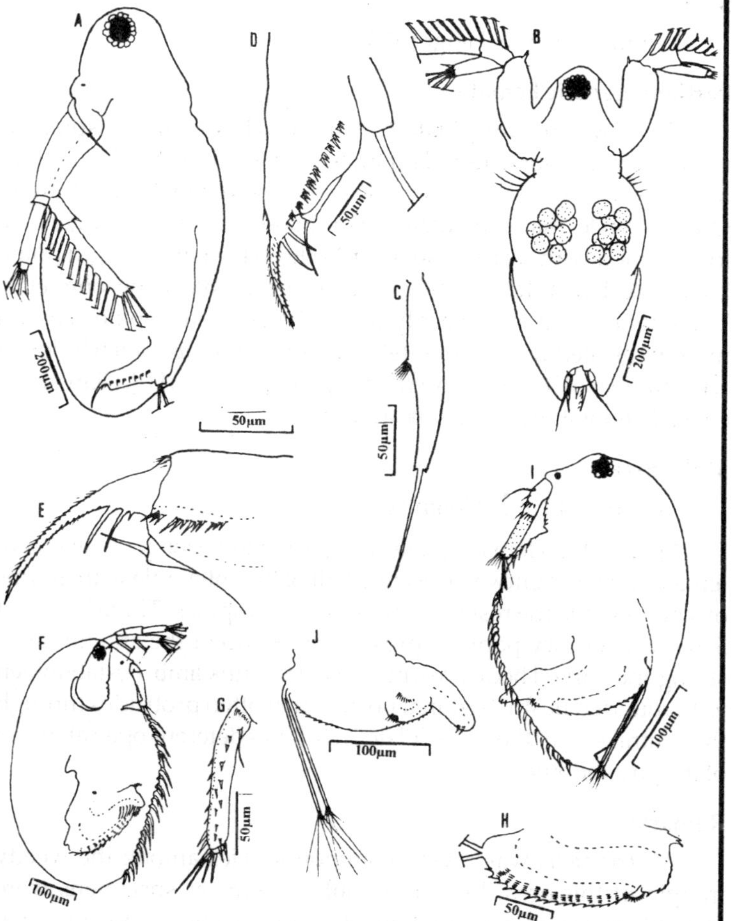

Plate 2.1

(A) *Pseudosida szalayi* Daday (female); (B) *P. szalayi* (dorsal view); (C) Antennule of *P. szalayi* (female); (D) Postabdomen of *P. szalayi* (female); (E) Postabdomen of *P. szalayi* (lateral view); (F) *Macrothrix triserialis* Brady (female); (G) Antennule of *M. triserialis* (female); (H) Postabdomen of *M. triserialis* (Female); (I) *M. triserialis* (male); (J) Postabdomen of *M. triserialis* (male).

Macrothrix triserialis (Brady) [Plate 2.1(F–J)]

Size

Length: 0.56 mm, Height: 0.34 mm

Parthenogenetic Female

Body oval in shape [Plate 2.1(F)]. Head large with conspicuous ridge on its anterior side and separates it from the rest of the body by means of a cervical depression. Rostrum pointed labrum straight. Ocellus smaller than eye situated in the middle of the head. Antennule long, armed with series of small spinule on the entire length [Plate 2.1(G)]. Antenna short with 3 spines on the middle of the long setae. Valve reticulated and sharply pointed ending in a spine. Ventral margin serrated with moderately long setae. Postabdomen bi-lobed with number of backwardly curved spines [Plate 2.1(H)]. Claw short, curved and without a true basal spine.

Male: Size

Length: 0.40 mm, Height: 0.25 mm.

Smaller than female [Plate 2.1(I)]. Antennule longer than that of female and bend outward. Second half of the antennule with groups of setules while the rest without setules and spines. Three long and 5–6 short sensory papillae present on the tip of the antennule. A strong and curved hook present on the first trunk limb. Postabdomen with rounded fleshy terminal portion that often protrudes through the ventral side of the valve [Plate 2.1(J)]. Vas deferens open at the tip of the postabdomen.

Remarks

M. triserialis was a very common species among the weedy margin of this area. Male rare, only two males specimens were present among a large numbers of parthenogenetic and ephippial females. All major characters agree with the descriptions given by Smirnov (1992), Michael and Sharma (1988), Dumont *et al.* (2002).

Alona cheni Sinev [Plate 2.2(A–L)]

Size

Length: 38 mm, Height: 25 mm.

Parthenogenetic Female

Oval in shape, length about 1.6 times of the maximum height [Plate 2.2(A)]. Valve ornamented with striations and reticulations.

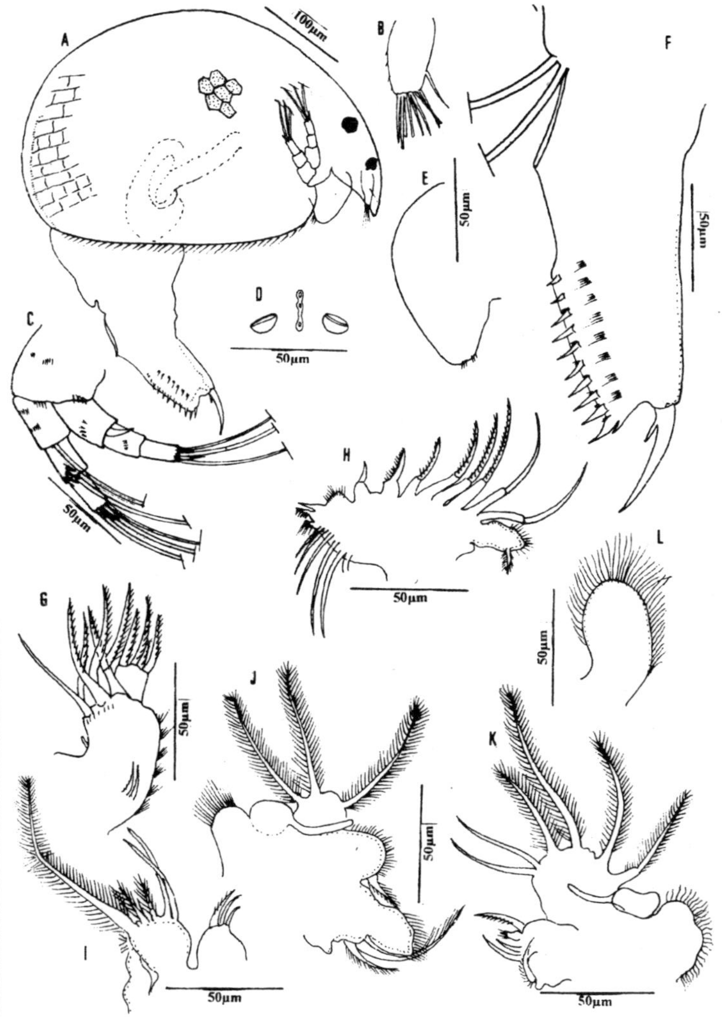

Plate 2.2

(A) *Alona cheni* Sinev (female); (B) Antennule of *A. cheni* (female); (C) Antenna of *A. cheni* (female); (D) Head pores of *A. cheni* (female); (E) Plate of labrum of *A. cheni* (female); (F) Postabdomen of *A. cheni* (female); (G) First trunk limb of *A. cheni* (female); (H) Second trunk limb of *A. cheni* (female); (I) Third trunk limb; (J) Forth trunk limb of *A. cheni*; (K) Fifth trunk limb of *A. cheni*; (L) Sixth trunk limb of *A. cheni*.

Dorsal margin of the valve arched and ventral margin almost straight with plumose setae. Rostrum directed downward and short. Antennule short and not reaching the tip of the rostrum [Plate 2.2(B)]. Lateral sensory hair of the antennule situated more closer to the tip than to the center. Setations on the antennae 0-0-3/1-1-3 [Plate 2.2(C)]. Ocellus smaller than eye and situated half way between the tip of the rostrum and eye. Headshield with three connected head pores, lateral pores large and pocket shaped [Plate 2.2(D)]. Plate of labrum with convex anterior margin and rounded tip possessing two clusters of setules [Plate 2.2(E)] and the tip possesses two groups of setules. The outer distal lob (ODL) of the trunk limb first with single long setae while the inner distal lobe (IDL) of the trunk limb first with 3 setae [Plate 2.2(G)]. Trunk limb second with 8 scrapers and first two scrapers were thin, long and non-ciliated [Plate 2.2(H)]. Exopodite of the third limb with seven setae, of which, 5th one was the longest [Plate 2.2(I)]. Epipodite oval in shape without bract. Exopodite of trunk limb four with six setae, first and second setae were thin and without setules [Plate 2.2(K)]. The epipodite oval in shape with bract. Exopodite of fifth leg with two lobes and epipodite oval in shape with bract [Plate 2.2(J)]. The sixth leg was very much reduced, lobe type and all margins were ciliated [Plate 2.2(L)]. Postabdomen slightly elongated with 10–11 anal denticles [Plate 2.2(F)]. The dorsal margin of the postabdomen more or less straight with an acute angle between distal and dorsal margin. Postabdominal claw elongated and curved with single basal spine.

Male

Not observed.

Remarks

Formerly *Alona costata* Sars was considered as one of the cosmopolitan species but during the revision of this group Snive (1999) confirmed that *A. costata* is not a cosmopolitan species. Originally the *A. costata* was described by Sars (1862) from Norway. This species is now distributed from Norway to Siberia and Mongolia (Smirnov, 1974). The present species *Alona cheni* was first reported from Allahabad, India (Sinev, 1999). In general appearance the *A. cheni* and *A. setigera* were the closely related species but they were different in numbers of soft parts characters, shape, number of head pores, structure, shape of labral plate and epipodite structure of trunk limbs. So, to resolving the problems of 'species complex' the soft parts details are more relevant.

Disparalona adiyodii sp. Nov. [Plate 2.3(A–I) and Plate 2.4(A–B)]

Size

Length: 0.47 mm SD ± 0.02, Height: 0.28 mm SD ± 0.04 (n = 5)

Type Locality

Muriyad Wetlands, Trichur Disrict, Kerala, India.

Type Material

17 mature parthenogenetic females and 8 immature females were collected on 5 October 2002.

Holotype

One mature female mounted in glycerin and sealed with Bee wax and the slide was kept in the museum of Christ College and will be deposited in the museum of Zoological Survey of India, Chennai.

Paratype

Three additional slides of parthenogenetic females mounted on the same medium, kept in the Zoological museum of Christ College, of this, two of them will be deposited in ZSI, Chennai.

Etymology

The species is named after Professor Dr. K.G. Adiyody in appreciation of his encouragement and in recognition of his important contributions to invertebrate endocrinology.

Parthenogenetic Female

Brownish in colour, maximum height just before the middle [Plate 2.3(A)]. Dorsal margin curved and ventral margin almost straight. Posteroventral corner rounded without spines. Rostrum pointed and directed backward. Antennule small not reaching even in the middle of the rostrum [Plate 2.3(C)]. Lateral sensory hair situated in the middle. Ocellus smaller than eye and situated closer to eye. Setation on the antenna 0-0-3/1-1-3 and the spination is 1-0-1/0-0-1 [Plate 2.3(D)]. Valve ornamented with longitudinal lines

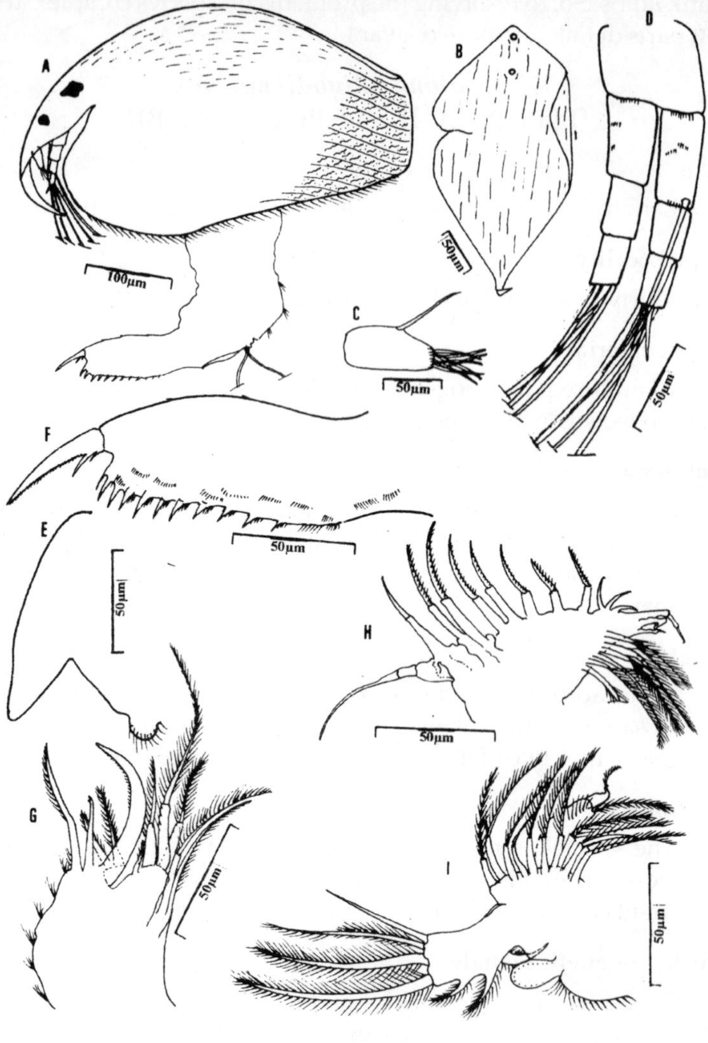

Plate 2.3

(A) *Disparalona adiyodi* (sp. nov) female; (B) Head shield with head pores of *D. adiyodi*; (C) Antennule of *D. adiyodii* (female); (D) Antenna of *D. adiyodii* (female); (E) Plate of labrum of *D. adiyodii* (female); (F) Postabdomen of *D. adiyodi* (female); (G) First trunk limb of *D. adiyodii* (female); (H) Second trunk limb of *D. adiyodii* (female); (I) Third trunk limb of *D. adiyodii* (female).

and fine serrations. The longitudinal lines on the valve were interconnected forming honeycomb like pattern. Head shield with two main head pores and no lateral pores [Plate 2.3(B)]. Plate of labrum slightly pointed on ventral side [Plate 2.3(E)]

First Leg

Inner distal lobe (IDL) with three setae, of these, two are curved and hook-like. Inner side of the hook armed with spinules and teeth [Plate 2.3(G)]

Second Leg

Exopod with long single setae and the endopod with eight scrapers. First scraper without spinules on marginal side and the smaller than near by second and third one. All other even scrapers situated on long base and uni-laterally setulated. First setae of the gnathobase (GT) long and setulated on the both margin. The filter comb composed of seven long un-segmented, bilaterally setulated setae. [Plate 2.3(H)].

Third Leg

Exopodite with seven setae, four of them were on the posterior side and three of them on dorsal side. In this posterior setae, the first one was thin and partially pubescent, while rest three of them were long and bilaterally pubescent [Plate 2.3(I)].

Fourth Leg

Exodite with very large six feathered setae. The first anterior setae partially pubescent and rest of them long and bilaterally pubescent. Gnathobase less developed [Plate 2.4(A)].

Fifth Leg

Exopodite large, rounded and bears four teathered setae. Of this three of them on the posterior side and one on anterior side. The anterior one is short and hairy [Plate 2.4(B)].

Postabdomen moderately elongated with slightly convex ventral side [Plate 2.3(F)]. Dorsal side almost straight with 11–12 anal denticles. The lateral side with 9–10 cluster of spinules. Claw large and setulated on concave margin with two basal spine.

The first basal spine somewhat bigger than others.

Table 2.1: The Differences Between *D. rostrata* and *D. adiyodii*

Sl.No.		*D. rostrata*	*D. adiyodyii*
1.	Head pores	Two main head pores with two inter pores	Two main head pores and no inter pore
2.	Setaion on antenna	0-0-3/0-1-3	0-0-3/1-1-3
3	IDL of first leg	Two long and one short setae	Two strong hook like setae and one long seta
4.	Rostrum	Pointed and directed downward	Pointed but directed posteriorly
5.	Valve ornamentation	Longitudinal lines and interconnections	Interconnected longitudinal lines with small undulated striations between them
6.	Teeth on valve	Two or three teeth on postero-ventral corner	No teeth
7.	Labral plate	Labral plate small with rounded apex	Labral plate large with almost conical apex
8.	Postabdomen	Postabdomen elongated with rounded distal angle	Postabdomen elongated with acute distal angle

Male

Not observed.

Remarks

Fryer (1968) created the genus *Disparalona* to include some chydorids with elongated forms and subsequently transferred *Alonella acutirostris* to *Disparalona* fryer (1971). He attributed some specific characters for this genus such as sweeping setae on third trunk limb, one or more denticles on postero-ventral comer of the valve. The peculiar features of structure of thoracic limbs of *D. rostrata* were noted by Smirnov (1966c) as well. Fryer also noted the weak development of labral plate in this genus. Michael and Frey (1984) closely examined a number species of *Alonella* and *Disparalona* and finally concluded that the sweeping setae is not a specific character to the genus *Disparalona,* the sweeping setae was nothing but the proximal member of the gnathobasic filter comb. However, Venkataraman (1994) reported *D. rostrata* (Koch) from Tripura (India) and he collected several parthenogenetic females. He clearly agrees with the original description of *D. rostrata.* The *D. rostrata* were originally reported from Europe and the animal had some spines on the postero-ventral corner of the valve and the shape of the postabdomen was somewhat different from the animal described by Venkataraman. Further more, the shape and length of the rostrum were also different. Idris (1983) reported the same species from Malaysia, according to him, the presence and absence of the spines on the postero-ventral corner of the carapace and the number of basal spines on the claw are not major characters. From the above-mentioned controversial observations, the subject deserves some more detailed studies to ascertain the status of the genus *Disparalona*.

Alonella clathratula Sars [Plate 2.4(C–G)]

Size

Length: 0.32 mm, Height: 0.14 mm.

Parthenogenetic Female

Body oval in shape [Plate 2.4(C)] Posterior margin of the valve straight; a blunt protuberance on postero-ventral corner. Rostrum long pointed and directed backwardly. Plate of labrum with convex anterior margin and blunt apex with a notch. Valve obliquely striated and reticulated. Postabdomen short with blunted pre-anal corner. Claw moderately large with two un-equal basal spines.

Male

Size: Length: 0.23 mm, Height: 0.14 mm.

Body rectangular in shape, postero-ventral corner with a blunt protuberance as in female. Valve obliquely striated and reticulated. Maximum height little before the middle of the animal. Ocellus smaller than eye situated closer to the tip of the rostrum than the eye. Antennule short and not reaching the tip of the rostrum. Like female, the plate of the labrum with convex anterior margin and a blunt apex and a notch. The trunk limb first with a strong curved hook like structure [Plate 2.4(F)]. Postabdomen tapering distally with number of clusters of spinules on lateral margin [Plate 2.4(G)] and no true anal denticles. Claw somewhat large with two basal spines.

Remarks

A. clathratula can be confused with the related species *Alonella excisa* (Fischer) but the former can be distinguished by relatively large body size with length height ratio of 1:8. Further, the valve ornamentation was doted polygons with longitudinal striations and

Plate 2.4

(A) Forth trunk limb of *D. adiyodii* (female); (B) Fifth trunk limb of *D. adiyodii* (female); (C) *Alonella clathratula* Sars (female); (D) Postabdomen of *D. clathratula* (female); (E) *Alonella clathratula* (male); (F) First trunk limb of *Alonella clathratula* (male); (G) Postabdomen of *D. clathratula* (male); (H) *Ledigia australis ceylonica* (Daday) (female).

the posterior margin of the postabdomen is almost straight. Taxonomical details of present specimen agree with the descriptions given by Smirnov (1996). Males very rare, only three animals were collected along with large numbers of parthenogenetic females. Gamogenetic female not observed. The hook and setations of trunk limb was very clear even through the low power microscope. Characters of males like, the first leg details and shape and structure of antennule are very important to confirm the species status in Cladocera (Venkataraman, 1995). *A. clathratula* is recorded first time from India.

Leydigia australis ceylonica (Daday) [Plate 2.4(H) & Plate 2.5(A–E)]

Size

Length: 88 mm, Height: 0.44 mm.

Parthenogenetic Female

Body white in colour and transparent, valve ornamented with serrations, and no lines [Plate 2.4(H)]. Dorsal margin arched while the ventral margin more or less straight and fringed with long setae. Ocellus larger than eye and situated closer to the eye than to the tip of rostrum. Antennule large and not reaching the tip of the rostrum [Plate 2.5(A)]. The lateral sensory hair of the antennule situated on ventral side. Labrum large and the anterior margin ornamented with very fine and short hairs but the lower portion of the labrum naked [Plate 2.5(B)]. Head shield with prominent, characteristic three head pores. Postabdomen large with straight ventral margin and curved dorsal margin armed with groups of denticles [Plate 2.5(C)]. Each group consists of 4–6 denticles and the distal most denticles were large [Plate 2.5(D)]. Claw thin, elongated and setulated on concave margin [Plate 2.5(E)].

Male

Not observed.

Remarks

Michael and Sharma (1988) reported *L. australis ceylonica* with long ciliated labrum and valve ornamented with longitudinal lines. Smirnov (1974) described the species with very fine and small hairs on the anterior margin and longitudinal lines on valve. The specimens described in the present study have no longitudinal

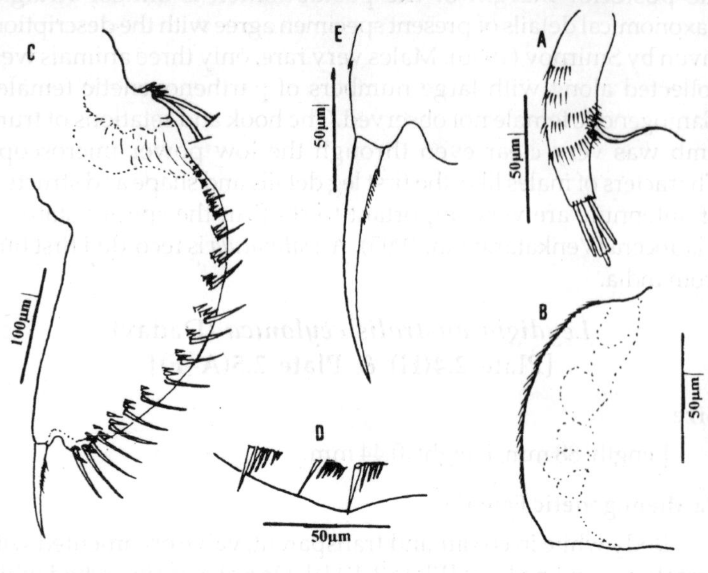

Plate 2.5

(A) Antennule of *Ledgia autralis cylonica* (Daday) (female); (B) Plate of labrum of *L. australis cylonica* (female); (C) Postabdomen of *L. australis cylonica* (female); (D) Marginal denticles of postabdomen; (E) Postabdominal claw of *L. australis cylonica* (female).

striations on the valve and the other details well agreed with Smirnov (1974).

Acknowledgments

We gratefully acknowledge the financial assistance provided by Kerala Research Programme on Local Level Development, CDS, Thiruvananthapuram. We thank Dr. C.K.G. Nayar, Director, Kerala Limnological Research Institute, Irinjalakuda, for his constructive comments and critical review of the manuscript. Thanks are also due to Rev. Fr. Thomas Chakaramakkal, the former principal, Christ College, Irinjalakuda for his encouragement and support

References

Brehm, V., 1953. Indisch Diaptomiden, Pseudodiaptomidae und Cladocera. *Ost. Zool. Zeit.*, 4: 241–345.

Dumont, H.J. Silva-Briano, M. and Subhashbabu. K.K., 2002. A re-evaluation of the *Macrothrix rosea-triserialis* group, with the description of two new species (Crustacea : Anomopoda : Macrothricidae), *Hydrobiologia*. 467: 1–44.

Fryer, G., 1968. Evolution and adaptive radiation in the Chydoridae (Crustacea : Cladocera): A study in comparative. functional morphology and ecology. *Phil. Roy. Soc., London. Ser. B. Biol. Sci.,* 254: 221–385.

Fryer, G., 1971. Allocation of *Alonella acutirostris* (Birge) (Cladocera : Chydoridae) to the genus *Disparalona*. *Crustaceana,* 21: 221–222.

Hann, B.J. and Chengalath, R., 1981. Redescription of *Alonella pulchella* Herrick, 1884 (Cladocera : Chydoridae) and description of the male. *Crustacea,* 41: 249–262.

Idris, B.A.G., 1983. *Freshwater Zooplankton of Malaysia* (Crustacea : Cladocera). Penerbit Universiti, Pertamian, Malaysia, 153 pp.

Michael, R.G. and Frey, D.G., 1984. Separation of *Disparalona leei* (Chien, 1970) in the North America from *D. rostrate* (Koch, 1841) in Europe (Cladocera, Chydoridae). *Hydrobiologia,* 114: 81–108.

Michael, R.G. and B.K. Sharma 1988. *Fauna of India and Adjacent Countries* (Indian Cladocera). Zoological Survey of India, 262 pp.

Ragunathan, R., 1989. Cladocera (Crustacea) from Wynadu district, Kerala. *Geobios New Report,* 8(2): 195–196.

Sars, G.O. 1862. Hr. studios medic. G.O. Sars fortsatte sit foredrag over de af ham I omegnen af Christiania iagttagne Crustacea Cladocera. F orhandlinger I Videnskabs-Selskabet I Christiania aar 1861: 250–302.

Sinev, A.Y., 1999. *Alona costata* Sars versus related palaeotropical species: the first example of close relations between species with a different number of main head pores among Chydoridae (Crustacea : Anomopoda) *Arthropoda Slecta,* 8(3): 131–148.

Smirnov, N.N., 1966c. *Alonella* and *Dunhevedia* (Chydoridae : Cladocera) morphology of trunk limbs. *Hydrobiology,* 33: 547–560.

Smirnov, N.N., 1974. *Chydoridae of the World. Fauna of USSR* (English translation of 1971) Jerusalem, 644 pp.

Smirnov, N.N., 1992. The Macrothricidae of the world. *Guide to the Identification of the Macro-invertebrates of the Continental Waters of the World*, SPB, The Hague, 143 pp.

Smirnov, N.N., 1996. The Chydorinae and Sayciinae (Chydoridae) of the world. *Guides to the Identification of the Micro-invertebrates of the Continental Waters of the world.* SPB, The Hague, 197 pp.

Subhashbabu, K.K. and C.K.G. Nayar, 1993. Observation on the Life-Cycle of *Ceriodaphnia cornuta* Sars. *J. Zool. Soc. Kerala.*, 3(1): 13–17.

Subhashbabu, K.K. and C.K.G. Nayar, 1997. Laboratory studies on the life-cycle of *Simocephalus serrulatus* Koch 1881 (Cladocera : Crustacea). *J. Bombay Nat. Hist. Soc.*, 94(2): 317–321.

Subhashbabu, K.K. and C.K.G. Nayar, 2004. Cladocera of Periyar Lake and adjacent sites, Thekkady, Kerala. *J. Bombay Nat. Hist. Soc.*, 101(3): 403–414.

Thomas, I.E., 1961b. Review of the genus *Pseudosida* Herrick, 1884 and *Latonopsis Sars* (*Cladocera*). *Ibid*, 3: 1–8.

Thressiamma, J. Mercy, T.V.A and Thamby, D.M., 1991. Production and population density of *Moina micrura* Kurz, cultured in differed media. *J. Zool. Soc., Kerala*, 1(1): 21–31.

Venkataraman, K., 1993. Sididae of Tamil Nadu (Crustacea : Cladocera). *J. Bombay Nat. Hist. Soc.*, 90(1): 50–57.

Venkataraman, K., 1994. Two new records of Cladocera of the genus *Echinisca Lievin* and *Disparalona* Fryer from India. *J. Bombay Nat. Hist. Soc.*, 91(2): 334–337.

Venkataraman, K., 1995. Cladoceran males from the Indian region. *J. Bombay Nat. Hist. Soc.*, 92(2): 378-385.

Chapter 3

Small-scale Mass Production Strategy for a Reduviid Predator *Rhynocoris longifrons* Stal (Heteroptera: Reduviidae)

K. Sahayaraj and C. Ravi

Crop Protection Research Centre, Department of Zoology, St. Xavier's College (Autonomous), Palayamkottai – 627 002, India

E-mail: ttn_ksraj@sancharnet.in

ABSTRACT

An effort was undertaken to mass rear a harpactorine reduviids, *Rhynocoris longifrons* Stal at different densities such as 1, 5, 10, 15 and 20 predators/1 L container in the laboratory in small scale by using different densities such as one and two alive *Corcyra cephalonica* Stainton (Pyralidae: Lepidoptera) larvae. Experimental results suggested that 10 and 20 predators/1 L container with two live *C. cephalonica* larvae was the ideal predator and prey proportion for small scale rearing. Heat killed *C. cephalonica* and alive *Spodoptera litura* Fabricius (Noctuidae: Lepidoptera) larvae further enhanced the growth, reduced the developmental period and increased the fecundity and longevity of both the reduviids. Life table studies also confirmed the same. *R. longifrons* (IV, V instars and adults) consumed. maximum *S. litura* than *Helicoverpa armigera* Hubner (Noctuidae: Lepidoptera) in the laboratory pot study.

Introduction

Use of synthetic pesticides is still considered to be the most reliable and economic way to control the pests. Farmers and agricultural workers are exposed directly when they mix and spray pesticides. The soil, water and air are also polluted and thereby cause severe health hazards to human beings, domestic animals and also various beneficial organisms such as microbes, arthropods etc. Biointensive Integrated Pest Management (BIPM) was initially considered as a strategy to manage a single pest, combining all the available BIPM components in a compatible manner and using pesticides as a last resort (Jayaraj *et al.*, 1994). BIPM has been practicing for the past few decades in most of the developed and developing countries. Over the past decade, BIPM has developed into a more holistic approach, in which the pest complex in a crop is considered as a whole, with ecological processes that already exist in the agro-ecosystem being encouraged to do most of the work (Fliert, 1998). The major components of BIPM are predators, parasitoids, botanicals, pathogens, microbes, intercrops and pesticides. Living organisms or its product used to control the pests are called as "biopesticides". Predators are organisms that kill and feed on their prey, require several preys to complete their development and they account for much of the reduction of pest insect populations in nature. Reduviids are generalist predators that mainly feed on lepidopteran, coleopteran, hemipteran and isopteran insects and are considered as a potential biological control agent against various economically important pests. Rearing and mass multiplication of the reduviids in the laboratory is an important requirement for successful biological control programme.

Attempts were made by several workers to mass multiply the reduviids in the laboratory (Lakkundi, 1989; Sahayaraj, 1991; Claver, 1998; Grundy and Maelzer, 2000; Sahayaraj and Paulraj, 2001 and Sahayaraj, 2002). But it is being a major constraint because of its high labour cost, laborious process (Ambrose, 1999) and nymphal cannibolism (Ambrose, 1999). The reduviids are reared in the laboratory mainly on the rice moth *Corcyra cephalonica* Staint on alive larvae (Lakkundi and Parshad, 1987; Sahayaraj, 1991, 2001, 2002b; Claver, 1998; Ambrose, 1999 and Sahayaraj and Paulraj, 2001) and other prey such as *Spodoptera litura* Fabricius (George, 2000a and Sahayaraj and Paulraj, 2001); *Earias vitella* Fabricius (George, 2000a); *Helicoverpa armigera* Hubner and *Nezara viridula* Linn.

(Grundy and Maelzer, 2000); frozen larvae of *C. cephalonica* (Sahayaraj and Jayalakshmi, 2002) and *H. armigera* (Grundy and Maelzer, 2000) and oligidic diet (Sahayaraj, 2004) were also tried for rearing the reduviids. In India, a few reduviids were mass reared in the laboratory and a little success was achieved. They are, *Pltemerus laevicollis* (Antony *et al.*, 1979); *Acanthaspis quinquespinosa* Fabricius (Lakkundi, 1989 and Sahayaraj, 1991); *Brassivola hystrix* Distant, *Coranus* sp., *Endochus parvispinus* Distant, *Irantha armipes* Stal, *Isyndus heros* (Fabricius) (Lakkundi, 1989); *Acanthaspis pedestris* Stal (Sahayaraj, 1991); *Cyndocoris gilvus* Burm (Venkatesan *et al.*, 1997); *R. kumarii* (Claver, 1998); *R. marginatus* (George, 2000b and Sahayaraj, 2002c); *Neohaematorrophus therasii* Ambrose and Livingstone (Sahayaraj, 2001).

Several workers attempted rearing of the predators by using frozen, heat killed and dead larvae. Yasuda and Wakamura (1991) tested the suitability of *S. litura* larvae to *Eocanthecona furcellata* (Wolff). Attempts were made on feeding dead larvae of *Bombyx mori* Linn. and frozen larvae of *Psorocampa denticulate* to *Apiomerus* sp. and *Montina confusa* (Zanuncio *et al.*, 1992); hot water killed larvae of *Tenebris molitor* (L.) and *H. armigera* to *P. plagipennis* (Grundy *et al.*, 2000) and frozen larvae of *C. cephalonica* to *R. marginatus* (Sahayaraj and Jeyalakshmi, 2002) and evaluated the biological parameter of the predators.

Rhynocoris longifrons Stal (Heteroptera: Reduviidae), a harpactorine reduviid predator, is generally found in agro ecosystems, semi arid zones and scrub jungles (Kumar, 1993 and Ambrose, 1999). The postembryonic development (Kumar, 1993); mating behaviour (Das, 1996); functional response (Kumar and Ambrose, 1996 and Claver *et al.*, 2002) and oviposition (Ambrose, 1999) were studied. Ambrose (2003) mentioned the biological control potential of this bug in his review paper. Ambrose (1980) and Kumar (1993) reared this bug on *C. compressus* and *O. obesus* respectively. However, the mass multiplication of *R. kumarii* and *R. longifrons* in the laboratory and the prey and predator density on its development, survival and life table characteristics are lacking. The present study has the following objectives:

1. To find out a small-scale rearing technique for *R. longifrons* in the laboratory condition using alive, frozen and heat killed *C. cephalonica* and alive *S. litura* larvae

2. To find out the impact of prey and predator density on the survival, development, reproduction and life table parameters of the two reduviids, and
3. To evaluate the predatory potential of *R. longifrons* on *S. litura* and *H. armigera* larvae (pot study) as a preliminary test for field release.

Materials and Methods

Collection and Maintenance of Reduviids

Nymphs and adults of *R. longifrons* were collected from the scrub jungles and bordering agroecosystems of Aralvaimozhi and Chunkankadai, Kanyakumari District, Tamil Nadu. They were maintained in the laboratory on the larvae of *C. cephalonica* at a temperature of 29 ± 2°C; 70–80 per cent RH and 13 L : 11 D hours photoperiod. The reduviids were reared in plastic boxes (IL capacity), filter paper was placed at the bottom of the box and chart paper folded into zigzag pattern (10 × 15 cm) was kept inside the boxes for the resting of the predator. Newly hatched nymphs (> 24 hrs) were used for the experimental purposes.

Rearing of the Host

The following ingredients were used as the culture medium for rearing *C. cephalonica:* Wheat flour (1 kg); Powdered Groundnut (200 g); Baker's yeast (5 g); Streptomycin (15 mg) (Sarabhai Piramal Pharmaceuticals Pvt. Ltd., Mumbai, India) and Multivitamin (500 mg) (Glaxo, India). The ingredients were mixed thoroughly and taken in a plastic trough (3 L capacity). Half cc of *C. cephalonica* egg was purchased from District Agriculture Office, Palayamkottai and sprinkled over the medium and the trough was covered with a white muslin cloth, tightly closed with rubber bands. The troughs were kept undisturbed in a wooden rack at ambient laboratory conditions (as mentioned earlier). After 25 days, the larvae (approximately 4.5 mm and 45 mg length and weight, respectively) were collected manually and their heads were crushed using a forceps to prevent the larvae from forming webs and provided to the reduviids.

Collection and Maintenance of *Spodopetra litura*

Egg mass and caterpillars of *S. litura* were collected from groundnut, castor and bhendi agroecosystems of Alankulam, Athioothu and Pavoorchatram, Tirunelveli District and Vallanad,

Theivacheyalpuram and Chakkamalpuram, Tuticorin District, Tamil Nadu, India. They were transported to the laboratory and reared in wide mouth plastic containers (2 L capacity, 20 cm diameter) on groundnut leaves with the earlier mentioned laboratory conditions. Laboratory emerged third and fourth instar larvae were used for rearing the reduviids.

Collection and Maintenance of *Helicoverpa armigera*

The caterpillars of *H. armigera* were collected from groundnut and bhendi agroecosystems of the same locality where *S. litura* caterpillars were collected. They were transported to the laboratory and reared individually in plastic containers (60 ml capacity) to avoid cannibolism in the earlier mentioned laboratory conditions on groundnut leaves and flowers. Laboratory emerged fourth instar larvae were used for the biological control potential evaluation studies.

Experimentation

Experiments were performed in four categories such as reduviids reared on alive (first), heat killed (second), frozen (third) larvae of *C. cephalonica* and alive larvae of *S. litura* (fourth category).

Alive *C. cephalonica* Larvae

R. longifrons were reared at five different densities *viz.*, 1, 5, 10, 15 and 20 predators/container with 1 prey/predator (A1, A2, A3, A4 and A5) and 2 preys/predator (A6, A7, A5, A9 and A10).

Heat Killed *C. cephalonica* Larvae

Water was heated at 60°C in a glass beaker and the IV and V instar *C. cephalonica* larvae were allowed to immerse in it, after thirty seconds, they were taken out and the water was removed using blotting paper and then provided as feed to the reduviids at 2 prey/predator/day (10 predators/container) *R. longifrons* with 60 replication in each category.

Frozen *C. cephalonica* Larvae

In another experiment, live fourth and fifth instar larvae of *C. cephalonica* were collected and taken in a petridish. They were kept in the refrigerator (4°C), undisturbed for four hours. Then they were taken out and kept at room temperature for five to ten minutes and provided as food to the reduviids (2 prey/predator) with 60 replication in each category.

Alive *S. litura* Larvae

In the fourth category, *R. longifrons* (10 predators/container) were provided with live *S. litura* (equal number of III and IV instar larvae at 2 prey/predator/day) with 60 replications. All the experiments were performed in 1L plastic containers and in all the experiments, the adults were maintained with the same feed till their death in the same container. Further more, equal number of male and female was maintained in all the experimental categories. The predators were fed once in three days and twice in a week during the nymphal instars and adults, respectively. The unfed if any and the exuvae were removed daily. The parameters such as individual nymphal developmental period, survival rate, predatory rate (no. of prey consumed/predator/day) and sex ratio (♀/♀+♂) were observed. Further more, pre-oviposition period, number of eggs laid by a female and hatching percentage of eggs was recorded for all the experimental categories.

Life Table

Life tables were constructed according to the methods recommended by Howe (1953), Watson (1964), Laughlin (1965) and Southwood (1978). For each female, oviposition and mortality were checked at two or three day intervals. Daily oviposition was calculated by dividing the total number of eggs laid over an interval by its length (number of days). Time of death was the beginning of the age interval in which the animal died. In life table statistics, the intrinsic rate of increase has been determined using the equation $\Sigma\, e^{-rm \times lxmx-1}$, where e is the base of natural logarithms, x is the age of the individuals in days, l_x is the number of individuals alive at age x as the proportion of 1 and m_x is the number of female offspring produced per female in the age interval x. The sum of products $l_x m_x$ is the net reproductive rate (R_0). The rate of multiplication of population for each generation has been measured in terms of females produced per generation. The precise value of cohort generation has been calculated as follows:

$$T_c = \frac{\sum l_x m_x}{R_0}$$

The arbitraty value of innate capacity for increase r_c has been calculated applying the equation

$$r_c = \frac{\log_e R_0}{T_c}$$

This is an appropriate rm value. The values of negative exponent of e^{-rmx} ascertained from this experiment often lay outside the range. For this reason both sides of the equation have been multiplied by a factor of $\Sigma\, e^{7-rm \times lxmx}$–1096.6 (Birch, 1948 and Watson, 1964). The two values of $\Sigma\, e^{7-rm \times lxmx}$ have been then plotted on the horizontal axis against their respective arbitrary rm on the vertical axis. Two points have been then joined to give a line, which is intersected by a vertical line drawn from the desired value of $e^{-rm \times lxmx}$ (1096.6). The point of intersection gives the value of rm accurate to three decimal places. The precise generation time (T) has been then calculated from the equation:

$$T = \frac{\log_e R_0}{rm}$$

The finite rate of increase (λ) has been calculated as e^{rm}. This λ represents the number of individuals added to the population per female per day (Siddiqui *et al.*, 1973). The weekly multiplication of predator population has been calculated as (e^{rm7}). The doubling time has been calculated as log 2/log λ.

Statistical Analysis

Statistics such as ANOVA and post ANOVA, Least Significant Difference test (LSD), Correlation and Chi square test were worked out using SYSTAT and STATISTICA computer package and results were interpreted.

Results

Alive *Corcyra cephalonica* Larvae

Developmental Period

In *R. longifrons,* among the one-prey category, the total developmental period was significantly shorter in A4 (41.50 ± 0.96 days), A3 (41.90 ± 2.27 days) and A5 (42.61 ± 1.12 days). These groups were statistically significant at 5 per cent level when compared to A2 (49.42 ± 1.84 days) and A1 (52.90 ± 1.20 days) by LSD test. The total numphal period was reduced but not statistically significant when the predators were provided two-prey (Table 3.1). Among the two-prey categories, the shortest and longest developmental period was observed in A8 (40.29 ± 1.24 days) and A6 (51.04 ± 1.16 days) respectively and is statistically significant at 5 per cent level. Irrespective of the predator and prey densities, the second and fifth instar had the shortest and longest duration respectively.

Survival Rate and Sex Ratio

100 per cent survival of *R. longifrons* nymphs was observed in both A1 and A6 and is highly significant ($P < 0.05$) than the rest of the categories by LSD test. Except one and five predator densities, the survival rate was the highest in two-prey density than one-prey density. The minimum survival rate of the predator was observed in A3 (50.0 per cent). The survival rate recorded in A8 was 80.0 ± 4.50 per cent. Almost all the tested categories showed female biased sex ratio except A2 and A3 (0.5) and it was high in A10 (0.74) (Tables 3.1 and 3.2). Chi-square test showed a higher significance for the latter category. The correlation between predatory rate and sex ratio shows a higher significance ($P < 0.05$; $p = 0.0014$ and 0.00018 for 1 prey and 2 prey category, respectively).

Reproduction

In *R. longifrons*, the pre-oviposition period was shorter in A8 (14.06 ± 1.92 days) and longer in A9 (20.94 ± 3.42 days) and they are statistically significant ($P < 0.05$; $p = 0.009$). Mean number of eggs produced by *R. longifrons* on different predator and prey densities is presented in Tables 3.1 and 3.2. Maximum and minimum egg production was obtained in A10 (130.6 ± 4.5 eggs/female) and A1 (107.0 ± 5.48 eggs/female), respectively. The LSD analysis between the developmental days and egg production shows high significance ($t = -12.654$ for one prey category and $t = -15.606$ for two prey category) and a positive correlation exists between the prey density and eggs/female ($r = 0.85$). The hatching percentage was high in A8 (85.70 ± 8.13 per cent) and low (74.60 ± 5.48 per cent) in A4 (Tables 3.1 and 3.2).

Table 3.1: Biological Parameters and Predatory Rate of *R. longifrons* Fed with One Alive *C. cephalonica* Larvae

Parameters	*A1*	*A2*	*A3*	*A4*	*A5*
Total Nymphal Period (Days)	52.90 ± 1.20	49.42 ± 1.84	41.90 ± 2.27	41.50 ± 0.96	42.61 ± 1.12
Survival (%)	100.00 ± 0.00	80.00 ± 10.0	50.00 ± 5.00	62.22 ± 3.84	71.66 ± 6.96
Sex ratio	0.66	0.50	0.50	0.57	0.66
Pre-oviposition period (Days)	17.92 ± 2.46	16.46 ± 1.32	16.24 ± 3.12	15.54 ± 2.18	14.82 ± 1.96
Eggs/female (No.)	107.00 ± 5.48	102.50 ± 6.36	109.79 ± 9.76	116.00 ± 7.20	121.00 ± 4.50
Hatching percentage	79.43 ± 6.66	77.07 ± 8.48	74.70 ± 7.54	74.60 ± 5.48	76.70 ± 8.32
Predatory rate (Prey/Predator/Day)	0.98 ± 0.24	0.80 ± 0.42	0.92 ± 0.72	0.98 ± 0.66	0.85 ± 0.45

A1: 1 prey/1 predator; A2: 1 prey/5 predator; A3: 1 prey/10 predator; A4: 1 prey/15 predator and A5: 1 prey/20 predator.

Table 3.2: Biological Parameters and Predatory Rate of *R. longifrons* Fed with Two Alive *C. cephalonica* Larvae (mean ± SD)

Parameters	*A6*	*A7*	*A8*	*A9*	*A10*
Total nymphal period (Days)	51.04 ± 1.16	49.70 ± 1.24	40.29 ± 1.24	43.65 ± 1.56	41.83 ± 1.06
Predatory rate (Prey/Predator/Day)	1.60 ± 0.20	1.80 ± 0.25	1.64 ± 0.18	1.58 ± 0.24	1.51 ± 0.18
Survival (per cent)	100.00 ± 0.00	80.0 ± 5.00	60.00 ± 4.50	73.33 ± 6.66	78.33 ± 5.99
Sex ratio	0.66	0.60	0.72	0.69	0.74
Pre oviposition period (Days)	18.32 ± 2.84	16.18 ± 2.36	14.06 ± 1.92	20.94 ± 3.42	15.32 ± 1.96
Number of eggs/female	114.33 ± 5.42	116.50 ± 3.50	120.60 ± 4.54	127.00 ± 6.00	130.60 ± 4.50
Hatching percentage	83.09 ± 3.66	78.34 ± 6.36	85.70 ± 8.13	77.80 ± 4.36	80.28 ± 1.18
Predatory rate (Prey/Predator/Day)	1.60 ± 0.20	1.80 ± 0.25	1.64 ± 0.18	1.58 ± 0.24	1.51 ± 0.18

A6: 2 prey/1 predator; A7: 2 prey/5 predator; AS: 2 prey/10 predator; A9: 2 prey/15 predator and A10: 2 prey/20 predator.

Bioefficacy of Reduviids

Reduviids reared on high prey density consumed more number of preys than those reared on low prey density. In *R. longifrons,* it was higher in A7 (1.8 prey/predator/day) and minimum in A2 (0.80 prey/predator/day) and is highly significant ($P < 0.05$) (Tables 3.1 and 3.2).

Adult Longevity

Irrespective of predator densities and prey categories, the females lived longer than the males. The male adult longevity of *R. longifrons* was longer in A8 (101.66 ± 3.22 days) and shorter in A1 (71.64 ± 3.48 days) respectively and longest female longevity was observed in A8 (114.63 ± 5.84 days) and are highly significant ($P < 0.05$) (Figure 3.1 and 3.2).

Life Table Statistics

The life table statistics of *R. longifrons* fed with one and two alive *C. cephalonica* are shown in Tables 3.5 and 3.6. The prey density had a great influence on the net reproductive rate (NRR) and was higher in A9 (54.76) and minimum in A7 (41.8). The NRR for A8 was 47.5. As the predator density increased, the mean length of generation (Tc) decreased and it was very low in A10 (69.69). The calculated Tc and corrected rm for A8 were 75.48, and 0.051, respectively. The doubling time was very short in A10 (11.02) and similar doubling time was recorded for A3, A5 and A8 (12.33). The hypothetical F_2 female was maximum in A9 (2998.65), minimum, in A7 (1747.24) and moderate in A8 (2256.25).

HK, and FL of *C. cephalonica* and ASL (alive *S. litura*) Larvae

Biological Parameters

Based on the previous experiment *R. longifrons* reared with 10 predators and two-prey, was moderately superior to the other categories. *R. longifrons* developed faster in ASL (35.31 ± 1.86 days) than HK (36.58 ± 2.42 days) and FL (45.62 ± 1.92 days) (Table 3.5, Figure 3.1). The comparison between the ASL and HK were not statistically significant at 5 per cent level. The sex ratio was female biased in all the three groups and were, significant by chi-square test except FL. The fecundity was higher in ASL (149.72 ± 4.27 eggs/

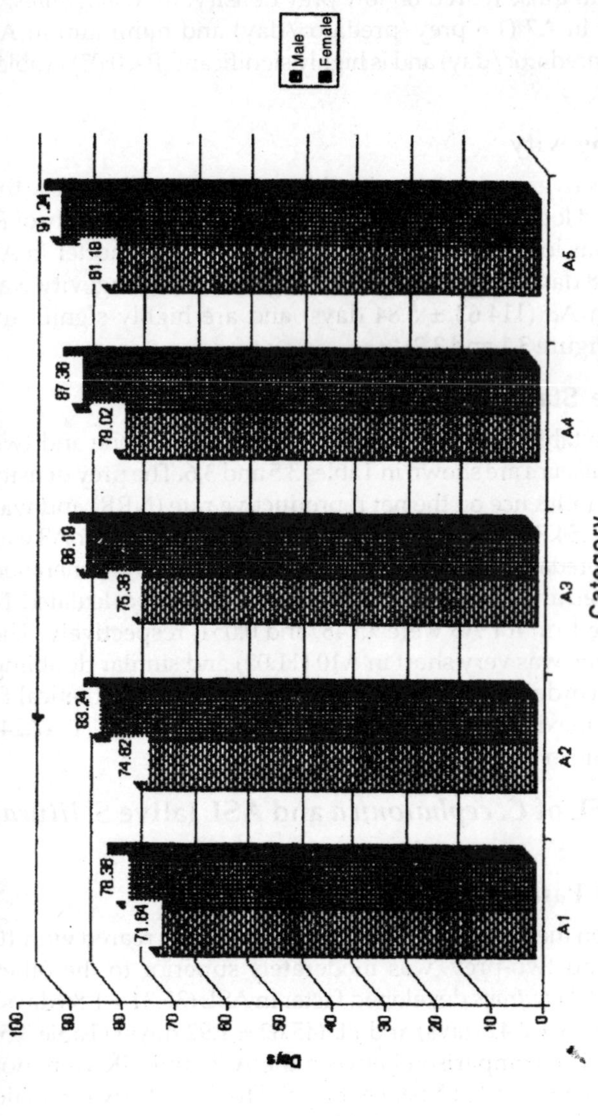

Figure 3.1: Adult Longevity (in days) of *R. longifrons* Fed with Only One Alive *C. cephalonica* Larvae

A1: 1 prey/1 predator; A2: 1 prey/5 predator; A3: 1 prey/10 predator; A4: 1 prey/15 predator and A5: 1 prey/20 predator

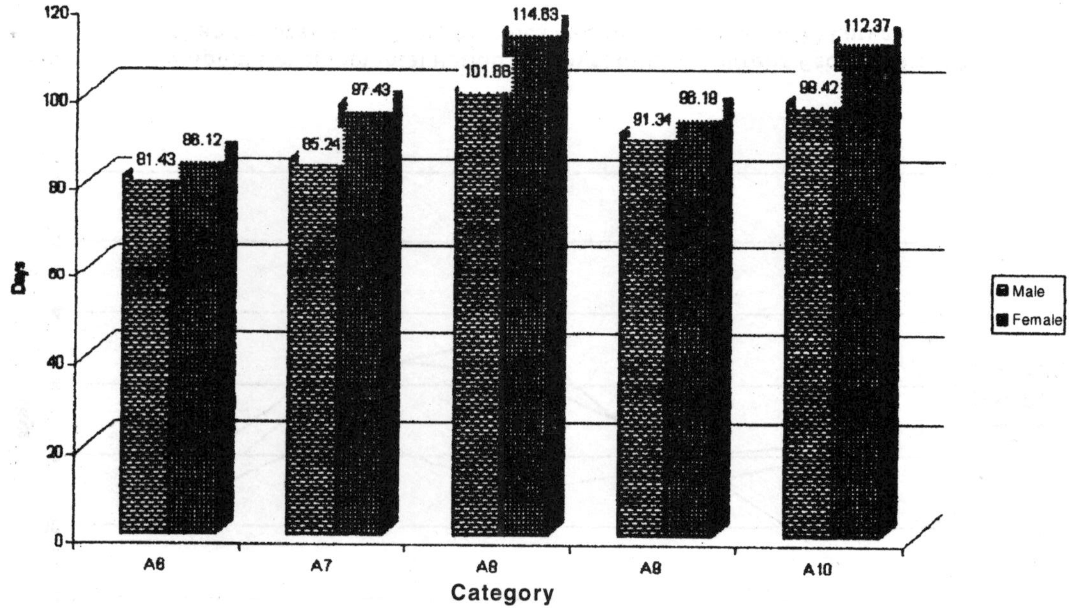

Figure 3.2: Adult Longevity (in days) of *R. longifrons* Fed with Two Alive *C. cephalonica* Larvae
A6: 2 prey/1 predator; A7: 2 prey/5 predator; A3: 2 prey/10 predator; A9: 2 prey/15 predator and A10: 2 prey/20 predator

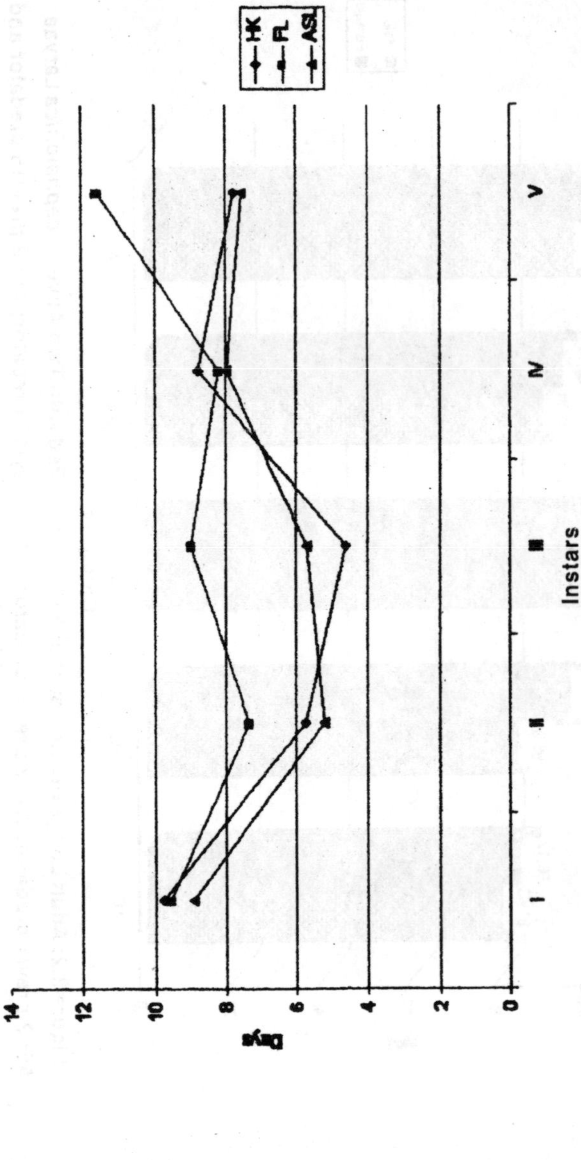

Figure 3.3: Nymphal Developmental Period (in Days) of *R. longifrons* Fed with Heat Killed (HK) and Frozen (FL) *C. cephalonica* and Alive *S. litura* (ASL) Larvae

Table 3.3: Life Table Parameters of *R. longifrons* Fed with One Alive *C. cephalonica* Larvae

Parameters	*A1*	*A2*	*A3*	*A4*	*A5*
Net reproductive rate (NRR)	50.25	52.75	44.65	43.44	50.33
Mean length of generation (Tc)	75.6	71.23	76.71	75.90	76.24
Innate capacity for natural increase (rc)	0.051	0.055	0.049	0.049	0.051
Corrected rm	0.055	0.059	0.057	0.056	0.057
Precise generation time (T)	71.21	67.21	66.64	67.34	68.74
Finite rate of increase (λ)	1.056	1.060	1.058	1.057	1.058
Doubling time (C)	12.75	12.04	12.33	12.54	12.33
Weekly multiplication time (erm^7)	1.469	1.540	1.490	1.479	1.058
Hypcthetical female in F_2 generation	2525.06	2782.56	1993.62	1887.03	2533.1

A1: 1 prey/1 predator; A2: 1 prey/5 predator, A3: 1 prey/10 predator; A4: 1 prey/15 predator and A5: 1 prey/20 predator.

Table 3.4: Life Table Parameters of *R. longifrons* Fed with Two Alive *C. cephalonica*, Larvae

Parameters	*A6*	*A7*	*A8*	*A9*	*A10*
Net reproductive rate (NRR)	54.0	41.8	47.5	54.76	48.43
Mean length of generation (Tc)	84.25	83.93	75.48	74.12	69.69
Innate capacity for natural increase (rc)	0.047	0.044	0.051	0.054	0.055
Corrected rm	0.054	0.050	0.057	0.061	0.063
Precise generation time (T)	73.87	74.65	67.73	65.62	61.58
Finite rate of increase (λ)	1.055	1.051	1.058	1.062	1.065
Doubling time (C)	12.97	13.93	12.33	11.53	11.02
Weekly multiplication time (erm^7)	1.459	1.419	1.490	1.062	1.554
Hypothetical female in F_2 generation	2916.0	1747.24	2256.25	2998.65	2345.46

A6: 2 prey/1 predator; A7: 2 prey/5 predator; A8: 2 prey/10 predator; A9: 2 prey/15 predator and A10: 2 prey/20 predator.

female), followed by HK (143.20 ± 5.81 eggs/female) and FL (109.34 ± 4.62 eggs/female) and HK and ASL are statistically significant than FL (P < 0.05). The pre-oviposition period was shorter in ASL (9.42 ± 0.96 days) and longer in FL (16.34 ± 2.62 days) and they were statistically significant (P < 0.05). The male and female longevity was also longer in ASL (114.36 ± 5.62 and 126.82 ± 4.16 days) for male and female respectively (Figure 3.4).

Life Table Statistics

In *R. longifrons*, the weekly multiplication was the same for both HK and ASL (1.521) (Table 3.5). Similar kind of observations was also recorded for ASL group that had high NRR (64.16), rm (0.054), precise generation time (69.39) and hypothetical F_2 generation (4116.5) (Table 3.5).

Table 3.5: Biological and Life Table Parameters of *R. longifrons* Fed with Heat Killed (HK) and Frozen (FL) *C. cephalonica* and Alive *S. litura* (ASL) Larvae

Parameters	*HK*	*FL*	*ASL*
Total nymphal period (Days)	36.58 ± 2.42	45.62 ± 1.92	35.31 ± 1.86
Survival (per cent)	90.67 ± 5.63	55.0 ± 5.00	90.00 ± 10.00
Sex ratio	0.66	0.56	0.77
Pre-oviposition period (Days)	10.32 ± 1.74	16.34 ± 2.62	9.42 ± 0.96
Eggs/female (No.)	143.20 ± 5.81	109.34 ± 4.62	149.72 ± 4.27
Hatching percentage	85.49 ± 6.82	81.65 ± 2.34	93.95 ± 4.03
Predatory rate (Prey/Predator/Day)	1.6	1.2	0.8
Net reproductive rate (NRR)	61.81	44.76	64.16
Mean length of generation (Tc)	77.80	76.75	77.69
Innate capacity for natural increase (rc)	0.053	0.049	0.0540
Corrected rm	0.060	0.055	0.060
Precise generation time (T)	68.73	69.10	69.35
Finite rate of increase (λ)	1.061	1.056	1.061
Doubling time (C)	11.71	12.75	11.71
Weekly multiplication time (erm^7)	1.521	1.469	1.521
Hypothetical female in F_2 generation	3820.47	2003.45	4116.5

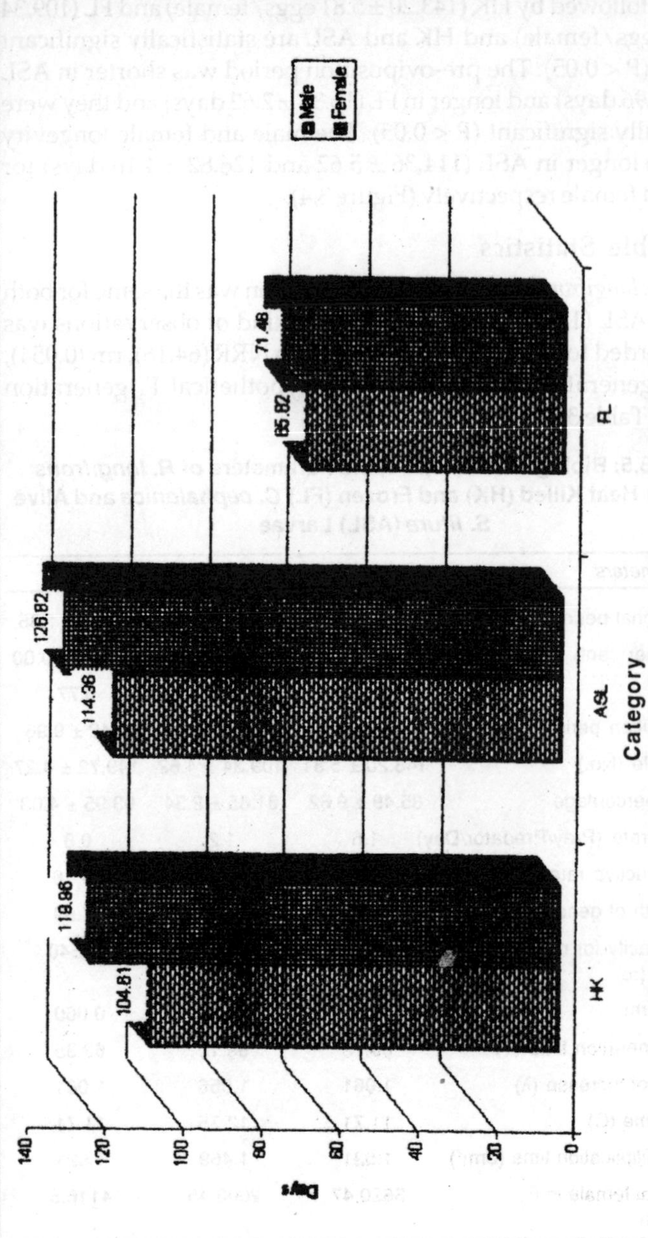

Figure 3.4: Adult Longevity (in Days) of *R. longifrons* Fed with Heat Killed (HK) and Frozen (FL) *C. cephalonica* and Alive *S. litura* (ASL) Larvae

Biological Control Potential

Pot studies revealed that at low prey density (one-prey), prey consumption given a uniform prey distribution was higher than that for contagious prey distribution ($P < 0.05$), but no similar prey consumption rates were observed at high prey density for the two prey spatial patterns ($P < 0.05$). *R. longifrons* preferred both *S. litura* and *H. armigera* and maximum consumption of 1.66 preys/predator/day for both pests. However, the maximum weight gain was observed in the adult that fed *S. lituna* (54.94 ± 5.00 mg) (Table 3.6).

Table 3.6: Predatory Rate (Prey/Predator/Day) and Amount of Food Consumption (AFC) [in mg] of *R. longiftons* on *S. litura* and *H. armigera* Larvae

Predator Life Stages	*Prey*	*No. of Prey*		*AFC*
		Provided	*Consumed*	
IV	*S. litura*	1.0	1.00	34.85 ± 3.21
V			1.00	34.97 ± 4.55
Adult			1.00	39.58 ± 5.22
IV		2.0	1.16	39.59 ± 3.85
V			1.32	44.81 ± 2.96
Adult		1.0	1.66	54.94 ± 5.11
IV	*H. armigera*		1.00	29.15 ± 2.59
V			1.00	33.92 ± 3.28
Adult			1:00	37.61 ± 4.25
IV		2.0	1.16	31.23 ± 1.98
V			1.32	36.41 ± 3.56
Adult			1.66	47.47 ± 5.15

Discussion

Developmental Period

The survival and growth of animals, their ability to pass through nymphal instars in life and the capacity to reproduce are all dependent on nutritional factors, but the actual nature of dietary demands in reduviids has not been analyzed, perhaps in *R. marginatus* (George *et al.*, 2002). In *N. therasii* and *A. pedestris*, studies of diet-based predator discrimination is the assumption that predator that have fed on a particular prey type should present a greater predation risk than predators that have fed on other prey (Sahayaraj, 2001).

The faster a predator completes its life cycle, the more generations it can produce in a specific period. Further more, the total developmental period of any insects, especially the predators mainly depend upon the prey species and its nutritional quality (Ambrose and Subbarasu, 1988; Ambrose *et al.*, 1990 and O'Neil and Widenmann, 1990). *C. cephalonica* serve as an important laboratory host for many predatory insects including reduviids (Lakkundi and Parshad, 1987 and Sahayaraj, 1998a, 2002b). It is a general concept that the amount of prey consumed is proportionate to the development, hence *R. longifrons* developed faster in the two-prey category as reported by Ambrose (1999). Ten and twenty predators with two *C. cephalonica* fed *R. longifrons* development was faster than the other groups. Rearing in groups also enhance the development of reduviids (Ambrose, 1999). Social behavior of insects is of prime importance in the augmentation of predatory insects. In nature, some of the harpactorine reduviids like *R. kumarii, R. marginatus, R. fuscipes* and *R. longifrons* live in groups (Ambrose, 1999). It is a kind of adaptive measure of reduviids to promote higher fecundity, hatchability and adult longevity for better success.

Alive *S. litura* and heat killed *C. cephalonica* further reduced the developmental period in both the reduviids and this preferred prey accelerated the food consumption and development of the insects. In the present study, live *S. litura* greatly reduced the developmental period of the reduviids. Ambrose (1999) pointed out that the most preferred and suitable prey for augmentation is based on the size and agility of the prey and the subsequent predatory stress developed by the predator to capture and satiate itself. George *et al.* (1998); George (1999, 2000a) and Sahayaraj and Paulraj (2001) also reported faster development of reduviids when they reared the reduviids on *S. litura*. George *et al.* (2002) reported that the biochemical constituents (total carbohydrate, protein and lipids) of *S. litura* were much greater than *C. cephalonica* and that might be the reason for the faster development of these reduviids when reared on *S. litura*. The heat-killed larvae also reduced the developmental period of *R. longifrons* than those fed with frozen and alive *C. cephalonica*. Similar finding was also reported by Sahayaraj and Jeyalakshmi (2002) in

R. marginatus where frozen *C. cephalonica* larvae postponed the developmental days. So the frozen larvae of *C. cephalonica* may be used to maintain the reduviids instead of using it for rapid multiplication purpose. It is ideal to group rear both the reduviids in the laboratory on live *S. litura* larvae.

Survival Rate and Sex Ratio

Cannibolism is a major constraint in-group rearing of reduviids (Sahayaraj, 1991; Ambrose, 1999 and George, 2000b) and that would cause an enormous amount of mortality and reduce the survival rate of the predators. Natural death of the predators may also occur due to diseases, scarcity of food and shelter (Cohen, 1993 and Ambrose, 1999). But in the present study, no disease was observed and although the shelter was given in the form of a zigzag chart, still mortality occurred and the reason for that may be natural death and cannibolism. Cannibolism might be a stronger reason for the mortality as reported by George (2000b) because 100 per cent survival of *R. longifrons* was observed when they were reared individually. Cannibolism occurs in predators mainly due to the scarcity of the prey. But the present study reveals that, irrespective of the predator and prey densities, some preys remained in the rearing arena. Hence, the mortality of the predators was not due to the non-availability of the preys.

In both the reduviids, a female biased sex ratio was observed in all the experimental categories except A2 and A3. Sahayaraj (1991); Lakkundi (1989); Claver (1998) and Ambrose (1999) have reported that the harpactorine reduviids are generally female biased in optimal conditions. In the laboratory, good condition was maintained like cleanliness and physical parameters for the culture of both the reduviids and that might be the reason for the female based sex ratio. Production of more number of females is a desirable characteristic of a best biological control agent.

Reproduction

Information on the fecundity is a very important criterion in selecting a predator as a biological control agent. The predator should have a reproductive potential close to that of its potential prey, and it should be able to increase its population with minimum time lag, when the prey population increase or decrease in the field. The relation between nutrition and reproduction has not been extensively studied among reduviids as in other insects. As observed for the development, the reproduction also depends upon the quality and quantity of the food provided (O'Neil and Widenmann, 1990). In *R. longifrons*, A10 had the maximum fecundity (130.6 ± 4.5 eggs/female) in the prey predator experiment and in general *S. litura* category recorded the maximum fecundity (149.72 ± 4.27 eggs/female). George (2000a) reported 132.78 eggs for a related harpactorine species (*R. marginatus*) reared on *C. cephalonica*. General hypothesis is that number of prey ingested (amount) also is of high value in improving fecundity and the total egg production being dependent on the quantity of nutrients taken in by the insects. Further more, enhanced quantity will promote maturation of more eggs. This hypothesis is not true in the present study. For instance, the fecundity (number of eggs laid/female) of A5, A8, A9 and A10 of *R. longifrons* has no correlation with the predatory rate. Moreover, a close relationship between food supply and reproduction in the female has been recorded here.

In the present study, pre-oviposition period was shorter when the reduviids were reared on alive *S. litura* (9.42 ± 0.96 day). The difference in pre-oviposition period occurs with varied prey type (Parajulee and Philips, 1993; Venkatesan *et al.*, 1997). The pre-oviposition of the predators also depends upon the prey and rearing condition. Group rearing shortened the pre-oviposition period (Ambrose, 1999). Hatching is basically a temperature and humidity dependent event and here all the eggs were placed in the same environment but difference in the observed hatching percentage suggests that the prey type also influence the hatching rate apart from the physical factors as reported by Fuller (1988). Maximum hatching was observed in *S. litura* fed category (93.95 ± 4.03 per cent). Similar increase in hatching percentage for *R. marginatus* eggs reared on *S. litura* was reported by George (2000c) and Sahayaraj and Paulraj (2001).

Predatory Rate

Predation is an important character of the carnivorous insects who meets their nutritional requirements by exploiting the prey's body content. The predatory rate of both the reduviids was high in the two-prey groups than the one-prey group. The predatory rate increased as the increased prey number is a general concept in the prey predator interaction. The present study coincides with the conclusions of Sahayaraj (1991, 2001) and Kumaraswami (1991) and Ambrose (1999). Claver (1998) also reported positive correlation between the number of prey and the predatory rate of *R. kumarii.*

Adult Longevity

The longevity of any individual is a genetic factor that can be influenced by the biotic and abiotic factors. The longer a predator can survive in the field, the more effective it could be a biological control agent. Longevity of female is of particular importance since it has a significant role in maintaining a high density of predator population in the field. Ambrose (1999) stated that eventhough longevity is a pre-determined one; the prey type and the amount of the prey can alter the adult longevity. He further reported that, in harpactorine reduviids, usually the ovipositing female live-longer than the males. A starved animal cannot survive longer, whatever be the genetical fitness the organisms have, it has to feed for long life. In the present study, the reduviids provided with two-prey lived longer than one-prey group. And the longevity was further increased in HK and ASL groups, in particular, the females in all the groups. In harpactorine reduviids, generally females live longer than males (Ambrose, 1999; Sahayaraj, 1991 and George *et al.*, 1998) and this might be due to the higher consumption of the food by the females to meet its extra energy requirement for oviposition. According to Parajulee and Phillips (1993) and Venkatesan *et al.* (1997), the longevity of ovipositing females vary with species.

Life Table

The life table is a useful method to study the life expectancy and net reproductive rate (Southwood, 1978). Life table of predatory reduviids are available only for few species (Venkatesan *et al.*, 1997; George, 1999, 2000 a; George *et al.*, 1998; Claver, 1998; Sahayaraj and Paulraj, 2001 and Sahayaraj and Jeyalakshmi, 2002). Life table studies give a clear-cut idea about the precise generation time, doubling time and hypothetical F_2 females. So the availability of predators for the future for a stipulated period could be easily predicted. *R. longifrons* and *R. kumarii* reared on *S. litura* had the maximum NRR (64.16 and 72.58 respectively) and the doubling time also was shorter in this category (11.71). In the prey-predator experiment, in *R. longifrons* A9 had the maximum NRR and minimum doubling time (11.53) with 2998.65 hypothetical female. In this reduviid, the heat killed larvae group had a better NRR, doubling time and hypothetical F_2 female than the alive and frozen *C. cephalonica* group. High NRR and intrinsic rate of increase have been reported for temperate bugs such as *Oncopeltus faxiatus* (Klausner *et al.*, 1980); *Pilegodorus guildnii* (Panizzi and Slansky, 1985) and *Clavigralla tomentosicollis* (Iheagwam, 1982) when reared on their preferred prey. In reduviids also, the *S. litura* larvae increased the NRR, life span and decreased the doubling time when compared to other prey species (George *et al.*, 1998). A maximum of 2256.25 *R. longifrons* females can be expected in the F_2 generation when they were reared with 10 and 20 predators per container, respectively with two alive *C. cephalonica*. The heat-killed larvae increased the NRR of *R. marginatus* and decreased the doubling time (Sahayaraj and Jeyalakshmi, 2002). *S. litura* had a great superiority than the other prey species for reduviids (George *et al.*, 1998; George, 1999; 2000a; Claver, 1998; Venkatesan *et al.*, 1997 and Sahayaraj and Paulraj, 2001) and the present findings also confirmed the same.

Biological Control Potential

R. longifrons nymphs and adults feeds both *S. litura* and *H. armigera* larvae and the predatory potential remained the same for both the prey *R. longifrons* preferred *S. litura* than *H. armigera* larvae. In both the case, the adults consumed more quantity than the nymphs. Ambrose (1999) reported that larger predator predates more and the weight gain would be more than the smaller predators. Sahayaraj (l999a) has reported the predatory efficiency of *R. marginatus* on three major groundnut pests and opined that the reduviid preferred the soft bodied prey than the prey with hard cuticle. Grundy and Maelzer (2000) also reported the predation of *P. plagipennis* on *H. amigera* in the laboratory.

Conclusion

It is concluded from this study that the quality and quantity of the food have been found to have influence on reduviids. *R. longifrons* can be reared in the laboratory with 10 individuals/1L container with 1 alive *S. litura* larvae/predator/day. Heat killed larvae of *C. cephalonica* can also be used during the non-availability of *S. litura*. Frozen *C. cephalonica* larvae can be used to maintain the predator culture during hard and critical times. Moreover, in the post study, *R. longifrons* life stages (IV, V instars and adults) consumed more *S. litura* larvae than *H. armigera* larvae and the weight gain were also high in that category. The predator reared on a particular prey type with higher reproductive potential and adult longevity and low developmental time is indirectly considered the most suited prey for augmentation of reduviids.

Acknowledgement

Senior author thanks Department of Biotechnology, New Delhi for the financial assistance provided to him. We are grateful to the management and head of the department for the laboratory facilities.

References

Ambrose, D.P., 1980. Bioecology, ecophysiology and ethology of reduviids (Heteroptera) of the scrub jungles of Tamil Nadu, India. *Ph.D. Thesis*, University of Madras, Madras, India.

Ambrose, D.P., 1999. *Assassin Bugs*. Science Publishers, Inc., Enfield, New Hampshire, U.S.A.

Ambrose, D.P. 2003. Bio control potential of assassin bugs. *J. Exp. Zool.*, India, 6(1): 1–44.

Ambrose, D.P. and Subbarasu, P.A., 1988. Prey influence on the development, reproduction and size of the assassin bug *Acanthaspis pedestris. Environ. and Ecol.*, 6(4): 948–955.

Ambrose, D.P., Saju, T. and Sahayaraj, K., 1990. Prey influence on the development, reproduction and size of the assassin bug *Rhynocoris marginatus. Ibid.*, 8(1): 280–287.

Antony, M., Daniel, J., Kurian, C. and Pillai, G.B., 1979. Attempts in introduction and colonization of the exotic reduviid predator *Platymeris laevicollis* Distant for the biological suppression of the coconut rhinocerous beetle, *Oryctes rhinoceros. Proc. Plant. Crops. Symp.*, 11: 445–454.

Claver, M.A., 1998. Biological control potential evaluation of the predator, *Rhynocoris kumarii* Ambrose and Livingstone (Hemiptera : Reduviidae) against three selected cotton insect pests. *Ph.D. Thesis*, M.S. University, Tirunelveli, India.

Claver, M.A., Ramasubbu, G., Ravichandran, B. and Ambrose, D.P. 2002. Searching behaviour and functional response of *Rhynocoris longifrons* (Stal) (Heteroptera : Reduviidae), a key predator of pod bug, *Clavigralla gibbosa* Spinola. *Entomon*, 27(4): 339–346.

Cohen, A.C., 1993. Organisation of digestion and preliminary characterization of salivary trypsin like enzymes in a predaceous heteropteran, *Zelus renardii*. *J. Insect Physiol.*, 39: 823–829.

Das, S.S.M., 1996. Biology and behaviour of chosen predatory hemipterans. *Ph.D. Thesis*, Madurai Kamaraj University, Madurai, India.

Fliert, V.E., 1998. Integrated Pest Management: Springboard to Sustainable Agriculture. In: *Critical Issues in Insect Pest Management*, (Eds.) G. S. Dhaliwal and E.A. Heinrichs. Commonwealth Publishers, New Delhi, India, pp. 250–266.

Fuller, B.W., 1988. Predation by *Calleida decora*, F. (Coleoptera : Carabidae) on velvet bean caterpillar (Lepidoptera : Noctuidae) in soybean. *J. Econ. Ent.*, 81: 127–129.

George, P.J.E., 1999. Biology and life table of *Rhynocoris marginatus* (Fabricius) (Heteroptera : Reduviidae) on three lepidopteran insect pests. *J. Biol. Cont.*, 13: 33–38.

George, P.J.E. 2000a. The intrinsic rate of natural increase of a harpactorine reduviid *Rhynocoris kumarii* Ambrose and Livingstone to three lepidopteran insect pests. *Entomon.*, 25(4): 281–286.

George, P.J.E. 2000b. Nymphal cannibalistic bahaviour of *Rhynocoris kumarii* Ambrose and Livingstone (Heteroptera : Reduviidae) to prey deprivation and mass rearing conditions. *J. Biol. Cont.*, 14(1): 1–3.

George, P.J.E. 2000c. Impact of antennectomy on the predatory efficiency of *Rhoyocoris marginatus* (Fab.) (Heteroptera : Reduviidae). *Adv. in Biosci.*, 19(11): 133–140.

George, P.J.E., Kannagi, J. and Ambrose, D.P. 2002. Nutritional influence of prey on the biology and biochemistry in *Rhynocoris marginatus* (Fabricius) (Heteroptera : Reduviidae). *J. Biol. Cont.*, 16(1): 1–4.

George, P.J.E., Seenivasagan, R. and Kannan, S., 1998. Influence of prey species on the development and reproduction of *Acanthaspis siva* Distant (Heteroptera : Reduviidae). *Entomon.*, 23(4): 313–319.

Grundy, P. and Maelzer, D. 2000. Predation by the assassin bug *Pristhesancus plagipennis* (Walker) (Hemiptera : Reduviidae) of *Helicoverpa armigera* (Hubner) (Lepidoptera : Noctuiidae) and *Nezara viridula* (L.) (Hemiptera : Pentatomidae) in the laboratory. *Aus. J. Ent.*, 39: 280–282.

Grundy, P.R., Maelzer, D.A., Bruce, A. and Hassan, E. 2000. A mass-rearing method for the assassin bug *Pristhesancus plagipennis* (Walker) (Hemiptera : Reduviidae). *Biol. Cont.*, 10: 243–250.

Howe, R.N., 1953. The rapid determination of intrinsic rate of increase of an insect population. *Ann. Appl. Biol.*, 40: 134–155.

Iheagwam, E.U., 1982. Effects on host plants (legume) on life and population parameters of *Clavigrella tomemtosicollis* (Hemiptera : Coreidae). *Rev. Zool. Africane.*, 96: 912–920.

Jayaraj, S., Ananthakrishnan, T.N. and Veeresh, G.K., 1994. Biological pest control in India: Progress and perspectives. RGICS Project No. 2. Rajiv Gandhi Institute for Contemporary Studies, New Delhi, India, 101 pp.

Klausner, E., Miller, E.R., and Dingle, H., 1980. Nerium oleander as an alternate host plant for South Florida milkweed bug *Oncopeltus fasciatus*. *Econ. Entomol.*, 5: 137–142.

Kumar, S.P., 1993. Biology and behaviour of chosen assassin bugs (Insecta : Heteroptera : Reduviidae). *Ph.D. Thesis*, Madurai Kamaraj University, Madurai, India.

Kumar, S.P. and Ambrose, D.P., 1996. Functional response of two reduviid predators *Rhynocoris longifrons* Stal and *Coranus obscurus* Kirby (Insecta : Heteroptera : Reduviidae) on *Odontotermes obesus* Rambur. In: *Biological and Cultural Control of Insect Pests: An Indian Scenario*, (Ed.) D.P. Ambrose, Adeline Publishers, Tirunelveli, India, pp. 321–327.

Lakkundi, N.H., 1989. Assessment of reduviids for their predation and possibilities of their utilization in biological control. *Ph.D. Thesis*, IARI, New Delhi, India.

Lakkundi, N.H. and Parshad, B., 1987. A technique for mass multiplication of predator with sucking type of mouth parts with special reference to reduviids. *J. Soil Biol. and Ecol.*, 7: 65–69.

Laughlin, R., 1965. Capacity for increase. A useful Population Statistics. *Journal of Animal Ecology*, 34: 77–91.

O'Neil, R.J. and Wiedenmann, R.N., 1990. Body weight of *Podisus maculiventris* (Say)–under various feeding regimes. *Can. Ent.*, 122: 285–294.

Panizzi, A.R. and Slansky, F., 1985. Legume host impact on performance of adult *Piezodorus guildnii* (Westwood) (Hemiptera : Pentatomidae). *Environ. Entomol.*, 14: 237–242.

Parajulee, M.N. and Philips, T.W., 1993. Effects of prey species on the development and reproduction of the predator *Lyctocores campestris* (Heteroptera : Anthocoridae). *Ibid.*, 25(5): 1035–1042.

Sahayaraj, K., 1991. Bioecology, Ecophysiology and Ethology of chosen predatory hemipterans and their potential in biological control (Insecta : Heteroptera : Reduviidae). Ph.D. Thesis, Madurai Kamaraj University, Madurai, India.

Sahayaraj, K., 1998. Laboratory rearing of predaceous bugs with special reference to reduviids (Insecta : Hemiptera : Reduviidae). *Zoo's Print*, 13(5): 17–18.

Sahayaraj, K., 1999. Effect of prey and their age on the feeding preference of *Rhynocoris marginatus* (Fab.). *Int. Arachis Newsletter*, 19: 39–41.

Sahayaraj, K. 2002a. Biopesticide: An Indian scenario. *Agrobios*, 1(7): 26–29.

Sahayaraj, K. 2002b. Small scale laboratory rearing of a reduviid predator, *Rhynocoris marginatus* Fab. (Hemiptera : Reduviidae) on *Corcyra cephalonica* Stainton larvae by larval card method. *J. Cent. Euro. Agri.*, 3(2): 137–147.

Sahayaraj, K. 2002c. Field bioefficacy of a reduviid predator *Rhynocoris marginatus* and plant products against *Aproaerema modicella* and *Spodoptera liltura* of groundnut. *Indian J. of Ent.*, 64(3): 292–300.

Sahayaraj, K. 2004. Reduviids in Biological Control; In: *Indian Insect Predators in Biological Control*, (Ed.) Sahayaraj, K. Daya Publishing House, New Delhi, India, pp. 134–176.

Sahayaraj, K. and Jeyalakshmi, T. 2002. Mass rearing of *Rhynocoris marginatus* Fab. on live and frozen larvae of *Corcyra cephalonica* Stainton. *Entomol. Croat.*, 6(1–2): 35–49.

Sahayaraj, K. and Paulraj, M.G. 2001. Rearing and life table of reduviid predator *Rhynocoris marginatus* Fab. (Heteroptera : Reduviidae) on *Spodoptera litura* Fab. (Lep., Noctuidae) larvae. *J. Appl. Ent.*, 125: 321–325.

Southwood, T.R.E., 1978. *Ecological Methods: With Particular Reference to the Study of Insect Population.* Chapman and Hall, London.

Venkatesan, S., Sennivasagan, R. and Karuppasamy, G., 1997. Influence of prey species on feeding response, development and reproduction of the reduviid *Cydnocoris gilvus* Burm. (Reduviidae, Heteroptera). *Entomon.*, 22(1): 21–27.

Watson, T.F., 1964. Influence of host plant condition on population increase of *Tetrancychus telerium* (L.) (Acaina : Tetranychidae). *Hilgardia*, 35: 273–322.

Yasuda, T. and Wakamura, S. 1991. Rearing of predatory stink bug *Eocanthecona furcellata* Wolff on frozen larvae of *Spodoptera litura* (Fab.). *Appl. Entomol. Zoologi.*, 27: 303–305.

Zanuncio, J.C. Alves, J.B.C. Sartorio, R.C. and Leite, J.E.M., 1992. Methods for rearing larval predators from the Hemiptera. *Anais-da-Sociedade-Entomologica-do-Brasil*, 21(2): 245–251.

Chapter 4

Studies on the Biodiversity and Population Density of Freshwater Pulmonate Malacofauna of Damoh and Sagar Districts of M.P. (India)

M. Bhide, Mohd. A. Khan, U. Dubey, P. Gupta, P. Thakur, P. Nema and S. Jain*

Department of Zoology, Dr. Hari Singh Gour University, Sagar – 470 003, M.P.

ABSTRACT

The present investigation was done on the 2264 specimens of pulmonate molluscs which were collected seasonally from the freshwaters of Damoh and Sagar districts of M.P., India for two consecutive calendar years *i.e.* 2000–2001 to 2001–2002 by scoop nets or by hand picking from the sites of molluscan habitats. After collection Pulmonate malacofauna was identified and classified in to 6 families, 12 genera and 21 species.

The most common pulmonates were *Lymnaea auricularia, Lymnaea acuminata, Lymnaea luteola* and *Indoplanorbis exustus* while *Gyraulus* species were uncommon and procured only from lentic water bodies of Damoh and Sagar districts. All the specimens of family–Ariophantidae were collected from stagnant water bodies and this family seems to be very large

* Corresponding author: E-mail: manglabhide@yahoo.com.

and also exhibited the high population density among surveyed gastropods. Member of family Lymnaeidae and Planorbidae were abundant which are known intermediate hosts for the life history of liver- and blood flukes while most of the non-pathogenic snails provided palatable dishes to natives solving to the some extent problem of protein malnutrition.

Keywords: *Biodiversity and population density of pulmonate malacofauna.*

Introduction

Almost all gastropods are aquatic and great majority of freshwater form occur in shallow bank of rivers. The biodiversity of freshwater gastropod fauna is precious for research of the corresponding area which provide information about the diversified environment of the concerning area. The freshwater pulmonate gastropod fauna constituted significant aspect of ecosystem. The different topographical areas give rise to very rich diversified fauna. Many species of freshwater pulmonate fauna having socio-economic importance, as intermediate hosts of number of disease causing parasitic helminthes, and also as exploitable food resource of men and duckeries as well.

Study of the Biodiversity of freshwater Pulmonate malacofauna is the valuable and research of the corresponding area which gives us an idea about the diversified environment of the concerning areas. The freshwater *Pulmonate* malacofauna constitutes an important part of ecosystem. They have significant participation with the ecological communities. Study of Biodiversity of fauna and their identification is one of the interesting fields of biological researchers which give us an idea about the environment of any particular area. The different topographical areas give rise to very rich diversified fauna (Odum, 1987).

Many scientists and researchers have done work on the terrestrial and aquatic biodiversity, diversified flora and fauna and even variation in species as surveyed by Tonapi (1980), Rao (1989), Sedgwick (1990) Crossley (1993), Hobdy (1993), Caldwell *et al.* (1995), Adoni (1995), Webb *et al.* (1995), Bour *et al.* (1996), Sharma and Sahai (1996), Emberton (1997), Minelli *et al.* (1997), Khan (1999), Khan (2002), Mohanta and Mahanta (2002), Bhide (2003), Bhide *et al.* (2003) and Bhide *et al.* (2004) but the work on freshwater biodiversity and population density of *Pulmonate* malacofauna has not been yet done in this specific regions of Madhya Pradesh (India) so this investigation was done to know more about these aspects of *Pulmonate* malacofauna of Damoh and Sagar districts of India.

Materials and Methods

Procurement of Pulmonate Malacofauna

Freshwater Pulmonate malacofauna is generally found adhering to the aquatic vegetation, floating objects or stones in stagnant waters *e.g.* ponds and lakes and also procured from freshwater rivers *e.g.* Sunar, Kopra, Bebas and Dhasan. They are also common in paddy fields and irrigation canals. So for the present investigation we have collected the maximum pulmonate specimens from their natural habitats of Damoh and Sagar districts for two consecutive calendar year (2000–2001 to 2001–2002). The procured pulmonate malacofauna was identified and classified according to the method and Key provided by Tonapi (1980), Rao (1989) and Hall (1996) summarized in Table 4.1 and analyzed in Table 4.2. The population density of the identified fauna was studied and summarized in Graphs 4.1–4.5 (method adapted after Odum, 1987).

Table 4.1: Classification, Population Density and Analysis of Data of Surveyed Pulmonate Malacofauna

Sl.No.	*Types of Malacofauna*	*Number of Individuals*	*Percentage*
	Class–Gastropoda		
	Subclass–Pulmonata		
	Order–Basommatophora		
1.	Family–Lymnaeidae		
	1. Genus–*Lymnaea*		
	Subgenus–Pseudosuccinea		
	Species		
	(i) *Lymnaea* (*Pseudosuccinea*) *acuminata*	135	20.36%
	(ii) *Lymnaea* (*Pseudosuccinea*) *acuminata rufescens*	90	13.57%
	(iii) *Lymnaea* (*Pseudosuccinea*) *biacuminata*	198	29.86%
	(iv) *Lymnaea* (*Pseudosuccinea*) *luteola*	110	16.59%

Contd...

Table 4.1–Contd...

Sl.No.	Types of Malacofauna	Number of Individuals	Percentage
	Subgenus–*Radix*		
	(v) *Lymnaea* (*Radix*) *auricularia*	55	8.29%
	Subgenus–Lymnaea		
	(vi) *Lymnaea* (*Lymnaea*) *stagnalis*	75	1.3%
	Total	**663**	**29.28%**
2.	Family–Planorbidae		
	Subfamily–Bulininae		
	1. Genus–*Indoplanorbis*		
	Species		
	(i) *Indoplanorbis exustus*	101	30.15%
	Subfamily–Planorbinae s.str.		
	2. Genus–*Gyraulus*		
	Species		
	(i) *Gyraulus convexiusculus*	75	22.39%
	(ii) *Gyraulus labiatus*	94	28.06%
	(iii) *Gyraulus rotula*	40	11.93%
	(iv) *Gyraulus auricularia*	25	7.46%
	Total	**335**	**14.80%**
3.	Family–Ariophantidae		
	1. Genus–*Macrochlamys*		
	Species		
	(i) *Macrochlamys pedina*	145	17.51%
	(ii) *Macrochlamys infausta*	114	13.77%
	2. Genus–*Ariophanta*		
	Species		
	(i) *Ariophanta bistrialis*	145	17.51%
	3. Genus–*Sitala*		
	Species		
	(i) *Sitala denselirata*	175	21.14%

Contd...

Table 4.1–Contd...

Sl.No.	Types of Malacofauna	Number of Individuals	Percentage
	4. Genus–*Kaliella*		
	Species		
	(i) *Kaliella bullula*	130	15.70%
	5. Genus–*Xesta*		
	Subgenus–*Fretum*		
	Species		
	(i) *Xesta* (*Fretum*) *semirugata*	119	14.37%
	Total	**828**	**36.57%**
4.	Family–Helicidae		
	1. Genus–*Eulota*		
	Species		
	(i) *Eulota Saclpturita*	164	7.25%
	Total	**164**	**7.25%**
5.	Family–Subulinidae		
	1. Genus–*Subuliana*		
	Species		
	(i) *Subuliana octona*	120	52.00%
	2. Genus–*Opeas*		
	Species		
	(i) *Opeas gracile*	110	48.00%
	Total	**230**	**10.16%**
6.	Family–Glessulidae		
	1. Genus–*Glessula*		
	Subgenus–*Glessula*		
	Species		
	(i) *Glessula* (*Glessula*) *notigena*	44	1.94%
	Total	**44**	**1.94%**

Analysis of Pulmonate Malacofauna

1. Family: 06; 2. Genera: 12; 3. Species: 21.

Following articles were used in the collection of Pulmonate malacofauna:

1. Hand net or a scoop net
2. A small piece of cloth
3. A few cloth or polythene bags
4. Specimen bottles (1 liter capacity)
5. A thermometer for the record of temperature of the concerning water bodies
6. A pair of blunt forceps
7. Enamel trays
8. pH meter for the record of pH of the concerning water bodies
9. Necessary chemicals *e.g.* fixatives and preservatives etc.
10. Field note book and data sheets
11. Appropriate labels etc.
12. A good camera to click the photograph of diversified pulmonate Gastropod fauna.

This investigation was carried out on the pulmonate specimens collected from the different desired sites of the stagnant and running freshwater bodies of Damoh and Sagar districts as follows:

Ponds and Lakes

Sl.No.	*Name of Ponds and Lakes*	*Location*
1.	Belatal	Damoh
2.	Futeratal	Damoh
3.	Purana talab	Damoh
4.	Quidi pond	Damoh
5.	Ramji pond	Damoh
6.	Rajnagar pond	Damoh
7.	Sagar lake and Botanical ponds	Sagar

Rivers

Sl.No.	*Name of Rivers*	*Location*
1.	Kopra	Damoh
2.	Sunar	Damoh
3.	Bebas	Sagar
4.	Dhasan	Sagar

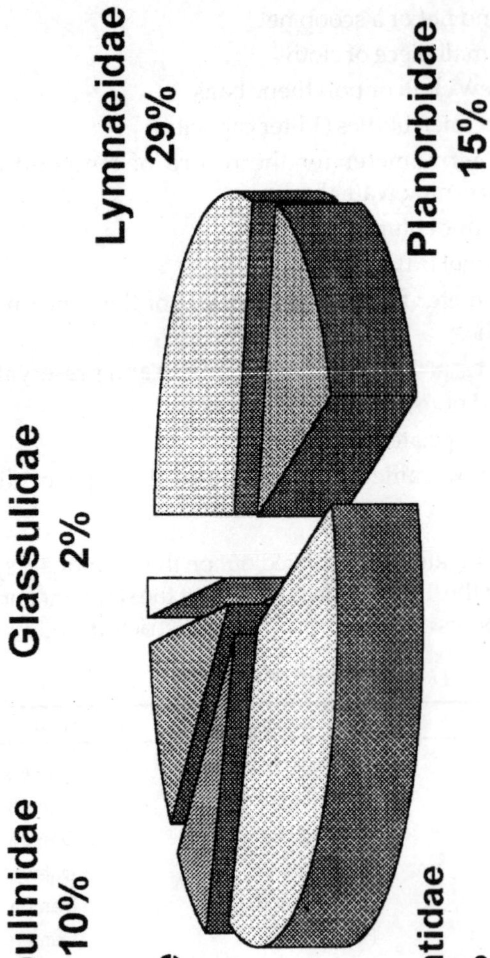

Graph 4.1: Population Density in Pulmonate Gastropod Molluscan Fauna

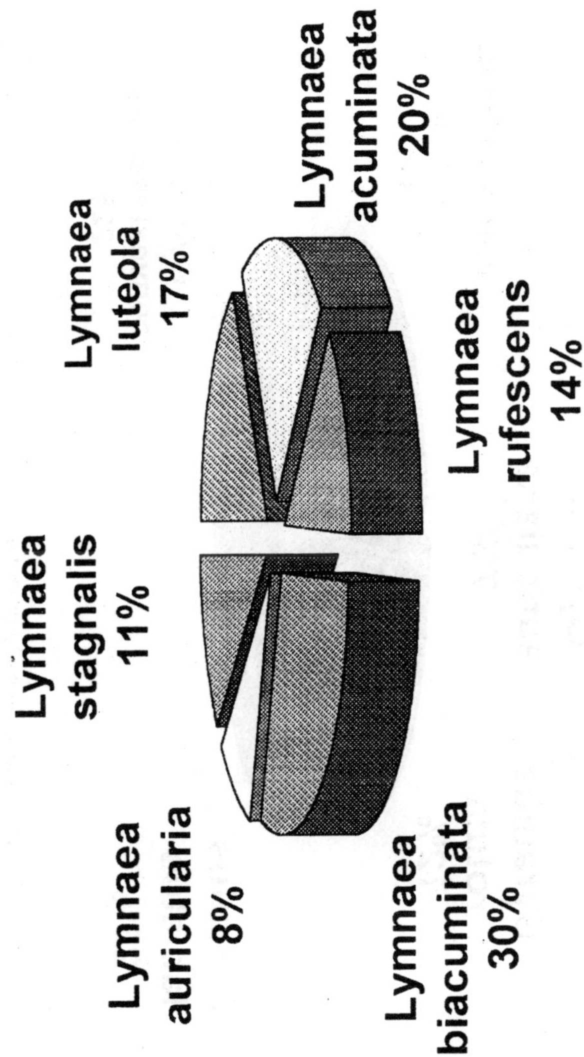

Graph 4.2: Population Density of Different Species of Lymnaea Belonging to Family–Lymnaeidae

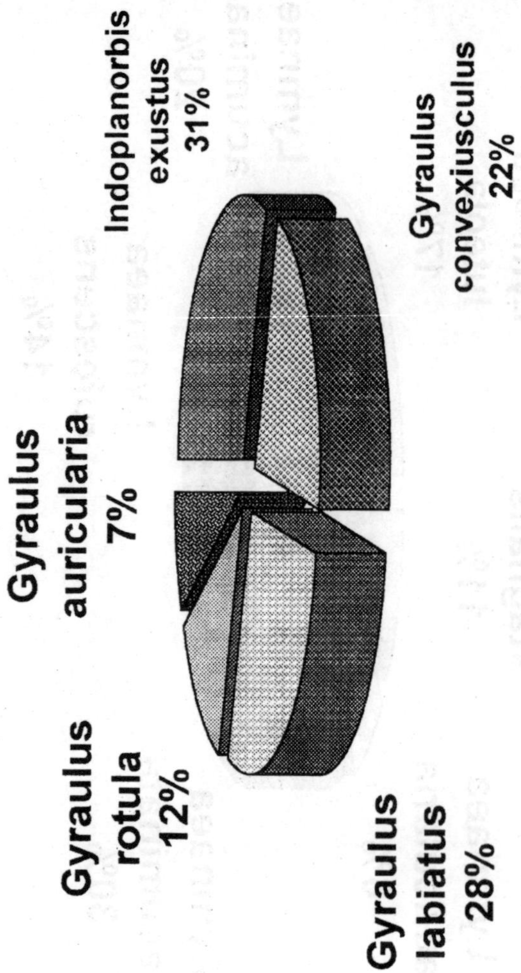

Graph 4.3: Population Density of Different Species of Family–Planorbidae

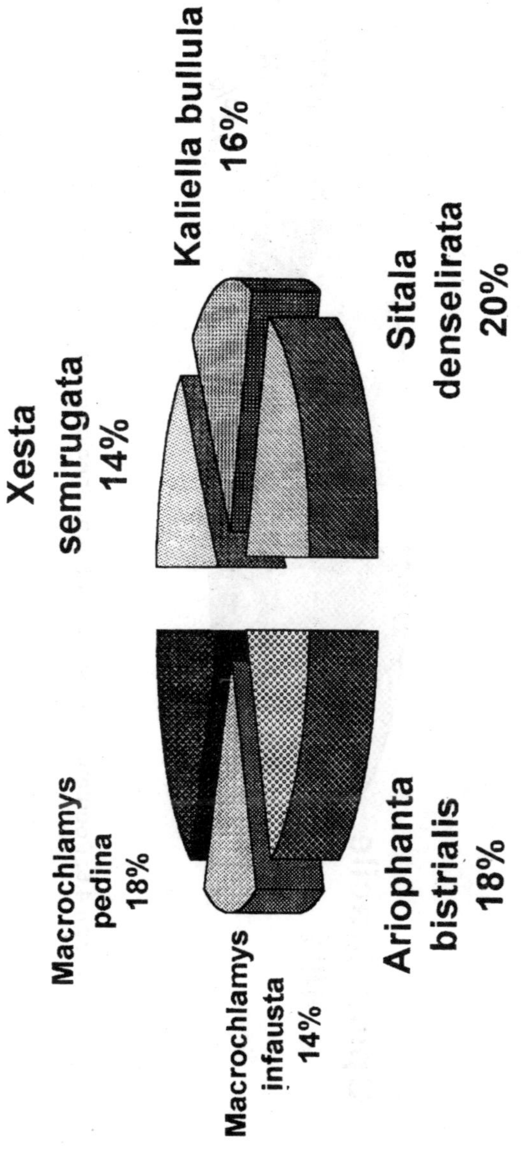

Graph 4.4: Population Density of Different Species of Family–Ariophantidae

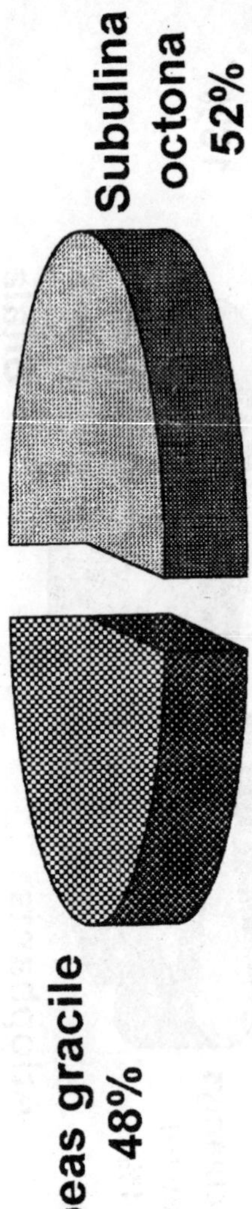

Graph 4.5: Population Density of Different Species of Family–Subulinidae

Table 4.2: Analysis of Data Surveyed on Pulmonate Malacofauna

Sl.No.	*Family*	*Specimens*	*Percentage*	*Genera*	*Species*
1.	Lymnaeidae	663	29.28%	01	06
2.	Planorbidae	335	14.80%	02	05
3.	Ariophantidae	828	36.57%	05	06
4	Helicidae	164	7.25%	01	01
5.	Subulinidae	230	10.16%	02	02
6.	Glessulidae	44	1.94%	0.1	01
	Total	**2264**	**100%**	**12**	**21**

Results and Discussion

The biodiversity of any area exhibited its environmental conditions (Odum, 1987; Adoni, 1995). Rich biodiversity and population density is the indication of restoration of good environment enrich with the palatable food of diversified fauna as reported by Crossley (19931. In woody debris in Southern forests, Emberton (1997) in South Eastern Madagascar, Webb *et al.* (1995) in Illinois spring (USA) while Hobdy (1993) noticed the loss of biodiversity on a small Hawaiian island due to the deterioration of the environmental conditions of the corresponding area.

All the specimens of Pulmonate were identified and classified up to species by the method adopted by Tonapi (1980), Rao (1989) and Hall (1996) because the species is the unit of biodiversity (Sedgwick, 1990 and Minelli *et al.* (1997). Some research work has been done on the biodiversity of Macroinvertebrate fauna by Caldwell *et al.* (1993), Baur *et al.* (1996), Sharma and Sahai (1996), Minelli *et al.* (1997) and have reported the presence of malacofauna along with some large sized invertebrate fauna. In the present investigation subclass–Pulmonata exhibited the abundance of numbers of only six families while no specimen of family "Ancylidae" and "Physidae" has been observed.

The temperature of all water bodies season to season varies. The minimum temperature was 12°C and maximum was 38–40°C. The pH of the water bodies also varies season to season; it ranges from pH 6.4 to pH 8.5.

All the collected Pulmonate malacofauna is classified as follows:

Phylum–Mollusca

Class–Gastropoda

Subclass–Pulmonata

Order–Basommatophora: Freshwater pulmonates in which eyes are located at the base of the tentacles and an external shell is present.

Six families have been surveyed and identified as follows:

Key to Families

Family–Lymnaeidae

Shell dextrally coiled, columellar axis typically twisted, tentacles triangular.

Family–Planorbidae

Shell generally discoidal or when sinistrally coiled, columellar axis not twisted extended and tentacles filiform.

Family–Ariophantidae

It is universally distributed family. The shell is translucent, sub-globose, often depressed, horny olivaceous with fine striae. The ground colour is sometimes superimposed by darker bands. The aperture is angular with their peristome. The umbilicus is relatively wide and conspicuous. The animals were capable of retraction into their shells. A caudal gland is usually present.

Family–Helicidae

This family is concerned with the highly evolved group of the pulmonates. The shell was conoid, broad with whorls increasing rapidly. The aperture was dilated and oblique to the main axis. The margin was fused by a thin callus. The umbilicus is moderately open and more or less perspective. The shells have fawn colour often with a belt of brown band.

Family–Subulinidae

The shell is uniformly coloured and no internal lamellae are present in the shell. The shell is variable in shape and size, small tower like as in Opeas or cylindrical pupiform like. The number of whorls differs in different genera.

Family–Glessulidae

These were common amphibious molluscs found in gardens and parks in wet humus of decaying leaves and litter in the rainy season. The shell was imperforate, elongated, copical or pupiform. The number of whorls may vary from seven to ten. The aperture was obliquely ovate with truncated columella. The shells were horny brown with fine crenulations. The molluscan fauna of this otherwise obscure group been thoroughly studied has found in lentic water bodied of Damoh and Sagar districts in the present investigation.

Family–Lymnaeidae

Genus–Lymnaea (Lamarck): Shell generally thin, whorls large, spire almost always exerted, columella spirally twisted.

Subgenus–Pseudosuccinea: Shell thin and spirally twisted.

Key to species: The outer lip was very much expanded and convex in outline. Spire acuminate.

Lymnaea (*Pseudosuccinea*) *acuminata* (Lamarck)

The nature of spiral course of the shell, the form of outer lip and a number of negative characters were used in the specific diagnosis of the shell. Found in all lentic water bodies of Damoh and Sagar districts (Figure 4.1).

Lymnaea (*Pseudosuccinea*) *acuminata rufescens* (Gray)

Shell with 3–4 whorls which gradually increase in size from apex to body whorl. Found in lentic water bodies of Damoh and Sagar districts (Figures 4.2–4.4).

Lymnaea (*Pseudosuccinea*) *biacuminata* (Annandale)

It is a phenotypic variant of *Lymnaea acuminata*. Found in all lentic water bodies of Damoh and Sagar districts (Figures 4.3–4.4).

Lymnaea (*Pseudosuccinea*) *luteola* (Lamarck)

Spire not very much pointed and the shell more ovate with a narrow ovate aperture. Found in lentic water bodies of Damoh and Sagar districts (Figures 4.3, 4.4 and 4.6).

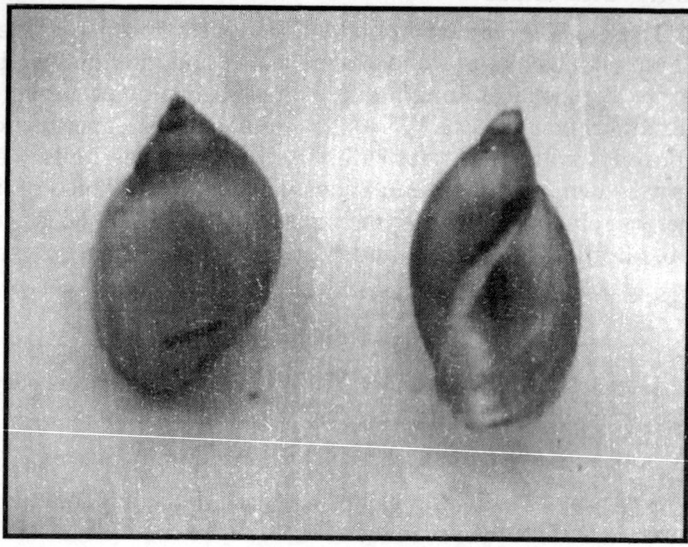

Figure 4.1: *Lymnaea (Pseudosuccinea) acuminata*

Figure 4.2: *Lymnaea (Pseudosuccinea) acuminata rufescens*

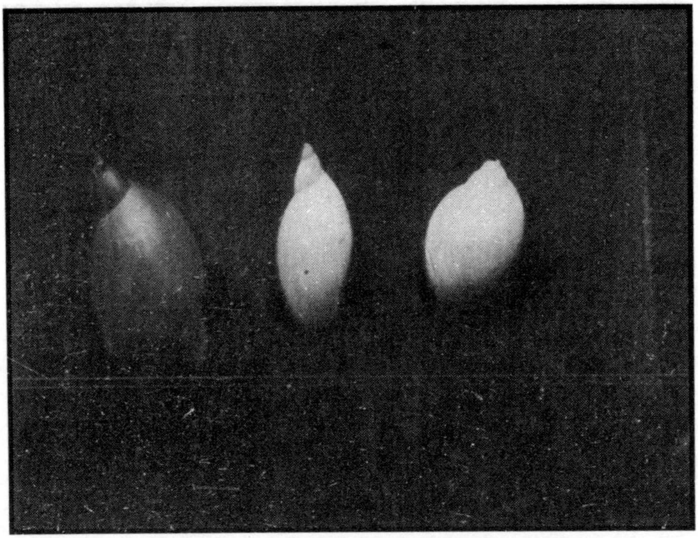

Figure 4.3: ***Lymnaea (Pseudosuccinea) acuminata rufescens, Lymnaea (Pseudosuccinea) biacuminata, Lymnaea (Pseudosuccinea) luteola***

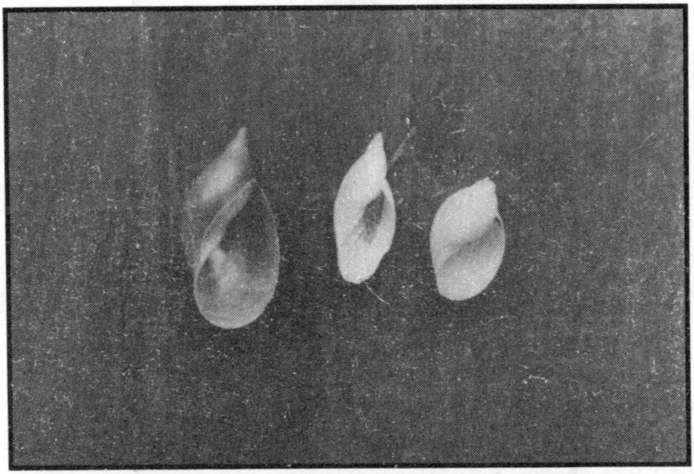

Figure 4.4: ***Lymnaea (Pseudosuccinea) acuminata rufescens, Lymnaea (Pseudosuccinea) biacuminata, Lymnaea (Pseudosuccinea) luteola***

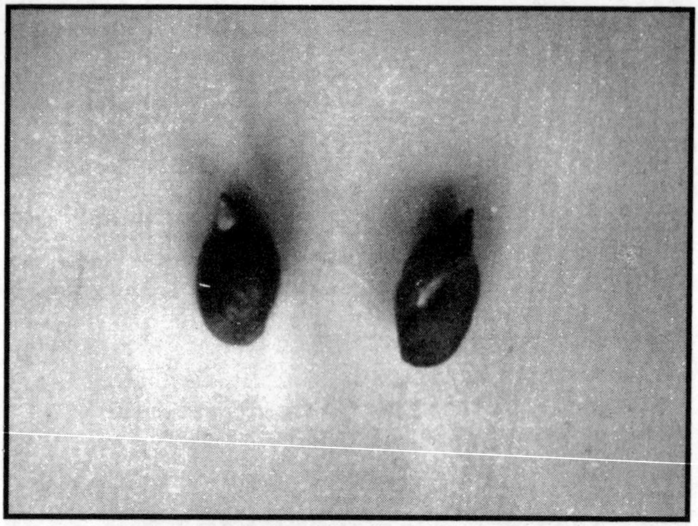

Figure 4.5: ***Lymnaea*** **(*Radix*)** ***auricularia***

Figure 4.6: ***Lymnaea*** **(*Pseudosuccinea*)** ***luteola,***
Lymnaea **(*Lymnaea*)** ***stagnalis***

Subgenus–*Radix*

Lymnaea (*Radix*) *auricularia* (Linnaeus)

Shell with four whorls. Outer chamber large and operculum margin triangular and showed zig-zag condition. Shell bigger, inflated, columella very much twisted, fold occluded umbilicus, spire shrunken into body whorl. Found in lentic water bodies of Damoh and Sagar districts (Figure 4.5).

Subgenus–*Lymnaea*

Lymnaea (*Lymnaea*) *stagnalis* (Linnaeus)

A very distinctly elongated form with six whorls, pointed spire, outer lip sharp and not expanded or retroverted, umbilicus closed. Found in lentic water bodies of Damoh and Sagar districts (Figures 4.6 and 4.7).

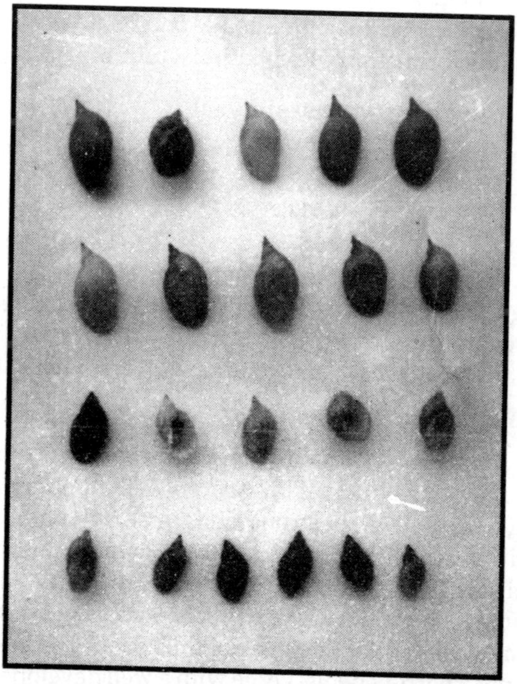

Figure 4.7: ***Lymnaea* (*Lymnaea*) *stagnalis***

Family–Planorbidae

Shell generally discoidal or when extended sinistrally coiled, columellar axis not twisted, animal with filiform tentacles.

Subfamily–Bulininae

Shell ovoid or elongated and acuminate as in *Camptoceras* or disc shaped as in *Indoplanorbis,* sinistral, when disc shaped distinguish from shells of other subfamilies by the shape of the last whorl, large and dilated forward to the aperture. Branchial process lobed, prostate gland compact and fan shaped, verge introverted and coiled inside the verge sac.

Key to Genera: Shell large, whorls rounded at periphery.

Genus: *Indoplanorbis* (Annandale and Prashad)

Shell large and thick, sutures deeply impressed, aperture ear shaped. Animal sinistral, foot relatively broad and leaf shaped, broadly rounded anteriorly and pointed posteriorly, branchial process lobed. Genital aperture slit like and relatively large, penis long and in the form of simple cylindrical tube but no penial style.

Indoplanorbis exustus (Deshayes)

Characters were same as given for genus. Found in lentic water bodies of Damoh and Sagar districts (Figure 4.8).

Subfamily–Planorbinae

Shell discoidal, whorls distinct from each other, regularly increasing in size, upper and lower size flattened. Prostate duct long and separate from sperm duct, prostate diverticula simple and placed all along the duct. Verge may be with or without style.

Genus: *Gyraulus* (Charpentier)

Shell small always less than 1 cm in diameter, thin, greatly depressed, pale translucent or transparent with three to five rapidly increasing whorls, without strong transverse ribs, with or without short, hair like projections of periostracum, with or without peripheral keel, body whorl at aperture slightly deflected, aperture oblique, simple lip, radula with a bicuspid laterals and several sharp cusps on marginals. Mantle edge not thickened. The vas deferens continue into a narrow penis, verge with a well developed stylet.

Figure 4.8: *Indoplanorbis exustus*

Gyraulus convexiusculus (Hutton)

Maximum diameter rarely exceeded 5 mm whorls 4 to 5, rounded, suture well defined, periphery sub angular, closely and obliquely striated and umbilicus wide. All the whorls were distinctly seen from above. Aperture was ovate or lunate.

Radula with a well developed central bearing sharply pointed cusps, inner laterals with two stout, sharply pointed cusps, outers with three cusps and marginals with 4 to 6 similar and slender cusps. Found in lentic water bodies of Damoh and Sagar districts (Figure 4.9).

Gyraulus labiatus (Benson)

Shell oblique and striated. Whorls were three and half, sutures impressed, aperture oblique heart shaped. Body whorl is with remarkable departure from the axis of previous whorls, whitish rib within the aperture. Found in lentic water bodies of Damoh and Sagar districts (Figure 4.10).

***Gyraulus rotula* (Benson)**

Whorls were three and half, obsolete spiral striations, sutures impressed, periostracum thin, brightly but sometimes lightly tinted greenish, sculptured with longitudinal striae, sometimes strong enough to be called sub costae. Found in lentic water bodies of Damoh and Sagar districts (Figure 4.10).

***Gyraulus velifer* (Annandale)**

Shell transparent with very fine close set regular transverse striae, whorls three and half, upper surface vary slightly convex and the suture impressed, periphery faintly carinate, aperture large, broad, moderately oblique, edge of the lip sharply defined without thickening. Found in lentic water bodies of Damoh and Sagar districts (Figure 4.9).

Family–Ariophantidae

This is universally distributed family. The shell was translucent, sub globose often depressed, horny olivaceous with fine striae. The ground colour was sometimes superimposed by darker bands. The aperture was angular with peristome. The umbilicus was relatively wide and conspicuous. The animal was capable of retraction into shell. A caudal gland was usually present.

Genus–Macrochlamys

The shell was thin, less transparent. Body whorl was dextral.

***Macrochlamys pedina* (Benson)**

Shell is bigger in size. Vertical lines on shell, operculum big size, 5–7 whorls which were dextral. Found in lentic and lotic water bodies of Damoh and Sagar districts (Figure 4.11).

***Macrochlamys infausta* (Blanford)**

Shell with 6–7 whorls. Operculum was curved at one point. Found in lentic and lotic water bodies of Damoh and Sagar districts (Figure 4.11).

Genus–Ariophanta

Shell concave at one side and convex at the other side. Shell was disc like in appearance. Shell having light band at the exterior side.

Figure 4.9: *Gyraulus convexiusculus* and *Gyraulus velifer*

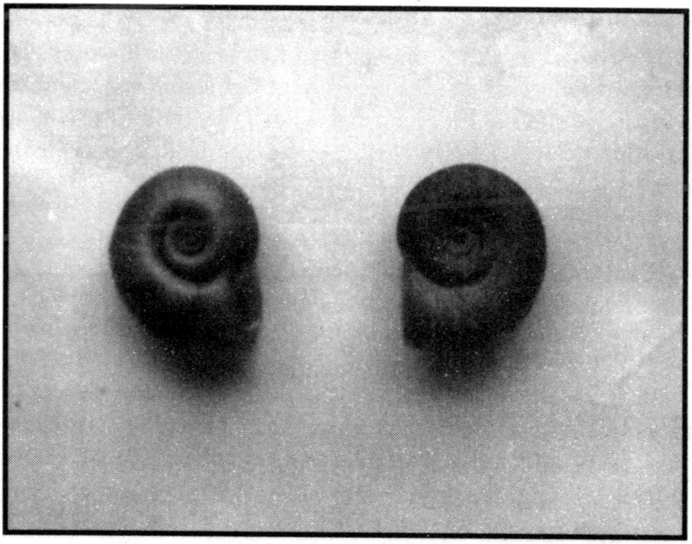

Figure 4.10: *Gyraulus labiatus* and *Gyraulus rotula*

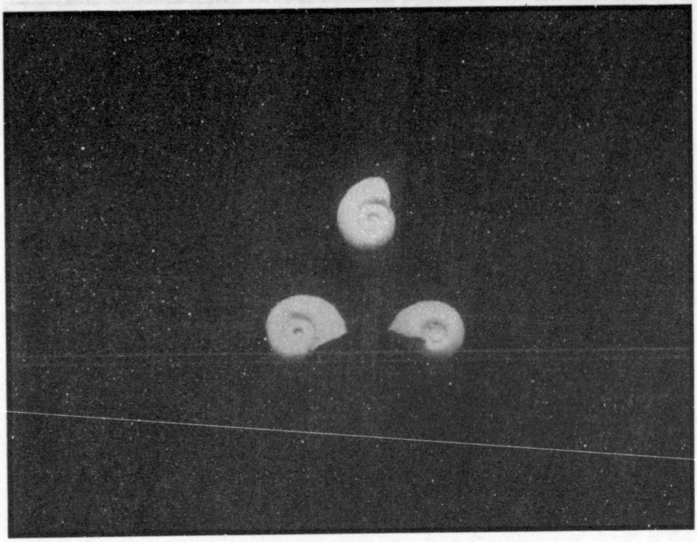

Figure 4.11: *Macrochlamys pedina* and *Macrochlamys infausta*

Ariophanta bistrialis (Beck)

Shell was convex at one side and concave at the other side. Dark bands were present at inner side. A light band is present from the centre of shell to the marginal whorl. Operculum was placed obliquely. Found in lentic water bodies of Damoh and Sagar districts (Figure 4.12).

Genus–Sitala

Shell was transparent, brownish in colour. One side was concave and other side was convex in appearance.

Sitala denselirata (Preston)

Shell was transparent, brownish in colour, one side was concave and other side was convex in appearance. Genus *Sitala* shell was found in all lentic water bodies of Damoh and Sagar districts (Figure 4.13).

Genus–*Kaliella*

Shell was concave at one side and convex at the other side. Whorl started with a pit. Operculum became oblique in condition.

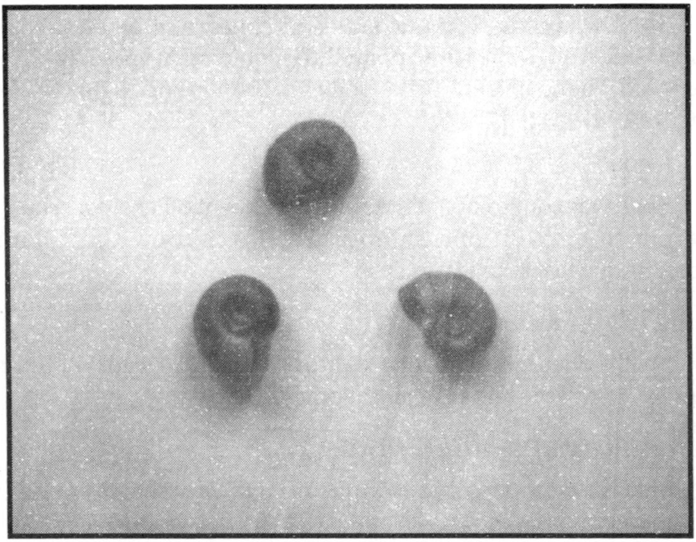

Figure 4.12: ***Ariophanta bistrialis***

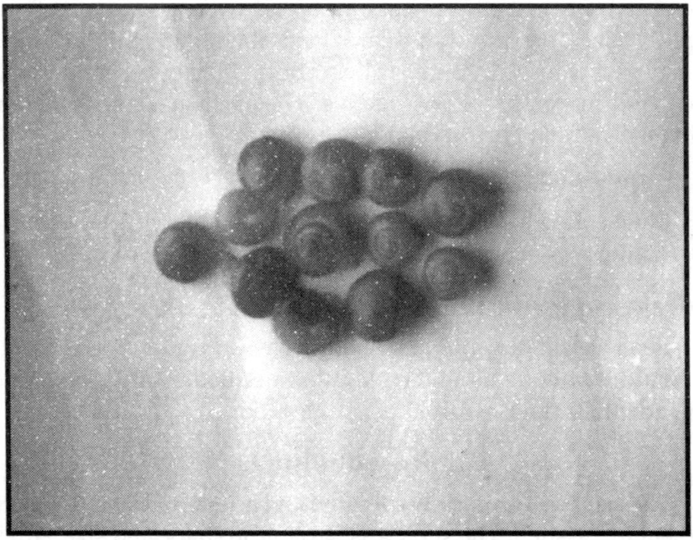

Figure 4.13: ***Sitala denselirata***

Kaliella bullula (Hutton)

Shell was convex at one side and concave at other side. At dorsal side from the starting point three equidistant lines have been observed. Found in lentic water bodies of Damoh and Sagar districts (Figure 4.14).

Genus–*Xesta*

Shell was convex at one side and concave at other side. Whorl's margin fused with one another. Shell was brown in colour. Operculum was obliquely placed.

Subgenus–*Fretum*

Shell was brightened, operculum obliquely placed. Shell was convex at one side and concave at other side.

Xesta (*Fretum*) *semirugata* (Beck)

Shell fused, operculum oblique margin very sharp. Found in lentic water bodies of Damoh and Sagar districts (Figure 4.15).

Family–Helicidae

This family is considered to be a highly evolved group of the pulmonate. The shell was conoid, broad with whorls increasing rapidly. The aperture was dilated and oblique to main axis. The margins were fused by a thin callus. The umbilicus was moderately open and more or less perspective. The shells have fawn colour often with a belt of brown band.

Genus–*Eulota*

Shell coiled fused and one side concave and one side convex, operculum was in oblique conditions.

Eulota scapturita (Benson)

Shell has pitate center, one side convex and other side concave, operculum oblique. Found in lentic water bodies of Damoh and Sagar districts (Figure 4.16).

Family–Subulinidae

The shell was uniformly coloured and no internal lamellae were present in the shell. The shell was variable in shape and size, small tower like as in *Opeas* or cylindrical pupiform like in *Zootecus* (Rao, 1989). The number of whorls differs in different genera. The columella

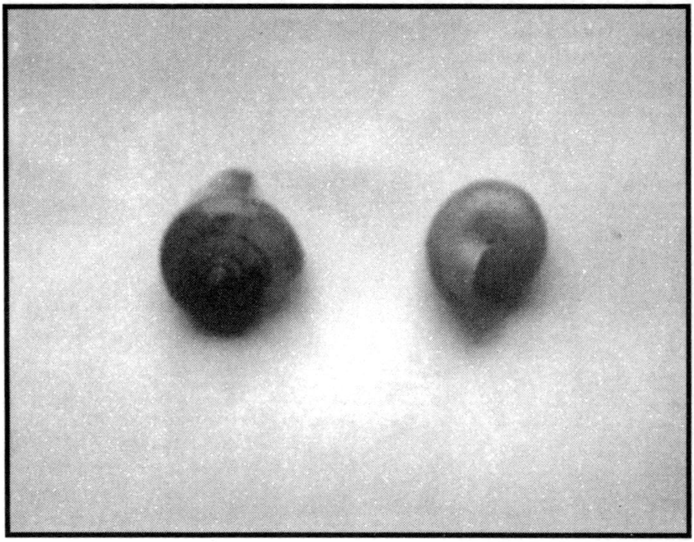

Figure 4.14: *Kaliella bullula*

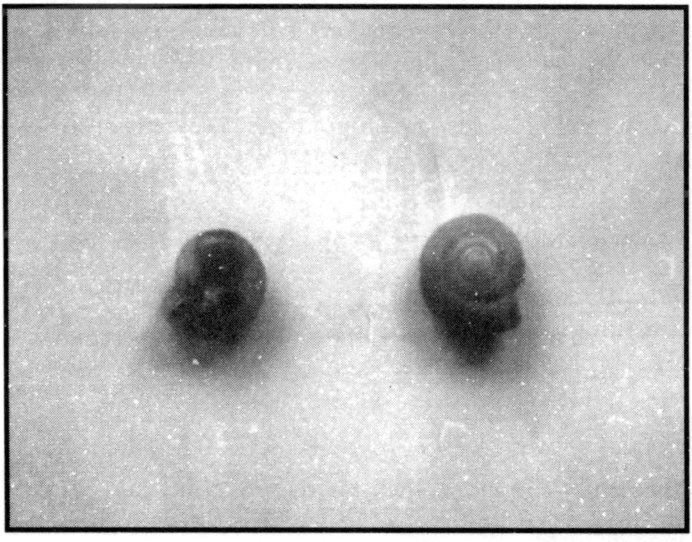

Figure 4.15: *Xesta* (*Fretum*) *semirugata*

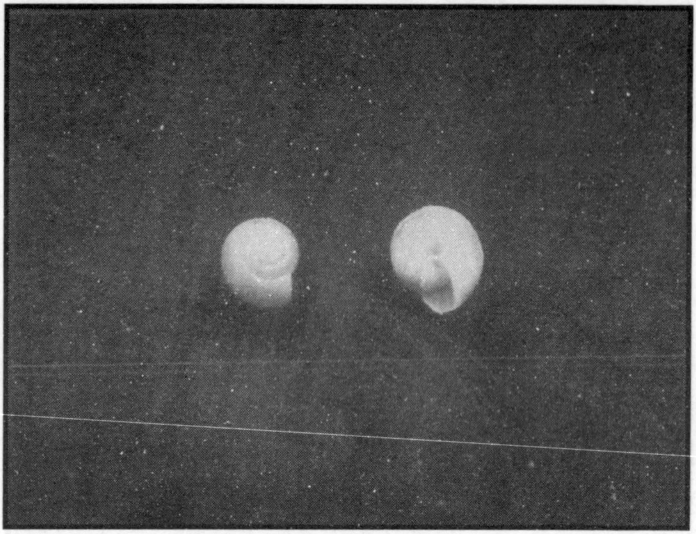

Figure 4.16: *Eulota scalpturita*

was not simple, it was either truncated or expanded at its lower margin. The aperture was semiovate in shape. The members of the family were found either in rivers *e.g.* in Kopra, Sunar, Bebas, and Dhasan or lying in vicinity of aquatic environment. In the wet rainy season several of these were found in the decaying slushy leaves rich in humus under luxuriant vegetations.

Genus–*Subulina*

Shell whorls were 10–12 in numbers.

Subulina octona

Body whorls 11 in number, operculum at one side rounded and at other side pointed or mango shaped. Found in lentic water bodies of Damoh and Sagar districts (Figure 4.17).

Genus–*Opeas*

Operculum rounded, shell whorls 8–10 in number.

Opeas gracile

Shell whorls were 10 in numbers, operculum was rounded. Found in lentic water bodies of Damoh and Sagar districts (Figure 4.18).

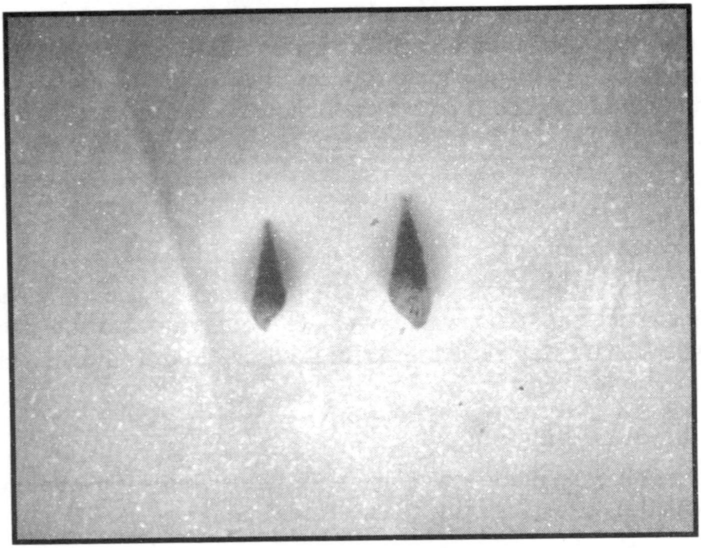

Figure 4.17: *Subulina octona*

Figure 4.18: *Opeas gracile*

Family–Glessulidae

These were common amphibious molluscs found in gardens and parks in wet humus decaying leaves and litter. The shell was imperforate, elongated and pupiform. The number of whorls may vary from 7–10. The aperture was obliquely ovate with truncated columella. The shells were horny brown with fine crenulations. The molluscan fauna of such group found in lentic water bodies of Damoh and Sagar districts has not been thoroughly studied.

Genus–*Glessula*

The shell was imperforate, elongated, conical or pupiform. The number of whorls might vary from 7–10. The aperture was obliquely ovate with truncated columella. The shells were horny brown with fine crenulations.

Subgenus–*Glessula*

Shell elongated, tower like with 9–10 whorls. Operculum pointed at both sides while convex in the middle.

Glessula (Glessula) notigena

Shell with 7--10 whorls smooth and conical. Operculum pointed at both the ends. Found in lentic water bodies of Damoh and Sagar districts (Figure 4.19). All the specimens of pulmonates procured in the present investigation from different freshwater bodies of Damoh and Sagar districts are mentioned in the freshwater fauna of India by Tonapi (1980) and Rao (1989).

After identification of diversified pulmonate molluscan fauna some significant points investigated are as follows:

1. Most of the investigated species of pulmonates *e.g.* of family Lymnaeidae, Planorbidae, Ariophantidae, Helicidae, Subulinidae and Glessulidae have been procured only from lentic freshwater bodies of Damoh and Sagar districts and a few of them have been procured from the lotic freshwater bodies.
2. Member of family–Lymnaeidae and Planorbidae were abundant and easily reared in the laboratory while members of other families have not survived in the laboratory and died within 24 hrs and their mortality rate was very high in the laboratory so it is very difficult to do

Figure 4.19: *Glessula* (*Glessula*) *notigena*

research on any aspect in the laboratory while the egg masses of Lymnaeidae and Planorbidae provided good ideal material for the students pursuing research in developmental biology.

3. Most of the members of family Lymnaeidae, Planorbidae, Ariophantidae (*Kaliella bullula, Xesta Fretum*) *semirugata*), Helicidae (*Eulota scalpturita*) were herbivorous and act as pestiferous snails because they decline the productivity of the valuable vegetations of the corresponding areas while those species who fed on organic matter *e.g.* family Subulinidae and Glessulidae does not act as pestiferous snails due to the utilization of organic debris.
4. Member of some families act as intermediate host, in the life history of trematode parasites like liver flukes so they act as bioindicators and also showed the probable presence of parasitic trematodes and causes infections *e.g.*, Schistosomiasis and Bilharizasis and *Lymnaea auricularia, Indoplanorbis exustus, Gyraulus convexiusculus* etc. act as an intermediate hosts hence are known as pathogenic snails.

5. The members of family Glessulidae were nonpathogenic and their number was less than fifty in the total collection so they need conservation from the ecological point of view.
6. Member of family Lymnaeidae survived in most polluted water due to the presence of hemoglobin in the plasma.
7. Members of family Planorbidae are artistic and very good looking hence used in India for decorative purposes.
8. All nonpathogenic pulmonate snails have nutritious value due to protein rich contents hence used as palatable dishes by the natives.
9. The powder of the shells of the snails is used for medicinal purposes and also one of the substitutes of poultry and duckeries.

The data showed that freshwater Pulmonate is the diversified group found in each lentic and some lotic water bodies. The present investigation was done on the 2264 specimens of Pulmonates which were collected seasonally from the freshwater bodies of Damoh and Sagar districts. About 663 specimens collected belonged to family–Lymnaeidae, 335 specimens to family–Planorbidae, 828 to family–Ariophantidae, 164 to family–Helicidae, 230 to family–Subulinidae and just 44 specimen belonged to family–Glessulidae as earlier reported by Khan (2002).

Total collection of specimens of sub class-pulmonata is classified into 6 families, 12 genera and 21 species. The present investigation was done on diversified pulmonate molluscan fauna of freshwater bodies of Damoh and Sagar districts which indicated the presence of good environmental conditions of the corresponding freshwater bodies. Generally mixed or multispecies population was observed in the present investigation. Although seasonal significant variation in the population density of the species between seasons and between two consecutive calendar years (from 2000–2001 to 2001–2002) were noticed, the pattern was almost similar during both years. Two peak periods of abundance were recorded, first in pre monsoon (February–March–April) and second in monsoon (June–July–August) which could be due to abundance of newborn or juvenile individuals as reported by Khan (2001) in the small pond of Kolkata and Khan (2002) in the same region of Madhya Pradesh.

The most common Pulmonates were *Lymnaea auricularia, Lymnaea acuminata, Lymnaea luteola* and *Indoplanorbis exustus,* while *Gyraulus auricularia* species was uncommon and only procured from lentic water bodies of Damoh and Sagar districts confirming the earlier reports of Khan (2002), Bhide (2003), Bhide *et al.* (2003, 2004).

Member of Family Lymnaeidae and Planorbidae

Member a family Lymnaeidae and Planorbidae which are pathogenic and provided intermediate host for the life history of liver and blood flukes as reported by Adoni (1995), Mohanta and Mahanta (2002) were abundant while most of the non pathogenic snails provided palatable dishes to natives so at some extent solved the problem of protein malnutrition hence the present investigation suggested that the shell fisheries might be the alternative or substitute of pisciculture which would solve the problem of protein malnutrition in developing country like India.

Acknowledgements

The authors wish to express their thanks for Professor S. Sahai, HOD, Department of Zoology, Dr. H.S. Gour University, Sagar (M.P.) for providing the necessary laboratory facilities and one of the author (Dr. Mangla Bhide*) would like to express her sense of gratitude to UGC for providing the financial assistance.

References

Adoni, A.D., 1995. *Workbook on Limnology*. Pratibha Publisher, Sagar, India, pp. 164–166.

Baur B., Joshi, J., Schmid, B., Hanggii, A., Borcard, D. and Stary, J., 1996. Variation in species richness of plants and diverse groups of invertebrates. *Revue Suisse de Zoologic*, 103(4): 801–833.

Bhide, M. 2003. Biodiversity of freshwater malacofauna. *Him. J. Env. Zool.*, 17(2): 159–163.

Bhide, M., Khan, M.A., Dubey, U., Gupta, P. and Nema, P., 2003. Biodiversity and population density of freshwater gastropod fauna of Damoh (M.P.). *J. Ecophysiol. Occup. Hlth.*, 3: 179–190.

Bhide, M., Khan, M.A., Dubey, U., Gupta, P. and Nema, P., 2004. Biodiversity and population density of freshwater macrofauna. In: *91st Proc. Ind. Sci, Cong.* Abst. No. 8: 7–8.

Caldwell, R.S., Minn, Mc. J.W. and Crossley, D.A. Jr., 1993. Macroinvertebrate and their relationship to coarse woody debris with reference to land snails. In: *Proceeding of the Workshop on Coarse Woody Debris in Southern Forest: Effect on Biodiversity*, held in Athens Georgia, pp. 146.

Crossley, A.A. Jr., 1993. Biodiversity and course woody debris in Southern forests. In: *Proceeding of the Workshop on Coarse Woody Debris in Southern Forest: Effect on Biodiversity* held in Athens Georgia, pp. 146.

Emberton, K.C., 1997. Diversities and distribution of 80 land snail species in South Eastern most Madagascar rain forests with a report that lowlands are richer than highlands in endemic and rare species. *Biodiversity and Conservation*, 6(8): 1137–1154.

Hobdy, R., 1993. A case study: The loss of biodiversity on a small Hawaiian island. *Pacific Sci.*, 47(3): 201–210.

Hall, G.S., 1996. *Methods for the Examination of Organismal Diversity in Soil and Sediments*, 12th edition. Academic Press, New York.

Khan, A., 2001. Population Dynamics and Biology of Aquatic Gastropod *Gyraulus Convexisculus*. (Hutton) in a small pond of Calcutta. In: *14th Proc. Ind. Sci. Cong.*, New Delhi, Abst. No. 35: 34–35.

Khan, M.A., 2002. Studies on the biodiversity of freshwater molluscan fauna of Damoh and Sagar districts. *Ph.D. Thesis*, Dr. H.S. Gour University, Sagar (M.P.), India.

Minelli, A., Foddai, D, Cloridge, M.F., Dawah, H.A. and Wilson, M.R., 1997. The species in terrestrial non-insect invertebrate (Earthworm, Arachnids, Mariposa, Woodlice and Snails). *Species the Unit of Biodiversity*, pp. 309–324.

Mohanta, K. and Mahanta, M.C. 2002. Studies on the freshwater molluscs of Chotanagpur Plateau (Jharkhand state). Family: Corbiculidae. 89 *Proc. Ind. Sci. Cong.*, held at Lucknow, Abst. No. 46: 38, 324.

Odum, E.P., 1987. *Fundamentals of Ecology*. W.B. Saunders Company, Ltd., Tokyo, Japan.

Rao, S.N.V., 1989. *Handbook, Freshwater Molluscs of India*. Ed. by Director, Zoological Survey of India, Kolkata.

Sedgwick, A., 1990. *A Student's Textbook of Zoology*, Vol 1. Low Price Publication, Delhi, pp. 316–416.

Sharma, N. and Sahai, Y.N., 1996. Abundance and seasonal distribution of the macrobenthos group mollusca and Oligochaeta. In: *National Conference on Environmental Pollution and their Chain Reactions* held at Dr. H.S. Gour V.V., Sagar, 16–18 February 1996, Ab. No. 92, 40.

Tonapi, G.T., 1980. *Freshwater Animals of India: An Ecological Approach.* Oxford and IBH Publishing Co., New Delhi.

Webb, D.W., Wetezi, M.J., Reed, P.C., Phillippe, L.R., Horris, M.A. and Ferrington, L.C., 1995. Aquatic biodiversity in Illinois Springs (USA). *J. Kansas Entomological Society*, 68(2): 93–107.

Chapter 5

Taxonomic Richness and Peculiarities in the Himalaya Fish Fauna (Indian Subcontinent)

P. Nautiyal*, R. Nautiyal and H.S. Rawat*****

**Department of Zoology, H.N.B Garhwal University, Srinagar – 246 172*

***Department of Zoology, Govt. P.G. College, Dak Pathar, Dehradun – 248 125*

****Department of Zoology, Govt. P.G. College, Rishikesh*

ABSTRACT

Present investigations resulted in 426 native species of, 43 families and 13 orders from the highland areas of India, Nepal and Bhutan. The Himalayan region harboured 268 species of 107 genera, 33 families and 11 orders compared with 236 in the Indian Himalaya, suggesting restricted distribution of 22 species to the Central Himalaya. Only 90 species of, 21 families and 9 orders were common to these three components of the Himalaya. Consequently, a large number of species were restricted in Himalaya. Among the west, central and east Himalaya 42, 22 and 24 taxa, respectively, were restricted. The taxa restricted to individual Himalayan components amounted to 66 species in the Indian Himalaya and 87 species in the Himalaya. Only family Chaudhuriidae appears to be restricted in Eastern Himalaya. Further, a sizeable share (38 per cent) of species in one family (Cyprinidae) and domination of Ostariophysian element (86 per cent) suggests recent richness with respect to species.

Keywords: *Himalaya, Fish fauna, Richness, Similarities, Peculiarities.*

Introduction

Geographically, India can be divided into three main regions: the Peninsula, extra Peninsula and the Gangetic Plain (Wadia, 1983). The mountain regions especially the eastern Himalaya and the Western Ghats are considered to be hot spots of biodiversity. The Peninsula is strewn with large number of hilly ranges, besides the Western and Eastern Ghats (Figure 5.1). The latter have shown to be particularly rich in fish fauna (Easa and Shaji 1997; Kharat *et al.*, 2003). Among the mountain chains the Himalaya is the largest. It extends c.1,500 miles (2,410 km) E from the Indus River in Pakistan through India, the Tibet region of China, Nepal, East India, and Bhutan to the southern bend of the Brahmaputra River in SE Tibet (Columbia Encyclopedia 2001). The Indian Himalayan Region (IHR; Rao, 1997) is spread between 21°57′–37°5′N latitudes and 72°40′–97°25′E longitudes. It has an area of 5,31,250 km^2 spread over 12 states and constitutes about 16.16 per cent of India's total geographical area (Nautiyal, 2004).

In the Himalayan realm studies have graduated from the fish fauna of small geographical and administrative units (Dutta and Kaur, 1999; Dutta, 2003) to medium (Heckel, 1938; Jerdon, 1849; Hora, 1921; Hora and Mukherji, 1935; Johal and Tandon, 1979; Singh *et al.*, 1983; Hussain, 1995; Hussain and Tilak, 1995) and larger entities (Menon, 1962; Joshi and Joshi, 2004) of the country. It is obvious from the studies quoted above that voluminous information exists on the fish fauna of the various parts of the Himalaya. However, very few authors have attempted to study taxonomic richness with a geographical orientation, so necessary to understand diversity and distribution of the species (Rajbanshi, 1982, 2001, 2002; Shrestha, 1995; Petr, 1999, 2002, Petr and Swar, 2002; Joshi and Joshi, 2004; Nautiyal, 2004). Notable shifts or reductions occur in the distributional ranges of species that either become invasive or extinct, a basis for determining their status with respect to IUCN categories.

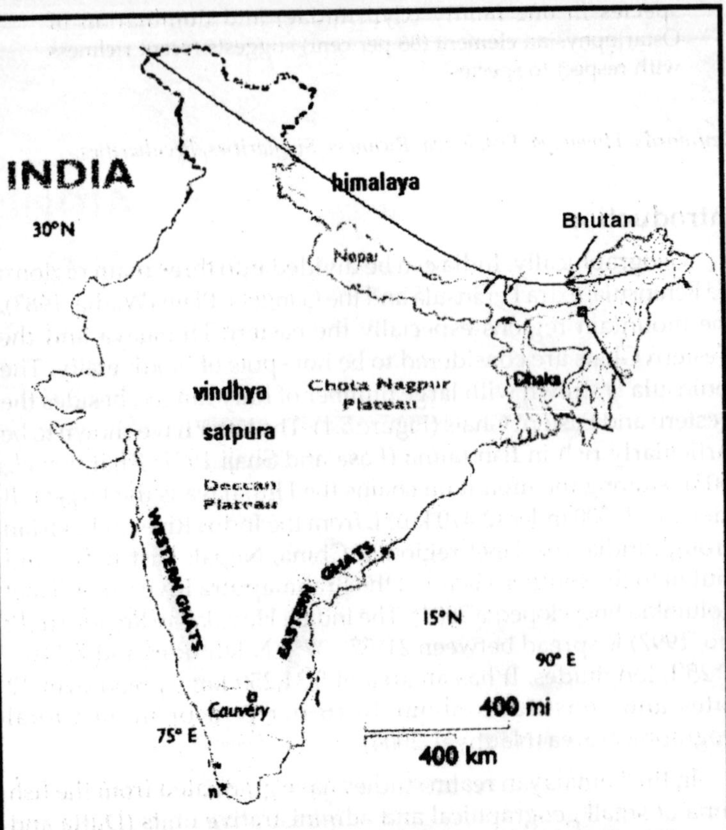

Figure 5.1: Topographic Map of India Showing Location of Various Mountain Chains and Hill Ranges. The entire Peninsula and north east regions have hilly topography.

Materials and Methods

Published works were consulted for preparing a comprehensive list of the highland fish fauna (Singh *et al.*, 1983; Dhendup and Boyd, 1994; Shrestha, 1995; Hussain, 1995; Hussain and Tilak, 1995; Dutta and Kaur, 1999; Kullander *et al.*, 1999; Johal, 2001; Nautiyal, 2001; Dutta *et al.*, 2002a; Dutta *et al.*, 2002 b; Dutta *et al.*, 2003; Dutta and Fayaz, 2003; Dutta, 2003; Vass and Raina, 2002; Joshi and Joshi, 2004; Biswas, 2005; Johal, 2005; Joshi, 2005). Considering that the nomenclature of fishes has undergone wide changes in last few decades, especially cyprinids and multiplicity of names for many of them, the nomenclature and synonyms suggested for fishes of India by Jayaram (1999) were treated as standard. In the process name of the species mentioned in the original publication of the above said authors has changed for quite a number of fish species.

In case of schizothoracids nomenclatural problems needed special attention. Jayaram (1999), most probably on the basis of Misra (1962) and Tilak (1987), split the subfamily Schizothoracinae (=Oreinini) into *Schizothorax* Heckel (4 species) and *Schizothoraicthys* Misra (13 species). Contrary to it, Kullander *et al.* (1999) explained the fact that both *Racoma* and *Schizopyge* have priority over *Schizothoraicthys* which did not occur to Misra or Tilak and place all 5 species found in Kashmir under *Schizothorax* and considered others as their synonyms. Jayaram (1999) has listed 17 species considering *S. richardsonii* = *S. plagiostomus* (to which I do not agree; see Pandey and Nautiyal, 1997; neither do Kullander *et al.*, 1999 and more and more evidence genetical and osteological evidence is accumulating B. Bist, Personal communication). Jayaram makes no mention of *S. intermedius*, *S. punctatus*, and *S. molesworthii* not even in the synonyms. Obviously, these names have been in use. Kullander *et al.* (1999) considers *S. intermedius* similar to *S. curvifrons* though not as synonym, which included *S. curvifrons*, *S. hugelli*, *S. micropogon*. These names have been listed below *S. curvifrons* in the text but excluded, from numbering. Consensus emerged only for 8 out of 20 species obtained from the works of various authors mentioned above. Hence, only 8 were considered for the richness estimate.

This exercise was followed by counting the number of taxa at lower (species) and higher level (genera, families and orders) in the Himalaya (H). Taxonomic richness was examined in the western, central and eastern components of the Himalaya, each being geographical as well as geological entity, which is highly varied (alpine to subtropic) owing to northwest positioning at subtropical latitudes. The taxonomic richness in these individual components was also compared with CH and WG, as the latter have a mix of Peninsular and Indo-Chinese and Malayan elements (Jayaram, 1974; Mani, 1974). Similarity index was computed to compare the magnitude of taxonomic richness in these components of the Himalaya.

Results

The present study is significant, as a major part of the Himalaya has been considered including Nepal and Bhutan. Though the Himalaya extends to Pakistan and Afghanistan, attention has been focused to the Indian, Nepal and Bhutan part, primarily, because these countries account for a very large part of the Himalaya. Moreover, the fish fauna of these regions (Pakistan and Afghanistan) was similar to west Himalayan fauna to a great extent. The western Himalaya is drained by Indus and Ganges, while the central solely by the Ganges and the eastern by the Brahmaputra. The study of mountain fish fauna would hence be representative of the specialist Indian elements, the hill stream fish fauna so as to distinguish it from the fauna of the Indo-Gangetic Plains. They also help in monitoring short and long-term changes in richness and species composition at a geographical scale vis-à-vis distribution.

Taxonomic Richness

Species

The Himalaya is home to 268 native species. The exotics (introduced for culture and aquarium trade) in the Himalaya amount to 17 species. The Himalayan segments varied slightly in richness; 172, 180 and 156 (of which 4 were exclusive to Bhutan, where 41 species have been listed), in the western, central and eastern Himalaya, respectively. This large length of Indian Himalaya (nearly 1600 km), was represented by 236 species of 29 families (Figure 5.2).

Genera, Family and Orders

Within the Himalayan region, the number of higher taxa was least in the western Himalaya, 78 genera, 24 families and 9. The eastern Himalaya had more number of orders (11), genera (81) and families (29). The number of genera, families and orders increased to 90, 33 and 11, respectively in Central (Nepal) Himalaya.

Discussion

In Himalaya fish fauna has been largely studied with respect to administrative or small geographical units. There have been fewer studies with respect to habitats especially sanctuaries and biosphere reserves (Hussain and Tilak, 1995). The authors have, however, not considered them from the viewpoint of highland fishes as Johal (2001), who has studied the hillstream fishes of Himachal Pradesh

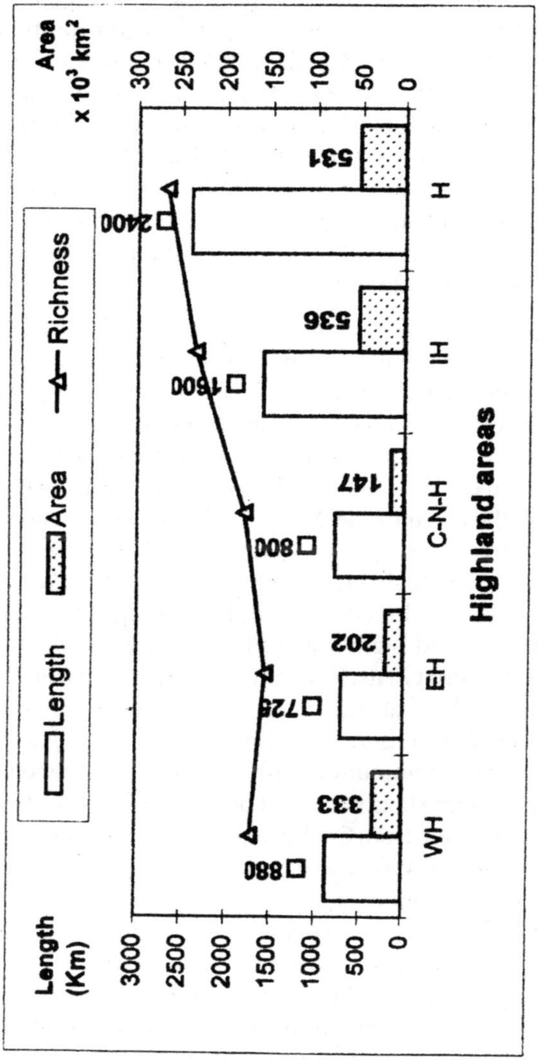

Figure 5.2: Species Richness (Line) Versus Length (Bars) End Area (Bars) of the Highlands. The species richness exhibited a tendency of increases with area and length, but was better related to length compared with area.

with habitat orientation. Again his view did not pertain to a geographical entity. Himachal Pradesh is a significant part of the Western Himalaya, important biogeographically. Mani (1974), refers to flora and fauna to west and east of Sutlej defile to be unique. Joshi and Joshi (2004), nevertheless, made a significant contribution by listing the fauna of western Himalaya Jammu and Kashmir to Uttaranchal. Much earlier, Menon (1962) listed 218 indigenous species in the Himalayan region of India, which is quite close to 236 species enumerated in present study. IGCMC lists 2546 species from the Indian region. Jayaram (1999) maintains that there are 2500 species of fishes from Indian region; 1570 marine and 930 freshwater. He described 852 teleost species falling under 272 genera, 71 families and 16 orders including common exotic species.

Joshi and Joshi (2004) have listed 256 species of fresh water teleost from Western Himalaya compared with 172 species of 78 genera, 24 families and 9 orders recorded during the present investigation. On comparing the list it was found that the differences in higher taxa have arisen primarily out of classification schemes. The scheme adopted is not mentioned whereas Jayaram (1999) has been followed in the present case. Moreover, the exotic species have been excluded from the present counts and hence the order Salmoniformes also. Joshi and Joshi (2004) have excluded Osteoglossiformes and Clupeiformes, but Dahanukar *et al.* (2004) have included them and many other visitor species in the list of fauna from the Western Ghats, thus considering them as highland fauna. Nautiyal and Singh (2005) have included visiting and inhabitants of elevated lands in the category of highlands. Nautiyal and Agarwal (separate communication) have discussed the concept of highland fauna. Some taxa in this list of 256 species according to present investigations are from Central and Eastern Himalaya. Even the species restricted to WG (*e.g. B. canarensis* as per the present investigation) have been included [Nautiyal and Singh (In Press) 2005]. These facts account for the high number (256 species[1] (on next page) belonging to 62 genera, 12 families and 05 orders, Cypriniformes, Siluriformes, Mastacembeliformes, Salmoniformes and Cyprinodontiformes) compared with almost same number of species (267) obtained from the present survey for the whole of Himalaya.

In the Western Himalaya–WH (represented by 3 states) 172 species were listed compared with 180 species in central or Nepal

Himalaya (C-N-H) arid 156 species in Eastern Himalaya (EH) and adjoining mountain ranges. More species are expected from the north east. It is notable that within the Himalaya, central or Nepal Himalaya was rich (discussed below) followed by WH and the EH (157). Earlier studies, Singh *et al.* (1983)–69 species from the Garhwal region: Hussain (1995)–126 species from the western Himalaya (U.P. now Uttaranchal state) are well within the present estimate (133 species) for Uttaranchal. In view of these observations it is necessary to review the richness in fish fauna of Nepal.

The first compilation of the reported fish fauna for the Central or Nepal Himalaya within the boundary of the Kingdom of Nepal for the period 1793–1982 was prepared, and it contained 171 fish species, of which 164 were indigenous and 7 exotic (Rajbanshi, 1982). Shrestha (1994) has reported a total of 188 fish species from Nepal, out of which 179 are indigenous and 9 exotic. Shrestha (1995) listed only 185 species, representing 11 orders. Shrestha (1995) records a total of 183 species, out of which 173 indigenous and 10 exotic fish species. Shrestha (2001) has reported 182 fish species belonging to 92 genera under 31 families and 11 orders, as compared to 186 species of 75 genera, 31 families and 11 orders reported earlier from Nepal. Though there is variation in the number of species (Shrestha, 1995; Shrestha, 1995) from Nepal, the current estimate of 180 native species is closest to the above estimates. Petr (2002) reports 156 species in the Nepal Himalaya, and 41 in Bhutan. The latter estimate is more conservative and may be more close to reality. The numbers appear

---

[1] This is almost at with par the current estimate of 267 species for the whole of Indian Himalaya. Probably some species can be synonimised. Joshi and Joshi have excluded some orders considering them to be fishes of terai flat land adjacent to the foothills where elevation may recede to 250 m asl. However, some of these orders have been reported from Dehradun, a flat land (Valley) sandwiched between low and high hills ranging from 300–600 masl. It is an upland and has been considered as mountain area much in the fashion same as Kashmir Valley and Central Highlands where elevations vary from > 1500 m to 100–500 m, respectively. Can we consider such fish as mountain fauna? Yes, because a major share of the Himalayan fauna is sub-Himalayan as already stated. The truly mountain forms like Oreinini (*Schizothorax*) inhabit the foothills (River Ganga Rishikesh–Hardwar) at 300–350 m elevation which explains that for them temperature is the barrier, not the morphometry (stream flowing through undulated or flat land). *S. niger*, primarily lacustrine in the Kashmir Valley does not inhabit gushing fast flowing water and thereby doesn't cease to be a mountain fish. Mountains are complex ecosystems and the subject probably needs more insight to develop the concept of mountain fish fauna.

to be more with respect to WH and EH. Mani (1974) suggested that EH was the point for exchange of Peninsular and eastern fauna. He was also of the view that this fauna extended to the limits of central/ west Himalaya.

The richness in Kashmir valley deserves special mention. Kullander *et al.* (1999) reported 14 native species from the immediate vicinity of dam construction site (Uri) on the River Jhelum in Kashmir valley. Earlier, Heckel (1838, 1844) had listed 16 species from the same river. Other reports include Hora *et al.* (1955) 17 species, Das and Nath (1971) 20 species from Poonch Valley, Nath (1986) 42 species from Kashmir valley. In the present evaluation more species could be listed from the Jammu and Kashmir region. The Kashmir valley in particular appears to be depauperate, while more fauna belongs to other sub Himalayan (subtropical) parts of Jammu and Kashmir.

However, the present data does generate a trend of increase in numbers within the Himalaya as it progresses from west to east, most probably due to latitudinal decrease of 6–7° within first 800 km. In addition to latitude, the elevation (msl) of J&K is much higher than the central and eastern sector of the Himalaya. Considerable part of this region is arid. Mesic parts of eastern USA support more richness per local site compared with arid west-central region (Matthews 1998). Sub-Himalayan conditions prevail in the rest of the Himalaya where harsh continental conditions give way to monsoonal climate, especially the northeast India. Lower latitude reduces the harshness of the environment. This allows their colonization by lowland species and their prevalence in the mountain ecosystems, as also opined by Kullander *et al.* (1999).

Kullander *et al.*, (1999) states, that the Kashmir Valley had low diversity at higher levels (4 families), which increased to 7 families of 3 orders in present case for the whole state and 2–5 species in several genera. However, richness was high at levels in the rest of Western Himalaya. Among the eastern, central and western Himalaya, the former two were essentially rich at higher taxonomic level (orders, families) whereas richness at species level was more or less similar with a slight exception of Central (Nepal) Himalaya. These differences can be explained on the basis of the numbers of species per family. A few families can incorporate most of the species-level diversity at a site, For example, in the study by Ibarra and

Stewart (1989), characoids made up 51 per cent of the species and 82 per cent of the individuals, and catfishes made up 39 per cent of the species and 15 per cent of the individuals collected. In the Himalaya the cyprinids (Family-Cyprinidae) were most dominant represented by 102 species amounting to 38 per cent of the total Himalayan fauna. However, the ostariophysians accounted for nearly 86 per cent of the fauna. This can be attributed to adaptive radiation in this group of fishes which coincided with the formation of Himalaya and availability of newly created habitats in this vast mountain area.

Similarities and Peculiarities

Only 90 species of, 23 families and 8 orders were common to all the three components of the Himalaya (Table 5.1). Within the Himalaya 41, 22 and 23 taxa, respectively, were restricted to the west, central and east Himalaya. Together they amounted to 64 species in the Indian Himalaya and 86 species in the Himalaya (Table 5.2). However, only 6 species were restricted to WH–CNH, 3 species to CNH–EH, while 2 species to WHI–EH (Table 5.3). Only family Chaudhuriidae appears to be restricted in EH. Engraulidae, Chacidae and Scianidae though appear to be restricted to C-NH, it is likely that they will be found in EH as more fauna is added from EH. Therefore, there was significant similarity among the highland areas with respect to families. This suggests that these areas were colonized simultaneously and some of them such as Cyprinidae, Balitoridae became speciose later on to create differences in the species richness and certain species evolved in isolation. High number of restricted species in the WH suggests more isolation in this part of the Himalaya. A deeper examination of the various elements in the fish fauna (Jayaram, 1974) can probably throw more light on these issues.

Species Richness–Area Relationship

At smaller scale, 92 species of 24 families from the 5 rivers in the Kerala part (1455 km^2) of the Nilgiri Biosphere Reserve (Easa and Shaji 1997) and 49 species in the Rajaji National Park (831.5 km^2) in Siwalik hills of Uttaranchal (Hussain and Tilak, 1995) have been reported. Thus, more area should mean more species as per the species richness-area relationship (Begon *et al.*, 1990), unless the area is an island or totally isolated. The Indian Himalayan Region

Table 5.1: Fish Fauna Common to the Three Components of the Himalaya in the Indian Subcontinent

Taxa

Subdivision–Osteoglossomorpha

(I) Order–Osteoglossiformes

Suborder–Notopteroidei

(1) Family–Notopteridae

1. *Chitala chitala* (Hamilton-Buchanan)
2. *Notoptereus notopterus* (Pallas)

Subdivision–Clupeiomorpha

(II) Order–Clupeiformes

(2) Family–Clupeidae

Subfamily–Alopsinae

3. *Gudusia chapra* (Day)

Subdivision–Euteleosti; Superorder–Ostariophysi

(III) Order–Cypriniformes

(3) Family–Cyprinidae

Subfamily–Rasborinae

4. *Salmostoma bacaila* (Hamilton-Buchanan)
5. *Aspidoparia morar* (Hamilton-Buchanan)
6. *Aspidoparia jaya* (Hamilton-Buchanan)
7. *Barilius bendelisis* (Buchanan) *Opsarius cocsa* (Hamilton) Kullander *et al.*, 1999
8. *B. barna* (Hamilton-Buchanan) *Opsarius* sp. *cf. barna* vide Kullander *et al.*, 1999
9. *B. vagra* (Hamilton-Buchanan)
10. *B. barila* (Hamilton-Buchanan)
11. *B. (Riamas) bola* (Hamilton-Buchanan) = *B. corbetti* Tilak & Hussain vide Gupta, 1983

    Tribe Danionini
12. *Chela laubuca* (Hamilton-Buchanan)
13. *Esomus danricus* (Hamilton-Buchanan)
14. *Danio devario* (Hamilton-Buchanan)
15. *D. aequipinnatus* (McClelland)
16. *Rasbora daniconius* (Hamilton-Buchanan)
17. *Amblypharyngodon mola* (Hamilton-Buchanan)

Contd...

Table 5.1–Contd...

Taxa
Subfamily–Cyprininae
Tribe–Cyprinini
Subtribe–Tores
18. *Tor putitora* (Hamilton-Buchanan)
19. *Tor tor* (Hamilton-Buchanan)
Tribe–Systomini
Subtribe–Osteobramae
20. *Osteobrama cotio cotio* (Hamilton-Buchanan)
Subtribe–Poropunti
21. *Chagunius chagunio* (Hamilton-Buchanan)
Taxa Subtribe–Systomi
22. *Puntius sarana sarana* (Hamilton-Buchanan), *P. chrysopoma* (Cuvier) in Rao *et al.*, 1991
23. *P. conchonius* (Hamilton-Buchanan)
24. *P. ticto* (Hamilton-Buchanan)
25. *P. chola* (Hamilton-Buchanan) syn *P. tetrarupagus* (McClelland) *P. titius* (Hamilton-Buchanan) in Rao *et al.*, 1991
26. *P. sophore* (Hamilton-Buchanan) syn *P. stigma* Day, *P. chrysopterus* Günther, syn *P. carletoni* (Flower)
Tribe–Labeonini
Subtribe–Labeones
27. *Cirrhinus reba* (Hamilton-Buchanan)
28. *C. mrigala* (Hamilton-Buchanan)
29. *Catla catla* (Hamilton-Buchanan)
30. *Labeo dero* syn *L. sindensis* (Hamilton-Buchanan), *Bangana dero*
31. *L. dyocheilus* (McClelland)
32. *L. gonius* (Hamilton-Buchanan)
33. *L. boga* (Hamilton-Buchanan)
34. *L. rohita* (Hamilton-Buchanan)
35. *L. calbasu* (Hamilton-Buchanan)
36. *L. pangusia* (Hamilton-Buchanan)
37. *L. bata* (Hamilton-Buchanan)

Contd...

Table 5.1–Contd...

Taxa
Subfamily–Schozothoracinae
38. *Schizothorax richardsonii* (Gray)
39. *S. labiatus** (McClelland & Griffith) syn *S. inetermedius* (McClelland & Griffith)
40. *Crossocheilus latius latius* (Hamilton-Buchanan)
41. *Garra gotyla gotyla* (Gray)
42. *G. lamta* (Hamilton-Buchanan)
(4) Family–Psilorhynchidae
43. *Psilorhynchus balitora* (Hamilton-Buchanan)
(5) Family–Balitoridae
Subfamily–Balitorinae
44. *Balitora brucei* (Gray) = *B. maculata*
Subfamily–Nemacheilinae
45. *Acanthocobitis botia* (Hamilton-Buchanan) syn *A. botia aeurus*
46. *Schistura rupecula rupecula* (McClelland)
47. *S. rupercula inglishi* (Hora)
(6) Family–Corbitidae
Subfamily–Botinae
48. *Botia Dario* (Hamilton-Buchanan)
Subfamily–Cobitinae
49. *Lepidocephalus guntea* (Hamilton-Buchanan)
(IV) Order–Siluriformes
(7) Family–Bagridae
50. *Rita rita* (Hamilton-Buchanan)
51. *Aorichthy seenghala* (Sykes)
52. *A. aor* (Hamilton-Buchanan)
53. *Mystus bleekri* (Day)
54. *M. vittatus* (Bloch)
55. *M. cavasius* (Hamilton-Buchanan)
56. *M. tengara* (Hamilton-Buchanan)
(8) Family–Siluridae
57. *Ompak bimaculatus* (Bloch)
58. *O. pabda* (Hamilton-Buchanan)
59. *Wallago attu* (Schneider)

Contd...

Table 5.1–Contd...

Taxa
(9) Family–Schilbedae
60. *Alia colia* (Hamilton-Buchanan)
61. *Clupisoma garua* (Hamilton-Buchanan)
62. *C. montana* (Hora)
63. *Silonia silondia* (Hamilton-Buchanan)
(10) Family–Amblyciptidae
64. *Amblyceps mangois* (Hamilton-Buchanan)
(11) Family–Sisoridae
65. *Bagarius bagarius* (Hamilton-Buchanan)
66. *B. yarrelli* Sykes
67. *G. cenia* (Hamilton-Buchanan)
68. *Glyptothorax pectinopterus* (McClelland)
69. *G. telchitta* (Hamilton-Buchanan)
70. *Euchiloglanis hodgarti* (Hora)
71. *Psudecheneis sulcatus* (McClelland)
(12) Family–Clariidae
72. *Clarius batrachus* (Linnaeus)
(13) Family–Heteropneustidae
73. *Heteropneustis fossilis* Muller
Superorder–Protacanthopterygii
(V) Order–Mugiliformes
(14) Family–Mugilidae
74. *Rhinomugil corsula* (Hamilton-Buchanan)
75. *Sicamugil cascasia* (Hamilton-Buchanan)
(15) Family–Belonidae
76. *Xenetodon cancila* (Hamilton-Buchanan)
(VI) Order–Cyprinidontiformes
(16) Family–Aplocheilidae
77. *Aplocheilus panchax* (Hamilton-Buchanan)
(VII) Order–Symbranchiformes
(17) Family–Symbranchidae
78. *Monopterus cuchia* (Hamilton-Buchanan)

Contd...

Table 5.1–Contd...

Taxa
(18) Family–Mastacembelidae
79. *M. pancalus* (Hamilton-Buchanan)
80. *Mastacembalus armatus* (Lacépède)
(VIII) Order–Perciformes
(19) Family–Ambassidae
81. *Chanda nama* (Hamilton-Buchanan)
(20) Family–Nandidae
82. *Badis badis*
83. *Nandus nandus*
(21) Family–Anabantidae
84. *Anabas testudineus* (Bloch)
(22) Family–Belontidae
85. *Colisa fasciata* (Schneider)
86. *C. lalia* (Hamilton-Buchanan)
Suborder–Channoidei
(23) Family–Channidae
87. *Channa striatus*
88. *C. punctatus*
89. *C. marulius*
90. *C. orientalis*

Abbreviations: WH: Western Himalaya; EH: Eastern Himalaya; IH: Indian Himalaya; C-NH: Central-Nepal Himalaya; H: Himalaya; B: Bhutan and rest; EH/Bo: Bhutan only in EH; #: The taxa indicated by this symbol are exotic species; *: Doubtful forms lacking consensus among the authors; # and * have not been included for richness count.

(IHR) with more area had more species. However, the richness in H, or IHR was not proportionate to their area. Area-wise the Indian Himalaya (531,250 km^2) is large but species do not occur in that proportion. Richness was also gauged with respect to length of the mountain chain (Figure 5.2). The length of each Himalayan segment; E.H (725 km Tista to Brahmaputra–156 species), C-N-H (Kali to Tista 800 km–180 species) and WH (Indus to Sutlej 560 km, Sutlej to Kali 320 km–172 species) was proportionate to richness in each segment. Probably considering the length is the best approach than area

Table 5.2: Fish Fauna Restricted to the Each Component (Part) of the Himalaya in the Indian Subcontinent (Abbreviations and other details as in Table 5.1)

Taxa	*WH*	*EH*	*CH–N*
(II) Subdivision–Clupeiomorpha			
Order–Clupeid formes			
(1) Family–Clupeidae			
Subfamily–Alopsinae			
1. *Hilsa* (*Tenualosa*) *ilisha* (Hamilton-Buchanan)		+	
2. *G. variegata* (Day)			+
3. *Goniolosa modestus* (Day)	+		
(2) Family–Engraulidae			
4. *Setipinna phasa* (Hamilton-Buchanan)			+
Subdivision–Euteleosti			
Superorder–Ostariophysl			
(III) Order–Cypriniformes			
(3) Family–Cyprinidae			
Subfamily–Rasborinae			
5. *Salmostoma punjabensis* (Day)	+		
6. *Barilius modestus* (Day)	+		
7. *S. acinaces* (Valenciennes)			+
8. *B. dimorphicus* Tilak & Hussain	+		
9. *B.* (*Riamas*) *guttatus* (Day)			+
Subfamily–Cyprininae			
Tribe–Cyprinini			
Subtribe–Cyprini			
#*C. carpio specularis* Lacépède		+	
#*C. carpio nudus* (Bloch)		+	
#*Carassius carassius* (Linnaeus)			+
Subtribe–Tores			
10. *Tor progenies* (McClelland)		+	
Tribe–Systomini			
Subtribe–Osteobramae			
11. *Osteobrama belangeri* (Valenciennes)		+	
12. *Cyclocheilichthys apogon* (Valenciennes)			+

Contd...

Table 5.2–Contd...

Taxa	WH	EH	CH–N
13. *Puntius terio* (Hamilton-Buchanan)	+		
14. *P. khohi* sp.nov. Dobriyal *et al.*, 2004	+		
15. *P. jayrami** Singh & Singh		+	
Tribe–Labeonini			
Subtribe–Labeones			
16. *Cirrhinus lata* in Dhendup & Boyd, 1994		+Bo15	
Barbus spp.*		+Bo16	
17. *Bangana diplostoma** (Heckel) Kullander *et al.*, 1999	+		
18. *Labeo caeruleus* (Day)			+
Subfamily–Schozothoracinae			
19. *S. kumaonensis* (Menon)	+		
20. *S. nepalensis*			+
21. *S. raraensis* (Terashima)			+
22. *S. macrophthalamus* (Terashima)			+
Schizotharichthys progastus (McClelland)		+	+
23. *Schizopygopsis stoliczkae* Steindachner	+		
24. *Ptychobarbus conirostris* Steindachner	+		
25. *Gymnocypris biswasi* Talwar	+		
26. *Schizocypris brucei* Regan	+		
Subfamily–Garinae			
27. *Crossocheilus latius diplocheilus C. diplochilus* (Heckel) Kullander *et al.* (1999) syn *C. latius punjabensis* (Mukerji)	+		
28. *Cyprinus gohama* (Hamilton-Buchanan) syn *C. latius*?	+		
29. *C. burmanicus* Hora		+	
30. *G. kemi* Hora		+	
(4) Family–Psilorhynchidae			
31. *Psilorhynchus sucatio* (Hamilton-Buchanan)			+
32. *P. pseudechensis* (Menon & Dutta)			+
(5) Family–Balitoridae			

Contd...

Table 5.2–Contd...

Taxa	*WH*	*EH*	*CH–N*
Subfamily–Balitorinae			
33. *Homaloptera bilineata* Blyth			+
Subfamily–Nemacheilinae			
34. *Acanthocobiti zoalternans* (Blyth) syn *A. botic aerus* (Day)		+	
35. *N. detamai* in Bhutia, 2002		+	
36. *N. gangeticus* (Menon)	+		
37. *N. kangree* (Menon)	+		
38. *Schistura punjabensis* (Hora)	+		
39. *S. nalbanti* (Menon)	+		
40. *S. prashari* (Hora)	+		
41. *S. Montana* (McClelland)	+		
42. *S.* sp cf. *afasciata* (Hora)	+		
43. *S. doonensis* (Tilak & Hussain)	+		
44. *S. gangeticus* (Menon)	+		
45. *S. kangjuphulensis* (Hora)		+	
46. *S. sikmaiensis* (Mirza & Nalbant)		+	
47. *S. horai* (Menon)			+
48. *S. devdevi* (Hora)		+	
49. *Triplophysa marmorata* (Heckel)	+		
50. *T. kashmirensis* (Hora)	+		
51. *T. yasinensis* (Alcock)	+		
52. *T. vittatus* (Heckel)	+		
53. *T. microps* Steindachner	+		
54. *T. stoliczkae* Steindachner	+		
55. *T. gracilus* (Day)	+		
(6) Family–Cobitidae			
Subfamily–Botinae			
56. *Botia birdi* Chaudhari	+		
57. *B. rostrata* Günther, *B. geto* (Hamilton-Buchanan) is juvenile	+		
58. *B. berdmori* (Blyth)		+	

Contd...

Table 5.2–Contd...

	Taxa	*WH*	*EH*	*CH–N*
	Subfamily–Cobitinae			
	59. *Lepidocephalus caudofurcatus* Tilak & Hussain	+		
(IV)	Order–Siluriformes			
(7)	Family–Bagridae			
	60. *Batasio tengara* (Hamilton-Buchanan)		+	
(8)	Family–Schibedae			
	61. *Pseudotropius murius batarensis* Shrestha			+
	62. *Clupisoma naziri* Mirza & Iqbal Awan	+		
(9)	Family–Sisoridae			
	63. *Gagata gagata* (Hamilton-Buchanan)			+
	64. *G. sexualis* Tilak			+
	65. *Nangra robusta* N. punctata is syn Mirza & Awan		+Bo32	
	66. *Conta conta* (Hamilton-Buchanan)		+	
	67. *Glyptosternon maculatum* (Regan)			+
	68. *Glyptothorax punjabensis* (Mirza & Kashmiri)	+		
	69. *G. madraspatanum* (Day)	+		
	70. *G. bervipinnis* (Hora)	+		
	71. *G. dakpathari* (Tilak & Hussain)	+		
	72. *G. garhwali* Tilak	+		
	73. *G. alaknandi* Tilak	+		
	74. *Coraglanis kishionougii* (Kimura)			+
	75. *Myresglanis blythi* (Day)			+
	76. *Exostoma labiatum* (McClelland)			+
(10)	Family–Chacidae			
	77. *Chacha chaca* (Hamilton-Buchanan)			+
(11)	Family–Olyridae			
	78. *Olyra kempi* Chaudhari		+	
	Superorder–Protacanthopterygii			
	Order–Salmoniformes			
	Family–Salmonidae			
	#*Oncorhynchus mykiss* (Walbaum, 1792)	+		
	#*O. masou rhodurus* (Jordon & McGregor)			+

Contd...

Table 5.2–Contd...

	Taxa	*WH*	*EH*	*CH–N*
(V)	Order–Mugiliformes			
(12)	Family–Mugilidae			
	79. *Mugil cascasia* (Hamilton-Buchanan)	+		
(VI)	Order–Cyprinidontiformes			
(13)	Family–Aplocheilidae			
	80. *Aplocheilus melastigma* in Biswas 2005		+	
(VII)	Order–Symbranchiformes			
(14)	Family–Symbranchidae			
	81. *Monopterus albus* (Zuiew)		+	
(15)	Family–Chaudhuriidae			
	82. *Chaudhuria indica* Yazdani		+	
	83. *C. khajuriai*		+	
(VIII)	Order–Perciformes			
(16)	Family–Scianidae			
	84. *Daysciaena albida*			+
(17)	Family–Belontidae			
	85. *Colisa sota* (Hamilton-Buchanan)		+	
Suborder–Channoidei				
(18)	Family–Channidae			
	86. *Channa barca*		+	

because the rivers have length unlike terrestrial habitat where land extends over an area. For rivers, length is always important. If the Punjab and Kumaun Himalaya (Wadia, 1983) are compared, the former has a large area but fewer species (over 100) while the latter even after excluding the area from Giri to Sutlej had 133 species. Probably, determining the rate of increase of species with respect to area would help in estimating the nature of species richness-area relationship in the highland areas.

Mountain Fauna in Adjoining Countries

The mountain fauna from adjoining countries has a large number of elements similar to the Himalayan fauna. The fauna of the Qinghai-Xizang Plateau includes 112 native and 17 introduced species. These native species are considered to be relief of mass

Table 5.3: Fish Fauna Common to Two Adjacent (WH–CNH, EH–CNH) and distant (WH–EH) Components of the Himalaya (Indian Subcontinent) [Abbreviations and details as in Table 5.1)

Taxa	*WH*	*C–NH*	*EH*	*WH–CNH*	*CNH–EH*	*WH–EH*
Subdivision–Osteoglossomorpha						
(II) Order–Angulliformes						
Suborder–Anguilloidei						
(1) Family–Anguiloidae						
1. *Anguilla bengalensis bengalensis* Gray & Hardwicke		+	+		+	
Subdivision–Euteleosti						
Superorder–Ostariophysi						
(III) Order–Cypriniformes						
(2) Family–Cyprinidae						
Subfamily–Rasborinae						
2. *Salmostoma phulo phula* (Hamilton-Buchanan)	+	+		+		
Tribe–Danionini						
3. *Chela cachius* (Hamilton-Buchanan)		+	+		+	
4. *Brachydanio rerio* (Hamilton-Buchanan)	+		+B6			+
Subfamily–Cyprininae						
Tribe–Cyprinini						
Subtribe–Cyprini						
#*Cyprinus carpio communis* Linnaeus		+	+		+	

Contd...

Table 5.3–Contd...

Taxa	WH	C–NH	EH	WH–CNH	CNH–EH	WH–EH
Subtribe–Tores						
Tor mosal (Hamilton-Buchanan)*		+	+		+	
(3) Family–Balitoridae						
Subfamily–Nemacheilinae						
5. *S. beavani* (Gunther)	+	+		+		
6. *S. savona* (Hamilton-Buchanan)	+	+		+		
(IV) Order–Siluriformes						
(4) Family–Schilbedae						
7. *Pseudoutropius atherinoides* (Bloch)	+	+		+		
(5) Family–Pangasiidae						
8. *Pungasius pangasius* (Hamilton-Buchanan)		+	+		+	
(6) Family–Sisoridae						
9. *L. ribeiroi* (Hora)	+	+		+		
10. *G. conirostrae conirostrae* (Steindachner)	+	+		+		
Superoder–Protacanthopterygii						
(V) Order–Perciformes						
(7) Family–Ambassidae						
11. *Parambassis baculis* (Hamilton-Buchanan)		+	+			+

extinctions in the Quaternary, when uplift of the Himalayas transformed the prevailing low-altitude tropical or sub-tropical conditions to the present cold, arid, high-altitude climate. Walker and Yang (1999) reviewed the limnology and fish/fishery literature for the three northwestern areas of China (Qinghai, Xinjiang and Xizang=Tibet), and supplemented it with their research and observations. All but four species belong to the High-Centro-Asiatic Fish Faunal Complex (Sensu Li 1981). *Glyptosternum maculatus, G. reticulates* and *Paraeuchiloglanis kamengensis* belong to the China-Indian Mountain Complex, and *Schistura naseeri* to the Southern Asian Complex. The fish fauna of different areas was as follows:

Northern Internal Drainages (Southern Xizang)

Triplophysa marmorata, T. stenura, T tibetana, Schizopygopsis younghusbandi.

Indus River Drainage

Schistura alepidota, S. naseeri, Triplophysa aliensis, T. gracilis, T. griffithi, T. marmorata, T. microps, T. stenura, T. stoliczkae, T. tenuicauda, T. yasinensis, Diptychus maculatus, Ptychobarbus conirostris, Schizothorax esocinus, S. labiatus, S. longipinnis, S. macropogon, S. nasus, S. plagiostomus, Glyptosternum maculates, G. reticulatum.

Yarlung Zangbo (Brahmaputra) River Drainage

Triplophysa microps, T. stenura, T. stewartii, T. tibetana, Gymnocypris xygymnocypris stewartii, Ptychobarbus dipogon, Schizopygopsis younghusbandi, Schizothorax waltoni, Paraeuchiloglanis kamengensis.

Nearly 75 per cent of Afghanistan is mountainous, with its north-east embracing the southwestern ranges of Hindu Kush. Northern Afghanistan is situated at the western end of the Himalayas-Karakoram-Pamir-Hindu Kush complex of high mountain ranges. Afghanistan rivers and streams contain a mixture of Oriental and Palaearctic species, with northern and southern species and high and low altitude-adapted species. For the Kabul River Coad (1981) has listed the following indigenous fish species: *Amblypharyngodon mola, Aspidoparia jaya, Barilius vagra, Cirrhinus reba, Danio devario, Esomus danricus, Labeo angra, L. dero, L. diplostomus, L. dyocheilus, L. gonius, L. pangusia, Puntius conchonicus, P. sarana, P. sophore, Salmostoma bacaila, Schizothorax barbatus, S. chrysochlora, S. edeniana, S. esocinus, S. labiatus, Tor putitora, Nemacheilus alepidotus*

(Kabul R. basin), *N. griffithi*, *N. sargadensis* (Kabul R. basin), *Rita rita*, *Ompok canio*, *O. pabda*, *Glyptosternum reticulatum* (Kabul R. basin), *G. jalalensis*, *Channa gachua* (Kabul R. basin), *C. punctatus* (Kabul R. basin). The following exotic fish were also noted for the Kabul River: *Carassius auratus*, *Ctenopharyngodon idella*, *Hypophthalmichthys molitrix*, *Gambusia affinis*.

Akhtar (1991a) listed 25 species (of these 4 introduced, *viz.* brown and rainbow trout (*Oncorhynchus mykiss*), common carp (*Cyprinus carpio*) and gold fish (*Carassius auratus*) of freshwater fish for the Northern areas as compared to 43 species (of which 4 have been introduced) for PoK (Pak occupied part of Jammu and Kashmir), in which he also included warm water species (Akhtar, 1991b). In the Northern Areas the Indus river receives numerous tributaries draining the Karakoram Range (length of streams and rivers is 2100 km). These regions too appear to be low in richness. Studies continue to throw more light on issues related to distribution and richness patterns.

Acknowledgements

We thank Prof. H.R. Singh, Vice Chancellor, University of Allahabad for going through the MS critically. The laboratory facilities and valuable discussions provided by Prof. (Ms.) A. Saklani-Chandola, Head, Department of Zoology, H.N.B. Garhwal University and academic support by Prof. P. Gaur, Head, Department of Zoology, University of Allahabad is acknowledged herewith. We are also thankful to Prof. B.D. Joshi, Dean Faculty of Science Gurukula Kangri University Hardwar, Dr. P.C. Joshi, Department of Zoology, Gurukula Kangri University, Hardwar and Prof. M.S. Johal, Department of Zoology, Panjab University, Chandigarh for providing relevant literature.

References

Akhtar, N., 1991a. The Northern Areas (Pakistan). Fisheries profile, feasible sites for trout culture and an overall sector development perspective. Report for Project PAK/91/008 Rome, FAO, 29 p.

Akhtar, N., 1991b. PoK Jammu and Kashmir. Fisheries profile, feasible sites for trout culture and an overall sectoral development perspective. Report for Project PAK/88/048 Rome, FAO, 25 p.

Begon, M., Harper, J.L. and Townsend, C.R., 1990. *Ecology, Individual, Populations* and *Communities*. 238, Main street Cambridge Massachusetts 02142 USA, 768 pp.

Biswas, S.P., 2005. Fisheries ecology of upland waters of north-east Himalaya, pages 164–176. In: *Biological Diversity: Structure and Function of Freshwater Environments*, (Eds.) Nautiyal, P., Bhatt, J.P., Gusain, O.P. and Dobriyal, A.K. Transmedia, Srinagar (In Press).

Coad, D., 1981. Fishes of Afghanistan an Annotated Check-list Publications in Zoology No. 14, National Museum of Canada, Ottawa, 26 p.

Dahanukar, N., Raut, R. and Bhat, A., 2004. Distribution, endemism and threat status of freshwater fishes in the Western Ghats of India. *Journal of Biogeography*, 31: 123–136.

Dhendup, T. and Boyd, C.E., 1994. Chemical features of water and soil farming areas of Bhutan. *Journal of Aquaculture in Tropics*, 9: 35–41.

Dubey, G.P., 1978. Survey of the waters of Bhutan. Physiography and fisheries potential report, FAO, Rome, 38 p.

Dutta, S.P.S., 2003. Fish fauna of Poonch District Jammu region, Jammu and Kashmir State. *Aquaculture*, 4: 241–246.

Dutta, S.P.S. and Kaur, H., 1999. Some new fish records for Jammu, J&K State. *Oriental Science*, 41: 45–47.

Dutta, S.P.S., Kaur, H., Gupta, S.C. and Bali, J.P.S., 2002a. Fish and fisheries of the river Chenab Jammu Province (J&K) p. 181–187. In: *Coldwater Fish Genetic Resources and their Conservation*. Nature Conservator, Publication, Muzzafarnagar.

Dutta, S.P.S., Gupta, S.C. and Salaria, S., 2002b. Ichthyofaunistic survey of Rajouri District (J&K State). *Aquaculture*, 3: 201–205.

Dutta, S.P.S., Kaur, H. and Zutshi, N., 2003. Ichthyofauna of river Tawi and its tributaries an important tributary of the river Chenab, J&K State. *Journal of Aquatic Biology*, 18: 61–68.

Dutta, S.P.S., and Fayaz, F.A., 2003. Ichthyofauna of Doda District, Jammu region (Jammu & Kashmir). *Journal of Aquatic Biology*, 18: 51–54.

Easa, P.S. and Shaji, C.P., 1997. Freshwater fish diversity in Kerala Part of Nilgiri Biosphere reserve. *Current Science*, 73(2): 180–182.

Heckel, J.J., 1838. Fische aus Caschmir. Carl Frieherm V. Hugel, Wein.

Hora, S.L., 1921. Fish and fisheries of Manipur with some observations on those of the Naga hills. *Records of the Indian Museum*, 2(1): 19–21.

Hora, S.L. and Mukerjee, D.D., 1935. Fish of the Eastern Doon United Provinces. *Records of the Indian Museum*, 38: 133–146.

Hora, S.L., Malik, G.M. and Khajuria, H., 1955. Some interesting features of the aquatic fauna of the Kashmir Valley. *Journal of the Bombay Natural History Society*, 53: 140–143.

Hussain, A., 1995. *Pisces*, pp. 117–150. In: Director Zoological Survey of India (Editor) Himalayan Ecosystem Series: Fauna of Western Himalaya Part I: Uttar Pradesh. Zoological Survey of India, Kolkata.

Hussain, A. and Tilak, R., 1995. *Fishes* (*Pisces*), pp. 115–193. In: Director Zoological Survey of India (Editor) Fauna of Conservation Area 5: Rajaji National Park. Zoological Survey of India, Kolkata.

Ibarra, M. and Stewart, D.J., 1989. Longitudinal zonation of sandy beach fishes in the Napo River Basin, eastern Eucador. *Copeia*, 1989: 364–381.

Jayaram, K.C., 1974. Ecology and distribution of freshwater fishes, amphibia and reptiles. pp. 517–584. In: *Ecology and Biogeography in India*, (Ed.) Mani, S., Dr. W. Junk b.v. Publishers, The Hague.

Jayaram, K.C., 1999. *The Freshwater Fishes of the Indian Region*. Narendra Publishing House, Delhi, India, pp. 551+xviii.

Jerdon, T.C., 1849. On the freshwater fishes of southern India. *Madras Journal of Literature and Science*, 15: 302–346.

Johal, M.S., 2001. Species richness of fish in relation to environmental factors. *Current Science*, 8(4): 499–501.

Johal, M.S., 2005. Ecology of hillstream fish, communities of Himachal Pradesh. pp. 36–57. In: *Biological Diversity: Structure and Function of Freshwater Environments*, (Eds.) Nautiyal, P., Bhatt, J.P., Gusain, O.P. and Dobriyal, A.K. Transmedia, Srinagar (In Press).

Johal, M.S. and Tandon, K.K., 1979. Monograph on the fishes of reorganized Punjab. *Punjab Fisheries Bulletin*, 3(2): 1–94.

Joshi, P.C. and Joshi, N., 2004. Aquatic resources and fish biodiversity of western Himalayan region. pp. 283–304, In: *Biodiversity and Conservation*. APH Publishing Corporation, New Delhi.

Joshi, B.D., Studies on the ecobiology of Tributaries of River Ganga between Devprayag and Rishikesh. pp. 19–35 In: *Biological Diversity: Structure and Function of Freshwater Environments*, (Eds.) Nautiyal, P., Bhatt, J.P., Gusain, O.P. and Dobriyal, A.K. Transmedia, Srinagar (In Press).

Kharat, S., Dahanukar, N., Raut, R., and Mahabaleshwarkar, M., 2003. Long-term changes in freshwater fish species composition in North Western Ghats, Pune District. *Current Science*, 84(6): 816–820.

Kullander, S.O., Fang, F., B. Delling and Åhlander, E., 1999. The fishes of the Kashmir Valley. In: *River Jhelum, Kashmir Valley: Impacts on the Aquatic Environment*, (Ed.) Nyman, L. The International consultancy Group of the National Board of Fisheries, Goteborgs Lanstryckeri AB. SWEDMAR, pp. 104–105.

Kulshrestha, S K., 2005. Biodiversity of aquatic environments in Central India, pp. 1–32. In: *Biodiversity and Ecology of Aquatic Environments*, (Eds) Singh, H.R. and P. Nautiyal. Narendra Publishing House, Delhi, India (In Press).

Li, Shizong, 1981. *Studies on Zoogeographical Divisions for Freshwater Fishes of China*, Beijing, 292 p.

Malviya, R.B., 1961. A list of fishes from Jabalpur, M.P. *Proceedings National Academy of Science*, India, 31(3): 347–354.

Mani, M.S., 1974. *Ecology and Biogeography in India*. Dr. W. Junk b.v. Publishers, The Hague, 725 pp.

Matthews, W.J., 1998. *Patterns in Freshwater Ecology*. International Thompson Publishing, pp. 756.

Menon, A.G.K., 1962. A distribution list of fishes of the Himalayas. *Journal of the Zoological Society of India*, Kolkata, 14(1): 23–32.

Misra, K.S., 1962. *An Aid to Identification of Commercial Fishes of India and Pakistan*. Records of the Indian Museum, 57: 1–320.

Nautiyal, P., 2001. Ichthyofauna, pp. 191–197. In: *Garhwal Himalaya: Nature, Culture and Society*, (Eds.) Kandari, O.P. and O.P. Gusain. Transmedia, Srinagar-Garhwal, Uttranchal, India.

Nautiyal, P. and Agarwal, A. (Separate Communication). Assessing diversity and peculiarities in the fish fauna from some highlands of the Indian subcontinent. *Proc. National Symposium on Reassessment of Fish Genetic Resources in India and Need to Evolve Sustainable Methodology for Conservation.* National Bureau of Fish Genetic Resources, 26–27 April 2005, Lucknow.

Nautiyal, P. and Singh, H.R., (2005). Fish diversity in the highlands of the Indian subcontinent (Oriental region). In: *Biodiversity and Ecology of Aquatic Environments*, (Eds.) Nautiyal, P. and H. R. Singh. Narendra Publishing House, Delhi [In Press].

Pandey, S.K. and Nautiyal, P., 1997. Statistical evaluation of some morphometric characters of taxonomic significance in *Schizothorax richardsonii* (Gray) and *Schizothorax plagiostomus. Indian Journal of Fisheries*, 44(1): 75–79.

Petr, T. (Ed.), 1999. *Fish and Fisheries at Higher Altitude, Asia*. FAO Fisheries Technical Paper No. 385, Rome FAO, 304 p.

Petr, T. 2002. Coldwater fish and fisheries in countries of the high mountain arc of Asia (Hindu Kush-Pamir-Karakoram-Himalayas): A review. In: *Coldwater Fisheries in the Trans-Himalayan Countries*, (Eds.) Petr, T. and D.B, Swar. FAO (http://www.fao.org/docrep/).

Petr, T. and Swar, D.B., 2002. *Coldwater Fisheries in the Trans-Himalayan Countries*. FAO Fisheries Technical Paper No. 431, Food and Agriculture Organization of the United Nations, Rome, (http://www.fao.org/docrep/).

Rajbanshi, K.G., 1982. *A General Bibliography on Fish and Fisheries of Nepal*. Royal Nepal Academy, Kamaladi, Kathmandu, Nepal.

Rajbanshi, K.J., 2002. Zoo-geographical distribution and the status of coldwater fish in Nepal. In: *Coldwater Fisheries in the Trans-Himalayan Countries*, (Eds.) Petr, T. and D.B, Swar. FAO. Paper presented in symposium on coldwater fish species in the trans-Himalayan region 10–14 July 2001, Kathmandu, Nepal, (http://www.fao.org/docrep/).

Shrestha, J., 1994. *Fishes, Fishing Implements and Methods of Nepal*. Smt. M.D. Gupta, Ladetpur Colony Lashkar, India, 145 pp.

Shrestha, J., 1995. *Enumeration of the Fishes of Nepal*. Technical Publication No. 10. Euroconsult, Arnhem, The Netherlands.

Shreshta, J., 2001. Taxonomic revision of fishes of Nepal. p. 171–180. Environment and Agriculture. In: *Biodiversity, Agriculture and Pollution in South Asia*, (Eds.) Jha, P.K., Baral, S.R., Karmacharya, S.B., Lekhak, H.D., Lacoul, P. and Baniya, C.B. (Editors), ECOS, Kathmandu.

Shrestha, T.K., 1995. *Fish Catching in the Himalayan Waters of Nepal*. Bimala Shrestha, Kuleswor, Kathmandu, Nepal. R.K. Printers Pvt. Ltd., Teku, Kathmandu, Nepal, 247 p.

Singh, H.R., Badola, S.P. and Dobriyal, A.K., 1983. Geographical distributional list of ichthyofauna of the Garhwal Himalaya with some new records. *Journal of the Bombay Natural History Society*, 84: 126–131.

Talwar, P.K. and Jhingran, A., 1991. *Inland Fishes of India and Adjacent Countries*. Oxford and IBH Publishing Co. Pvt. Ltd., New Delhi, Volumes I–II, 1157 pp.

Tilak, R., 1987. The fauna of India and adjacent countries. Pisces (Teleostomi). Subfamily: Scizothoracinae. Director Zoological Survey of India, Calcutta.

Vass, K.K. and Raina, H.S. (Editors), 2002. Highland Fisheries and Aquatic Resource Management. National Research Centre on Coldwater Fisheries (ICAR), Bhimtal.

Wadia, D.N., 1983. *Geology of India*, 4th edition. Tata-McGraw Hill Publishing Co., New Delhi, 508 pp.

Walker, K.F. and Yang, H.Z., 1999. *Fish and Fisheries in Western China*. FAO Fisheries Technical Paper No. 385, 237–278, Rome, FAO.

Internet Resources

Bartleby.Com, 2001–04. Columbia Encyclopedia, Sixth Edition 2001.htm New York: Columbia University Press, 2001–04, New York: http://www.bartleby.com/65/.

Khuswant Singh, 2002–2004 Khushwant Punjabi. Punjabilit.com http://india.punjabilit.com/land and resources.htm.

http://gbpihed.nic.in/envis/HTML/monograph2/IndianHimalaya.html.

http://www.indiaonestop.com/stategovetlist.htm

http://www.fao.org/docrep/.

Chapter 6

Heavy Metal Pollution in Freshwater and its Impact on Fisheries

R.K. Gupta, K.L. Jain, S.K. Gahlawat and N.K. Yadava

Department of Zoology and Aquaculture,
C.C.S. Haryana Agricultural University, Hisar – 125 004, India

Water pollution today has become world's most serious problem, as water although an essential commodity for sustaining life, is being maximally exploited by the industrialists, civic bodies vis-à-vis the water surface transporting units. Modern civilization with its rapidly growing industrial units and ever increasing population has led to an accelerated degradation of the freshwater resources. Today about 100 million people are known to severely suffer from these polluted waters, since a variety of causative agents are known to·occur for various physiological disorders and water born diseases.

Water a God's valuable gift to man is required in healthy state for every organism. Analysis of most of the natural water bodies revealed a critical loss of biodiversity, frequent occurrence of pathogens and a variety of other noxious problems with aquatic plants and animals due to man's careless activity. In India textile and dying industries alone contribute 10 to 15 per cent of about 1000 different chemicals in nature, leading to increased BOD and COD contents, thus simultaneously decreasing DO and creating anaerobic conditions in the rivers. In Delhi about 200 million litres of untreated water is pumped into Yamuna. The major industries causing water pollution are the paint and the detergent industries, metal and surgical industries, steel and pulp industries and various other chemical industries. Fly ash is the major contributor of heavy metal toxicity in natural waters. Over 100 million tones of fly ash are produced each year and much of it is transported in the atmosphere before deposition. It is a major source of contamination of heavy metal such as cadmium in the rivers and estuarine sediments in the vicinity of such power stations. Even most of the wetlands in India including Utter Pradesh today have become heavily polluted and unproductive.

The US Department of Health Education and Social Welfare has classified water pollutants in eight categories *viz.*,

1. Sewage wastes
2. Infectious agents
3. Plant nutrients
4. Particulates
5. Radioactive substances
6. Mineral and chemical substances
7. Heat
8. Organic chemical exotics.

Heavy metals are one major component among various effluents which has attained alarming status as water pollutants over the last few decades due to heavy industrialization. Toxic metals are added in aquatic system from industrial processes, domestic sewage discharge, steel dust, land runoff and fossil fuel burning. Traces of heavy metals such as Hg, Cd, Pb, As, Co, Mn, Fe and Cr have been identified as deleterious to aquatic ecosystem and human health. Excess inputs are of man made origin coming from industrial and mining effluents (Cd, Cu, Fe, Hg, Pb, Zn) besides (Cu, Hg, Zn), the burning of fuels and car exhausts (As, Ni, Pb, V, Zn). Since the rivers have become the natural dustbins for discharging all sorts of city, industrial and other wastes, these destroy natural harmony between biotic and abiotic components of the aquatic ecosystem. This envisages a great concern with the ecology of polluted waters and its perspectives with regard to its fitness for biotic community including fisheries.

Later being a natural food source for about half of the population of the world needs serious implications as the polluted food could draw not only nutritional complications in terms of diseases or other physiological disorders in humans and fish production, but also produce more hazards to the living world on account of a variety of indirect interacting factors.

Mercury is a unique pollutant because of its apparent indestructibility and ubiquitousness. It attracted worldwide attention as a potential pollutant because of a tragic episode which occurred in the Japanese villages surrounding the Minamata Bay. The victims were struck with a mysterious nervous illness commonly called "Minamata disease" which was later shown to be caused by consuming fish and shell fish that has lethal accumulation of methyl mercury. It resulted in 121 cases of poisoning and 41 death. Another similar incident of mercury poisoning was reported again in Japanese; villages near Niigata prefecture during 1969, which left 6 people dead with 47 poisoned cases.

The chlor-alkali industry in India has grown ten fold since 1947. Large quantities of caustic soda and soda ash produced are used in Paper and Rayon industries. Majority of the chlor-alkali factories adopt old process of using mercury as cathode. This has long since been discarded in Japan and in many other countries. The country imports around 200 tons of mercury every year of which 180 tons are used by the chlor-alkali industry alone and most of it finally goes out along with waste water and sludge, and ultimately enters into the environment.

Central Inland Fisheries Research Institute and National Environment Engineering Research Institute in 1981 reported that Hooghly estuary received 28.5 ppm of chromium from tannery and 0.007 ppm to 129.7 ppm of mercury from other various industries. Kaviraj (1989) also surveyed Hooghly estuary during 1977–81 and reported that an accumulation of 381.6 ppm of zinc and 9.6 ppm of chromium in the sediments, 120.4 ppm of zinc, 127.8 ppm copper and 20 ppm chromium in fishes and 112.2 ppm zinc, 10 ppm copper and 0.3 ppm chromium in shrimps.

The realization of the deleterious conditions of the rivers in the early nineteen century in Britain on account of disappearance of salmon from the Thomei and Mersey rivers prompted passing of certain acts to curb water pollution. In 1857, a royal commission on prevention of river pollution was set up. The federal water pollution central act of 1948 passed in America ascertained water quality and a new act passed in 1965 required the states to establish water quality standards. In India, the water (prevention and control of pollution) act came into existence in 1974 only and thereafter the environmental act was passed on 23 May, 1986. As per WHO, permissible limits for toxic metals in drinking water are: cadmium, 0.01, lead 0.1 and chromium 0.05 mg per litre. As per the analysis of ground water quality as a measure of impact of Bhilai steel plant effluents, it has been reported that the post monsoon ground water samples representing entire water shed area are in general high in toxic elements.

Two kind of heavy metal occur in a animal body, one termed as essential metals and others are non-essential metals. Non-essential metals are present in intracellular compartment and appear toxic to various animal function because of their binding to metallothione (MT). Experimental studies have proved that the role of heavy metal ions in living systems follows the pattern of natural availability. There is no life, that can survive without the participation of metal ions. Some of the major ions such as Na, K, Mg, C are essential to sustain life. In the same way some of the heavy metals are considered essential both for plants and animal nutrition, and they serve some useful biological functions in the body. Thus, Cu, Co, Mn, Mo, Se and Zn are essential to both plants and animals. However, some trace metals such as Cd, Hg, Pb and metalloids like As, Sb and Se are considered to be toxic, although some essential physiological roles have been inferred for As, Cd and Pb recently (Dara, 1993). Among the metals studied, Cr, Ni, Cd and Be have been found to be potentially carcinogenic (*i.e.* causing cancer). Even the heavy metal although essential, on "acute exposure" or "chronic exposure", severely affect different body organs and produces various types of diseases and deformities in the human body.

Copper necessary for the normal biological activity, excessive intake of however, can be toxic and can cause haemolysis and a disease of liver called Willson's disease. Generally liver, kidney, gills and intestine are the main site for MT formation in fishes which is either stored or excreted. Heavy metal residues, Nickle, chromium, cobalt in combined waste water from vegetable oil industries are generally below detectable limits (< 0.01 mg/l). These waters however, are reported to be iron rich (Av. 1.5 mg/litre), and implying very poor DO levels.

According to Haslan, 1978 the effects of sewage and industrial effluents on aquatic biotic system can be described as:

1. A decrease and loss of the species, most sensitive to pollution.
2. A decrease in species diversity at the sight of affluent industries.
3. An increase in any species actually favoured by the polluted waters.

Heavy metals are important, not only as a source of phytotoxicity in plants but also these enter into food chain causing hazardous impact on animals, including the humans. Chromium accumulation in an animal body occur mainly in liver, whereas lead and cadmium accumulation do not vary significantly in different tissues. The major symptoms of heavy metal toxicity in mammals are hypertension, cirrhosis of liver, lung cancer, nausea, vomiting, salivation, diarrhoea, abdominal cramps etc.

Body size is closely related to fish growth and metabolism and also an important attribute to metal accumulation in fishes. Liver and kidney mainly act for storage and transformation of such xenobiotics (Heath, 1995). Maiti and Banerjee (2002) studied bioaccumulation of certain heavy metals in three fish species such as *Labeo rohita, Catla catla* and *Oreochromis nilotica* in wastewater fed wetlands in Kolkata. They reported considerable role of the age of fish in metal accumulation in different fish species. The bioaccumulation of metals in tissues is further species specific.

The expression 'uptake efficiency' has been used to refer to how readily a fish removes a contaminant from the water by the gills. The metal uptake is however, greatly influenced by pH and temperature. Copper uptake by the gills of feathered minnows were found to the slower at low pH, cadmium appear to be taken up in direct correlation with pH and lead exhibit exactly the opposite *i.e.* the pH increases uptake (Camphell and Stokes, 1985). In terms of biomagnification however, different metals behave differently for instance, mercury biomagnifies easily, whereas cadmium and lead seemingly do not, nor do the body concentration of the later two metals increase with age and size of the fish, as dose mercury. The distribution of metals into body tissues and organs appear to vary greatly depending upon the fish species. Lead accumulation mainly occur in scales, gills, bone and liver, and in an estuarine fish-liver and muscles accumulated almost no lead (Spry and Weiner, 1991).

Relationship between mercury content and weight of the fish and fish length have been reported by many workers. They found the mercury content to increase with increase in the body weight. Llyod (1977) measured Hg levels in the liver and muscle tissues of rainbow trout and brown trout from trout parks stream of southern Missouri and found a relationship between concentration and size. Few studies showed decrease in the mercury level with increase in the size.

Tariq *et al.* (1979) reported increased accumulation of mercury in muscle and liver with many cases of increased level in muscle than liver. Out of several species of fishes analysed by Greig *et al.* (1975) most of the species were found to contain equal amount of mercury in muscle than in liver and a few cases of increased level in liver than in muscle.

Acute levels of metals in water, however, is said to stimulate mucus secretion which serve as a way of reducing their uptake in fishes (Part and Lock, 1983). Accumulation of metals further depend upon the environmental conditions and their availability in water and the sediments. Differences in bioaccumulation of metals in body tissues also depend upon the animals physiological activities. *O. nilotica* is said to have, increased accumulation of copper in liver, than in *L. rohita* or *C. catla.* The accumulation of metals by various intracellular compartments is further a function of relative rate of metal binding and release, as determined by their relative metal binding affinity and competition among different metals for the binding sites *i.e.* binding to metallothionein (MT). Low levels of cadmium are known to induce MT formation but with chronic exposures cadmium binding capacity of MT decreases. *Catla* is, however, said to possess a strong tendency to retain lead in the kidney. Organic mercury shows strongest affinity to bind with MT, so much so that it displaces copper and zinc from the molecule. Methyl mercury however, do not bind with MT. Eaton (1985) reported binding affinity of metal to MT in the order: Hg > Cd > Cu > Zn.

Cadmium toxicity depend upon intake of copper as copper displaces cadmium from MT molecules resulting in cadmium elimination. On the other hand lead is synergistic to copper uptake and elimination. There are also report for their vice-versa effects. Zinc uptake is inhibited by low level of cadmium but zinc seems to have no effect on uptake of cadmium. These interactions suggest further complicated role of essential and non-essential metals.

The Tilapia fishes (*O. mossambicus*) generally accumulate higher amount of metal residue as compared to other fishes and shows higher susceptibility to metal toxicity. The metal in combination have been found to exhibit higher toxicity effect. It has been reported that zinc and cadmium alone cause 32 per cent and 42 per cent reduction in growth respectively. Whereas these together at sublethal concentrations exert 65 per cent reduction. Likewise cadmium with copper cause 86 per cent reduction in growth, whereas copper alone is 47 per cent toxic. Even copper and zinc together could yield 73 per cent reduction in growth in tilapia fishes. Increase in lead concentration also cause higher mortality. In common carps, mortality or retarded growth occur at cadmium concentration ranging from 4–12 µg/l.

Bioconcentration factors have also evidenced greater bioaccumulation of metals, by molluscs than macrophytes in wastewater fed wetlands (Bhattacharya and Chakraborty, 2002).

Physiological Impact of Heavy Metals

Apart from acute toxicity, the heavy metals in water has proved dangerous and harmful because of their bioaccumulation and their impact on tissue degeneration. Beside these chemicals are also established to induce reproductive functions, thus influencing growth, survival and reproductive potential of animals/fishes. Loss of immunity, necrosis, inflammation, oedema are many other pathological conditions.

Chromic effect of copper adversely affected the growth and reproduction in fish. Precipitation of mucus on the gills of fishes exposed to copper leads to respiratory troubles. Fishes exposed to mercury exhibit neurological damage and histopathological changes in the visceral organs and gills. All metals are strongly chaleophilic (affinity for sulphur) in their mineralogy and chemistry and their affinity for sulphur is greater than oxygen. Hence they attack the sulphur bonds present in the enzymes and thus inhibit, various metabolic activities that are taking place in the organism. They also exert toxic effect by combining with one or more reactive groups essential for normal physiological function.

Most important are quantitative differences between the chemicals *i.e.* chromium is not nearly as harmful to gill tissue as is nickel, when fish are immersed in solution of comparable toxicity (Hughes *et al.*, 1979). Zinc at a dose level of 71.5 µg/l causes death due to hypoxia because of rapid drop of blood oxygen and pH. Cadmium causes both anaemia in a variety of fish species at low and high concentration as well as abnormally large number of malformed RBC's and altering the leukocyte differential count. Cadmium is further reported to cause inhibition of serum antibody titres in cunners, but a six fold stimulation in stripped bass (Robohm, 1996). Newman and Maclean (1974) has also reported a dose dependent 3–4 fold increase in neutrophils and a dose dependent decrease in lymphocytes in the cunners (Marine teleost fishes). A study conducted by Verma, (1999) from India has also evidenced significant increased in Hb content and RBC's count in case of Indian major carps, exposed to copper and zinc while WBC counts were decreased. But in the group exposed to cadmium a significant decrease in Hb content, RBC's count alongwith WBC count, as in the case of other heavy metals was observed. In case of copper and zinc the lymphocytes were decreased significantly, while the granulocyte and monocyte population tend to increase and with cadmium monocyte and lymphocytes decreased significantly. A significant decrease in the antibody titers was observed in the group exposed to copper and zinc. The serum protein levels were also decreased in all the cases of heavy metals. Likewise, chronic exposure of the cyprinid *Barbus* to lead for upto 60 days at 47 ug/l also resulted in severe reduction (12 to 31 per cent) in RBC's count, haematocrite, haemoglobin concentration and MCV (Tewari *et al.*, 1987).

Other major physiological aspects of metal toxicity is alterations in cellular enzyme activity. For example lead is said to act as a stimulator of alkaline phosphatase, but a strong inhibition of both acid phosphatase and xanthine oxidase. Cadmium along with mercury result in generally enzyme inhibition at high concentration. There was a clear cut increase in carbonic anhydrase activity in flounders exposed to cadmium (Gould, 1977). Shaffi (1993) also reported significant inhibition in certain metabolic enzymes such as lactate dehydrogenase, succinic dehydrogenase and malate dehydrogrenase in the freshwater fish species, following their 24 or 48 h exposure to mercury.

Prolonged cadmium exposure of male brook trout is reported to cause their hormone levels to continue to rise after spermiation at which time they normally decline and the testis further exhibit histopathological damage from cadmium possibly due to inhibition of the system responsible for normal metabolic breakdown of steroids. Hg both in organic and inorganic forms also cause histological lesions in the Leydig's cells of the freshwater catfish *Clarius* (Kirubagaran and Joy, 1992) and lead at low concentration is also said to cause marked suppression of spermatogenesis in trouts. The metals further interfere with the dynamic cell surface measurements due to complexion of ions particularly the bivalent cations like cadmium.

There are many other reports on the physiological and biochemical changes in animal tissue on account of heavy metal toxicity. The biochemical components like proteins, lipids, cholesterol and phospholipids in the liver of *Cyprinus carpio* were found significantly reduced following 60 days exposure to nickel (7.5 to 15.0 mg/l) and chromium (10.0 to 20.0 mg/l).

Heavy Metal Management

Today fishes have become the major diet and there have been attempts to devise ways to enhance fish production. The need of hour is to rationalize the use of industrial effluents on one hand and to minimize their interference with other biotic components including the fishes. Wastewater management technology has received considerable attention in recent years to meet the stringent regulation of water quality. Research in this field has resulted in development of various treatment techniques involving various chemicals and biological materials like bacteria, algae for the precipitation or elimination of various harmful toxicant. Organic loading in terms of biochemical oxygen demand (BOD) is found to be the limiting factor for ponds and rivers. Commonly around 9000 million cubic meters of waste water is discharged in India, of which nearly 32 per cent is currently treated with primary treatment systems. In eastern part of Kolkata alone a number of sewage fed ponds covering about 25000 hectare area are established which approximately produce about 7000 tonnes of fishes annually (Gupta and Mitra, 1997).

Among various biological material in use under wastewater management technology certain plant types are also being used for reducing heavy metal toxicity levels in water such as neem bark is used for the elimination of mercury, pine leaves for nickel, cypress leaves for lead and *Ipomea* leaves for cadmium. *Chlorella vulgaris* is reported to remove these metals upto 100 mg/l of the sewage waste (Shrivastava and Sohani, 2002).

Bioaccumulation method is effectively used for removal of toxic metals in wastewater treatment and industrial waste. This process is a component of commercial water treatment circuit and uses a combination of flocculation and settling as a mechanism by which metals are removed. Microorganisms and their extracellular products of metabolism play an active role in flocculation. Majority of the metals present in the system seem to be removed from solution through adsorption onto the flocs or onto the sludge by physicochemical interactions involving the ion-exchange types of mechanism (Dahiya, 1990).

Absorption and adsorption of metals by microorganisms involves more than biological activities. These mainly include–metal transport mechanism, which accumulate and store selective metals including Mg, Ca, Mn, Fe, Zn, Si, K, Cu, Co, Mo. The transport of metal in microbial cells has been studied in *Chlorella pyrenoidosa, Dunaliela tertiolecia,* Baker's yeast, *Neocosmospora vasinfectum* and *Escherichia coli* (Kelly *et al.*, 1979).

Failla *et al.* (1976) reported the accumulation and storage of Zn^{2+} by *Candida utilis.* Metal accumulation is affected by chemical factors such as cation soluble amino acids and physical factors including temperature, conductance, pH etc. Mostly, the binding efficiency slows in acidic pH. When pH goes to alkaline the binding efficiency decrease (Gloaguen and Morvan, 1997). In general, binding ability of the barks for the metal is, Pb > Cu > Cr Ni > Fe.

Krishnan *et al.* (2002) have further stated that heavy metals such as mercury, lead, cadmium and cobalt from wastewater can be effectively sorbed by sulphurized activated carbon. Sulphurized activated carbon was prepared from sugar industry waste, baggase pith, by single-step steam pyrolysis in the presence of SO_2 and H_2S

at 400°C. The maximum removal took place at the pH range of 4.0–8.0. Maximum removal of 98.2 per cent for Pb(II), 97.5, per cent for Hg (II), 91.5 per cent for Cd(II) and 73.6 per cent for Co(II) was observed at an initial concentration of 100.0 mg/L at 30°C and pH 6.0.

References

Bhattacharya and Chakraborty, S., 2002. In: *Ecology of Polluted Waters*, Vol. 1, (Ed.) Arvind Kumar. APH Publ. Corp., New Delhi, pp. 287.

Campbell, P.G.C. and Stokes, P.M., 1985. *Can. J. Dish. Aquat. Sci.*, 42: 2034.

Chatterjee, S. and Bhattacharya, S., 1984. *Toxicol. Lett.*, 22: 187.

Dahiya, J.S., 1990. *Indian J. Microbiology*, 30: 1–12.

Dara, S.S., 1993. S. Chand and Co. Ltd., Ram Nagar, New Delhi, 210 p.

Eaton, D.I., 1985. *Toxicol. Appl. Pharmacol.*, 78: 158.

Failla, M.L., Benedict, C.D. and Weinberg, E.D., 1976. *J. of General Microbiology*, 94: 23.

Gloaguen, V. and Morvan, H., 1997. *IJEH.*, A32: 901.

Gould, E., 1977. In: *Physiological Response of Marine Biota to Pollutants*, (Eds.) Vermberg, F.J. *et al.* Academic Press, New York, pp. 209.

Greig, R.A., Wengloff, D. and Shelpuk, C., 1975. *Pest. Monit. J.*, 9: 15.

Gupta, D.C., Peters E. and Yadava, R.N., 2002. In: *Ecology of Polluted Waters*, Vol. 1, (Ed.) Arvind Kumar. APH Publ. Corp., New Delhi, pp. 515.

Gupta, S.K. and Mitra, A., 1997. *J. Ind. Soc. Soil Sc.*, pp. 52.

Haslan, S.M., 1978. *River Plants*. Cambridge University Press, Cambridge, pp. 396.

Heath, A.G., 1995. CRC Press, NC. Florida, USA.

Hughes, G.M., Perry, S.F. and Brown, V.M., 1979. *Water Res.*, 13: 665.

Kaviraj, A., 1989. *Sci. Cult.*, 55(7): 257.

Kelly, D.P., Norris, P.R. and Brierley, C.L., 1979. *Symp. Soc. Gen. Microgeology*, 29: 263.

Kirubagaran, R. and Joy, K., 1992. *J. Fish Biol.*, 41: 305.

Krishnan, K., Vinod, V.P. and Anirudhan, T.S., 2002. In: *Ecology of Polluted Waters,* Vol. 1, (Ed.) Arvind Kumar. APH Publ. Corp., New Delhi, pp. 549.

Lloyd, E.T., 1977. *Environ. Res.,* 13: 62.

Maiti, P. and Banerjee, S., 2002. In: *Ecology of Polluted Waters,* Vol. 1, (Ed.) Arvind Kumar. APH Publ. Corp., New Delhi, pp. 217.

Newman, M.W. and MacLean, S.A., 1974. *NMFS SSRF,* 681, 27.

Norey, C.G., Brown, M.W., Cryer, A. and Kay, J., 1990. *Comp. Biochem. Physiol.,* 96C: 181.

Part, P. and Lock, R.A.C., 1983. *Comp. Biochem. Physiol.,* 76C: 259–263.

Robohn, R.A., 1986. *Vety. Immunol. Immunopathol.* 12: 251.

Shaffi, S., 1993. *Physiol. Res.,* 42: 7.

Shrivastava, V.S. and Sohani, D., 2002. In: *Ecology of Polluted Waters,* Vol. 1, (Ed.) Arvind Kumar. APH Publ. Corp., New Delhi, pp. 435.

Spry, D. and Wiener, J., 1991. *Environ. Pollut.,* 71: 243.

Tariq, W.K., M.D. George and R.S. Gupta, 1979. *Mar. Pollut. Bull.,* 10: 357.

Tewari, H., Gill, T. and Pant, J., 1987. *Bull. Environ. Contam. Toxicol.,* 38: 748.

Verma, A.R., 1999. *Research Thesis.* Central Institute of Fisheries Education, Versova, Mumbai.

Chapter 7

Effect of Dietary L-Thyroxine Supplementation on Growth of the Indian Major Carps Fry

A.K. Pandey*, P.K. Arvindakshan* and B.N. Singh**

**Central Institute of Freshwater Aquaculture, Kausalyaganga, Bhubaneswar – 751 002*

***National Bureau of Fish Genetic Resources, Canal Ring Road, Telibagh, Lucknow – 226 002*

ABSTRACT

In order to record the effect of dietary thyroxine (T_4) supplementation on growth, experiments were carried out on fly of the Indian major carps for 90 days under hatchery conditions. Basic ingredients of the diets were fish meal 5 per cent, soyabean oil-cake (SOC) 25 per cent, groundnut oil-cake (GOC) 35 per cent, rice bran 24.34 per cent, wheat flour 10 per cent and vitamin-mineral mixture 0.66 per cent. In the control diet, thyroxine was not added whereas in the experimental feeds, thyroxine was supplemented at the doses of 0.10, 0.03 and 0.05 ppm. Crude protein content of the control as well as the test feeds were kept at uniform level of 30.17 per cent. Feeding was done twice daily @ 3 per cent of biomass. The present study demonstrated that the fly maintained on 0.05 ppm thyroxine supplemented feed showed better growth as compared to those fed on the diet containing 0.03 and 0.10 ppm thyroxine.

***Keywords**: Dietary/Thyroxine, Growth, Weight gain, Fry, Indian major carps.*

Introduction

Role of thyroid hormones (T_3, T_4) in inducing metamorphosis among amphibians, especially in anurans, is well defined (Houndry, 1993; Weber *et al.*, 1994; Evans, 1998). These hormones have been detected in eggs (Tagawa and Hirano, 1987; Tam, 1994) and their concentrations vary during early development of the fish (de Jesus *et al.*, 1991; Yamano *et al.*, 1991). There are evidences that thyroid hormones enhance larval growth, development, metamorphosis and survival in the early stages of some teleosts (Donaldson *et al.*, 1979; Higgs *et al.*, 1982; Miwa *et al.*, 1988; Pandey, 1989; Hirata *et al.*, 1989; Lam, 1994; Inui *et al.*, 1995; de Jesus *et al.*, 1998; Park and Zhang, 1998; Solbakken *et al.*, 1999; Yoo *et al.*, 2000). There exist reports that exogenous thyroid hormone treatments either try feeding, immersion or injection stimulate growth, development and survival in many species including *Oreochromis mossambicus* (Lam, 1980; Reddy and Lam 1992a; Wan *et al.*, 1997), *Oreochromis niloticus* (Nacario, 1983), *Cyprinus carpio* (Lam and Sharma, 1985), *Colisa lalia* (Reddy and Lam, 1987), *Chrysophris major* (Woo *et al.*, 1991), *Carassius auratus* (Reddy and Lam, 1992b), *Ctenopharyngodon idella* (Azad, 1992), *Pagrosomus major* (Xu *et al.*, 1995), *Lates calcarifer* (Wan *et al.*, 1997) and *Paralichthys olivaceus* (Bao *et al.*, 1999). Improved larval survival has also been recorded even after maternal administration of thyroid hormones during breeding period (Brown *et al.*, 1989; Ayson and Lam, 1993). Thyroid hormones after getting associated with vitellogenin are accumulated in yolk of the developing oocytes (Monteverdi and Digiulio, 2000). While the growth-promoting effect of thyroxine has been recorded during the early stages of some freshwater teleosts (Donaldson *et al.*, 1979; Lam 1980), contradictory results have also been reported in some species (Chaudhary *et al.*, 1989; Donaldson and Dawn, 1993). Though thyroid hormones have potentials in enhancing growth and survival in fish, they are yet to be used in the aquaculture systems (Donaldson and Dawn, 1993; Brown and Kim, 1995). An attempt was made to record the effect of dietary L-thyroxine supplementation on growth of the fry of Indian major carps under hatchery conditions.

Materials and Methods

The fry of the Indian major carps (average weight: catla 0.51 ± 0.04 g, rohu 0.31 ± 0.03 g, mrigal 0.36 ± 0.02 g) were initially reared in plastic troughs of 400 litres capacity. Thereafter, they were divided into four equal groups–one group served as control, whereas the remaining three groups served as treatments. In each trough, 100 fry were stocked at a density of 6 per litre. The control diet (Feed T–1) comprised fish meal, rice bran, soyabean oil-cake (SOC), groundnut oil-cake (GOC), vitamin-mineral mixture (without thyroxine) whereas the three test diets were supplemented with thyroxine by spray-dried method at the dose of 0.10 ppm (Feed T–2), 0:05 ppm (Feed T–3), 0.03 ppm (Feed T–4) (Table 7.1). The thyroid hormone used in this experiment was *Eltroxin* (Glaxo, India) a commercial preparation of L-thyroxine sodium (T_4), each tablet containing 0.091 mg of the hormone. The feeds were given in the form of pellets of 2 mm. Feeding was done twice daily @ 3 per cent of biomass and the remaining food and faecal matter were collected regularly, dried in oven and stored for analysis following standard methods (AOAC, 1980). Regular sampling of the fish was done on every 15 days intervals and the feed was adjusted accordingly. The crude protein of the control as well as the test diets were kept at a uniform level of 30.17 per cent. Physico-chemical parameters of the water ranged between pH 7.3–7.9, temperature 28–30°C, dissolved oxygen 4.5–6.0 ppm, dissolved ammonia 0.05–0.10 ppm and alkalinity 122–131 ppm during the experimental period. Gentle aeration was provided during 90 days of the experiment.

Table 7.1: Proximate Composition of Diets Fed to the Fry of the Indian Major Carps

Ingredients (%)	*Feed T–1 (Control)*	*Feed T–2*	*Feed T–3*	*Feed T–4*
Fish meal	5	5	5	5
Soyabean oil-cake	25	25	25	25
Groundnut oil-cake	35	35	35	35
Rice bran	24.34	24.34	24.34	24.34
Wheat flour	10	10	10	10
Supplivite–M	0.66	0.66	0.66	0.66
Thyroxine (ppm)	–	0.10	0.05	0.03
Crude proteins	30.17	30.17	30.17	30.17

Weight gain percentage, feed conversion efficiency (FCE per cent), feed conversion ratio (FCR) and protein efficiency ratio (PER) of all the groups were calculated by using the following formulae

(Singh, 1990) which have been summarized for *Catla catla, Labeo rohita* and *Cirrhinus mrigala* in Tables 7.2–7.4, respectively. These values were evaluated for statistical significance using Students 't' test.

$$\text{Weight Gain Percentage} = \frac{\text{Final Weight} - \text{Initial Weight}}{\text{Total weight}} \times 100$$

$$\text{Feed Conversion Efficiency (FCE) (\%)} = \frac{\text{Gain in Wet Weight of the Fish}}{\text{Total Weight of Dry Food Consumed}} \times 100$$

$$\text{Feed Conversion Ratio (FCR)} = \frac{\text{Quantity of Feed Consumed}}{\text{Gain in Wet Weight of the Fish}}$$

$$\text{Protein Efficiency Ratio (PER)} = \frac{\text{Gain in Wet Weight of the Fish}}{\text{Quantity of Protein Fed}}$$

Table 7.2: Effect of Varying Concentrations of L-thyroxine on Growth and Feed Utilization of *Catla catla* Fry

Parameters	*Feed T–1 (Control)*	*Feed T–2*	*Feed T–3*	*Feed T–4*
Initial average weight (g)	0.52 ± 0.04	0.49 ± 0.02	0.51 ± 0.04	0.54 ± 0.05
Final average weight (g)	0.77 ± 0.08	1.00 ± 0.06	1.66 ± 0.08	1.46 ± 0.07
Increase in weight (g)	0.25	0.48	1.14	0.94
Weight gain %	48.07	92.30	219.23	180.76
FCE (%)	60.24	68.50	129.87	100.0
FCR	1.66	1.46	0.77	1.00
PER	0.25	0.35	0.58	0.44

Results and Discussion

Fry of the Indian major carps maintained on the diet supplemented with 0.05 ppm thyroxine (Feed T–3) recorded lowest feed conversion ratio (FCR) as well as maximum weight gain per cent, feed conversion efficiency (FCR) per cent and protein efficiency ratio (PER) followed by 0.03 ppm and 0.10 ppm, respectively. Though the fry fed on 0.10 ppm thyroxine-supplemented diet (Feed T–2) recorded significantly higher growth rate as compared to those fed on control diet, those fed on 0.05 ppm thyroxine-supplemented diet (Feed T–3) recorded maximum growth (Tables 7.2–7.4). Higher growth rates have also been observed in other fishes (*Oncorhynchus kisutch, Oreochromis mossambicus* and *Ctenopharyngodon idella* in response to thyroxine administration (Higgs *et al.*, 1976, 1979; Matty *et al.*, 1982; Azad, 1992). However, it is pertinent to remark that the higher doses of thyroxine (0.10 and 0.03 ppm) exhibited reduced growth in fry of the Indian major carps. There exist reports that the higher dosage of thyroxine or its prolonged treatment inhibit growth in *Sarotherodon mossambicus* (Lam, 1980), *Carassius auratus* (Reddy *et al.*, 1992b), *Ctenopharyngodon idella* (Azad, 1992) and *Etroplus suratensis* (Sambhu and Jayaprakas, 1997). Thus, 0.05 ppm L-thyroxine appears to be the optimum dose for eliciting maximum growth in India major carps (Tables 7.2–7.4). Thyroid hormones appears to promote growth in teleosts by increasing food intake and feed conversion efficiency (Higgs *et al.*, 1979; Muniandi, 1989; Woo *et al.*, 1991; Reddy and Lam., 1992b; Sambhu and Jayaprakas, 1997). In the present study also thyroxine supplementation in the diet resulted in improved feed.

Table 7.3: Effect of Varying Concentrations of L-thyroxine on Growth and Feed Utilization of *Labeo rohita* Fry

Parameters	*Feed T–1 (Control)*	*Feed T–2*	*Feed T–3*	*Feed T–4*
Initial average weight (g)	0.31 ± 0.03	0.40 ± 0.02	0.40 ± 0.03	0.34 ± 0.03
Final average weight (g)	2.55 ± 0.02	2.74 ± 0.03	3.44 ± 0.02	3.02 ± 0.02
Increase in weight (g)	2.24	2.34	3.04	2.68
Weight gain %	722.5	610.0	762.5	748.2
FCE (%)	34.24	42.38	76.30	47.49
FCR	2.76	2.11	1.16	1.98
PER	0.29	0.36	0.47	0.40

Table 7.4: Effect of Varying Concentrations of L-thyroxine on Growth and Feed Utilization of *Cirrhinus mrigala* Fry

Parameters	*Feed T–1 (Control)*	*Feed T–2*	*Feed T–3*	*Feed T–4*
Initial average weight (g)	0.29 ± 0.02	0.32 ± 0.03	0.30 ± 0.02	0.32 ± 0.02
Final average weight (g)	0.50 ± 0.04	50.72 ± 0.03	3.44 ± 0.02	3.02 ± 0.02
Increase in weight (g)	0.21	0.40	0.83	0.58
Weight gain per cent	72.41	125.00	276.60	181.25
FCE (%)	34.24	46.29	78.12	49.26
FCR	2.92	2.16,	1.28	2.03
PER	0.26	034	0.58	0.41

Acknowledgements

We are grateful to the Director, Central Institute of Freshwater Aquaculture, Bhubaneswar for extending necessary facilities to carry out this work. Thanks are due to Dr. P.V. Rangacharulu, Senior Scientist, Nutrition Division, CIFA for help in analysis of data. Financial support provided by the Department of Biotechnology (Government of India), New Delhi is thankfully acknowledged.

References

AOAC, 1980. *Official Methods of Analysis of the AOAC*. AOAC, Washington, D.C. (USA).

Ayson, F.G. and T.J. Lam, 1993. Thyroxine injection of female rabbitfish (*Signanus guttatus*) brood stock changes in thyroid hormone levels in plasma, eggs and yolk sac larvae and its effects on larval growth and survival. *Aquaculture,* 109: 83–93.

Azad, I.S., 1992. Dietary thyroxine vis-à-vis growth of grass carp (*Ctenopharyngodon idella*) fry. *J. Inland Fish. Soc., India,* 24(2): 28–35.

Bao, B., Z. Zhang, X. Gong and X. Wang, 1999. The effects of exogenous thyroxine (T_4) and cortisone on the growth, development and metamorphosis of *Paralichthys alivaceus* during early development stage. *J. Shanghai Fish. Univ.,* 8: 242–246.

Brown, C.L. and B.G. Kim, 1995. Combined application of cortisol and triiodothyronine in the culture of larval marine finfish. *Aquaculture*, 135: 79–86.

Brown, C.L., S.I. Dorosov, D.M. Cochran and H.A. Bern, 1989. Enhanced survival in striped bass fingerlings after maternal triiodothryonine treatment. *Fish Physiol. Biachem.*, 7: 295–299.

Chaudhary, M.A., K.P. Lone and A.J. Matty, 1989. Oral administration of thyroxine and triiodothryonine on growth and feed conversion of *Oreochromis mossambicus* and *Cyprinus carpio*. *Pak. J: Zool.*, 21: 307–318.

Donaldson, E.M. and N.E. Dawn, 1993. Hormonal enhancement of growth. In: *Recent Advances in Aquaculture, Vol. IV*, (Eds.) J.R. Muir and R.J. Roberts. pp. 109–120. Blackwell Sci. Pub., London.

Donaldson, E.M., U.H.M. Fagerlund, D.A. Hicks and J.R.L. McBride, 1979. Hormonal enhancement of growth. In: *Fish Physiology, Vol. VIII*, (Eds.) W.S. Hoar, D.J. Randall and J.R. Brett. pp. 455–597. Academic Press, New York.

Evans, D.H., 1998. *The Physiology of Fishes*, 2nd Edition. CRC Press, Boca Rarton and New York.

Higgs, D.A., E.M. Donaldson, H.M. Dye and J.R.L. McBride, 1976. Influence of bovine growth honnone, muscle composition and histological structures of the gonads, thyroid, pancreas and pituitary of coho salmon, *Oncorhynchus kisutch*. *J. Fish. Res. Bd., Can.*, 33: 1585–1603.

Higgs, D.A., U.H.M. Ferglund, J.R.L. McBride and E.G. Eales, 1979. Influence of orally administered L-thyroxine or 3,5,3'triiodo-L-thyroxine on growth, food consumption and food conversion of under yearling coho salmon. *Can. J. Zool.*, 57: 1974–1979.

Higgs, D.A., U.H.M. Ferglund, E.G. Eales and J.R.L. McBride, 1982. Application of thyroid and steroid hormones as anabolic agents in fish culture. *Comp. Biochem. Physiol.*, 73: 143–176.

Hirata, Y., H. Kurokura, S. Kasahara., 1989. Effects of thyroxine and thiourea on the development of larval red sea bream, *Pagrus major*. *Bull. Jap. Soc. Sci. Fish.*, 55: 1189–1195.

Houndry, J., 1993. Passage to terrestrial life in amphibians. II. Endocrine determinism. *Zool. Sci.*, 10: 887–902.

Inui, Y., K. Tamono and S. Miwa, 1995. The role of thyroid hormone in tissue development in metamorphosing flounder. *Aquaculture,* 135: 87–98.

de Jesus, E.G., T. Hirano and Y. Inui, 1991. Changes in cortisol and thyroid hormone concentrations during early development and metamorphosis in the Japanese flounder, *Paralichthys olivateus. Gen. Comp. Endocrinol.,* 82: 369–376.

de Jesus, E.G., J.D. Toledo and M.S. Simpas, 1998. Thyroid hormones promote early metamorphosis in grouper (*Epinephulus coioides*) larvae. *Gen Comp. Endocrinol.,* 112: 10–16.

Lam, T.J., 1980. Thyroxine enhances larval development and survival in *Sarotherodon (Tilapia) mossambicus,* Ruppell. *Aquaculture,* 21: 287–291.

Lam, T.J., 1994. Hormones and egg/larval quality in fish. *J. World Aquacult. Soc.,* 25: 2–12.

Lam, T.J. and R. Sharma, 1985. Effects of salinity and thyroxine on larval survival, growth and development in the carp, *Cyprinus carpio. Aquaculture,* 44: 201–212.

Matty, A.J., M.A. Chaudhary and K.P. Lone, 1982. The effect of thyroid hormones and temperature on protein and nucleic acid contents of liver and muscle of *Oreochromis mossambicus. Gen. Comp. Endocrinol.,* 47: 497–510.

Miwa, S., M. Tagawa, Y. Inui and T. Hirano, 1988. Thyroxine surge in metamorphosing flounder larvae. *Gen. Comp. Endocrinol.,* 70: 158–163.

Monteverdi, G.H. and R.T. Digiulio, 2000. Vitellogenin association and oocytic accumulation of thyroxine and 3,5,3'-triiodothyronine in gravid *Fundulus heteroclitus. Gen. Comp. Endocrinol.,* 120: 198–211.

Muniandi, S., 1989. The impact of L-thyroxine and thiourea on food consumption, assimilation, and conversion in fish, *Heteropneustes fossilis. ANJAL Journal,* 9: 21–24.

Nacario, J.F., 1983. The effect of thyroxine on the larvae and fry of *Sarotherodon noliticus* L. (= *Tilapia nilotica*). *Aquaculture,* 34: 73–83.

Pandey, A.K., 1989. Thyroid gland of the developing tilapia, *Oreochromis mossambicus*, with a note on the follicle formation. *J. Adv. Zool.*, 10: 142–144.

Park, I.M. and C.I. Zhang, 1998. The effect of thyroxine on the growth and yolk resorption of chum salmon (*Oncorhyanchus keta*) yolk sac larvae. *Israeli J. Aquacult.* (*Bamidgeh*): 50: 60–66.

Reddy, P.K. and Lam, T.J., 1987. Effects of salinity and thyroxine on larval survival and growth in the dwarf gourami, *Colisa lalia. J. Aquacult. Trop.*, 2: 79–87.

Reddy, P.K. and T.J. Lam, 1992a. Role of thyroid hormones in tilapia larvae (*Oreochromis mossambicus*). I. Effects of hormones and antithyroid drug on yolk absorption, growth and development. *Fish Physiol. Biochem.*, 9: 473–485.

Reddy, P.K. and T.J. Lam, 1992b. Effect of thyroid hormones on morphogenesis and growth of larvae and fry of telescopic-eye black goldfish, *Carassius auratus. Aquaculture*, 107: 383–394.

Sambhu, C. and V. Jayaprakas, 1997. Growth promoting potential of L-thyroxine in green chromide, *Etroplus suratersis* (Bloch) (Cichlidae–Pisces). *J. Aquacult. Trop.*, 12: 305–318.

Singh, B.N., 1990. Protein requirement of young silver carp, *Hypophthalmichthys molitrix* (*Val.*). *J. Freshwater Biol.*, 2: 89–95.

Solbakken, J.S., B. Notberg, K. Watanabe and K. Pittman, 1999. Thyroxine as a mediator of metamorphosis of Atlantic halibut, *Hippoglossus hippoglossus. Environ. Biol. Fish.*, 56: 53–65.

Tagawa, M. and T. Hitano, 1987. Presence of thyroxine in eggs and changes in its content during early development of chum salmon, *Oncorhynchus keta. Gen. Camp. Endocrinol.*, 68: 29–135.

Wail, L.S.C., P.W.S. Heng, L.W. Chan, C.C. Lee, K.K. Ho, W.K. Ngoh, J. Walford and T.J. Lam, 1997. Effect of feeding with micro spheres containing thyroid hormone on larval growth and development in tilapia, *Oreochromis mossambicus* and on larval growth in seabass, *Lates calcarifer. J. Aquacult. Trop.*, 12: 191–199.

Weber, G.M., E.S. Farrar, C.K.F. Tom and E.G. Grau, 1994. Changes in whole-body thyroxine and triiodothyronine concentrations and total content during early development and metamorphosis of the toad, *Bufa marinus. Gen. Comp. Endocrinol.*, 94: 62–71.

Woo, N.Y.S., A.S.B. Chung and N.G. Tb, 1991. Influence of oral administration of 3,5,3'-triiodo-L-thyroxine on growth, digestion, food conversion and metabolism in the under rearing red seabream, *Chrysophris major. J. Fish Biol.*, 39: 549–568.

Xu, B., Ge, G.C. and H.Z. Miao, 1995. The effects of recombinant eel growth hormone, methyltestosterone and L–thyroxine on the growth of red sea bream, *Pagrosomus major. Chin. J. Oceanol. Limnol.*, 13: 86–95.

Yamano, K., M. Tagawa, E.G. de-Jesus, T. Hirano, S. Miwa and Y. Inui, 1991. Changes in whole body concentrations of thyroid hormones and cortisol in metamorphosing conger eel. *J. Comp. Physiol.*, 161B: 371–315.

Yoo, J.H., T. Takeuchi and T. Seikai, 2000. Sensibility of the metamorphic events and morphogenesis of Japanese flounder, *Paralichthys olivaceus*, during larval development to thyroxine. *Fish. Sci.*, 66: 846–850.

Chapter 8

Nutritional Ecology of Carp (*Cyprinus carpio*, L.) Fingerling through Adult

P. Bambroo, Y.R. Malhotra** and K.K. Sharma***

**Department of Zoology, Government College for Women, Parade, Jammu (J&K)*

***P.G. Department of Zoology, University of Jammu (J&K)*

ABSTRACT

Foraging preference is tested in carp (*Cyprinus carpio*, L.) fingerlings through adult using Ivlev's index E. Fingerlings have been found to be planktivorous and column feeders. During early stages, these show strong preference for cladocerans (*chydorus*) over copepods. The advanced fingerlings show almost identical preference for both cladocerans and copepods. The sub-adults and adults are benthic feeders and exhibit preference for chironomids over oligochaetes. The zooplankton is the alternative food of adult carp showing plasticity in its feeding behaviour. The small gape (one-sixteenth of their standard length) does not allow the adult carp to show true macrophagy.

The plant matter is included in the diet by carp after attaining a size of 45–50 mm. The protsusible mouth in the adult also brings in some detritus, though passively, during its benthic foraging bout.

Introduction

Prey selectivity in larval fish, being of considerable relevance in aquaculture (Jhingran and Pullin, 1985 and Pillay, 1990), has been studied by many workers (Khadka and Rao, 1986; Chakrabarti and Jana, 1990; Mookerji and Rao, 1993 and Bambroo 1997). Such studies in respect of fingerling through adult stages of fish, also assume great significance to maximize the yield, but have not received the due attention of the fish biologists. This is why scanty information is available on the prey selectivity in advanced developmental stages of a fish, particularly in respect of carp.

It is a fact that carp is capable of digesting the artificial dry food from fingerling stage onwards, but unbalanced feeds given to a fish in ponds, *viz.*, oil seed cakes, feed mixtures etc. do not correspond to the nature of their feeding habits and obviously cannot be considered complete wholesome food. Any deficiency in one or the other nutritional requirement in the feed, is supplemented by carp from the physiologically complete natural food available in the pond. This is the basis of modern methods of feeding with natural foods constituting 30 per cent against 70 per cent artificial feed given (Martyshev, 1983). Considering this importance of natural food in carp ration, the present study on prey selection of carp fingerlings through adult stage has been undertaken.

Materials and Methods

The prey preference in the advanced stages of *C. carpio* was tested in five size classes (Table 8.1). Feeding trials on carp fingerlings were carried out in a pond at Kunjwani, Jammu. Happa (mosquito net enclosure) was pitched in the pond and left undisturbed for 1 hr. to ensure the availability of food organisms. Plankton sample was collected from the happa, using plankton net with 900 meshes/cm^2, and preserved in 8 per cent formaline for determination of percent species composition. This was followed by introduction of test fish into, the happa and farmer was allowed to feed. During the experimentation period 10 fingerlings were removed periodically and preserved in 8 per cent formaline.

For advanced stages of carp (Sub-adults and adults) larger sized happa (3 mts. × 3 mts. and 2 mts.) was specially designed to make it go deep into the bottom to facilitate the entry of *Zoobenthos* along

Table 8.1: Morphometery of Various Size Classes of *C. carpio* Fingerling through Adult

Size Class of Carp	Body Length (mm)	Gut Length (mm)	Upper Jaw Length (mm)	Mouth at 45° (mm)	Mouth at 90° (mm)	RLG	Ratio of Mouth Size to Body Size
I	33.6	36	2.5	19	35	1.07	1 : 17
	(30–35)	(30–45)	(2–3.2)				
II	48	66	4.5	34	63	1.37	1 : 14
	(45–50)	(58–76)	(3.6–5.2)				
III	74	120	6.5	49	91	1.62	1 : 15
	(70–80)	(110–134)	(5.6–7.0)				
IV	107	185	8.2	62	115	1.72	1 : 17
	(110–116)	(174–206)	(7.6–9.4)				
V	167	290	13	99	184	1.79	1 : 16
	(150–170)	(276–310)	(12–15.2)				

RLG: Relative length of gut.

Range in parenthesis.

with pond sediment, into the happa. The mesh size was no constraint for the entry of zooplankton either. The rest of the procedure remained same as described for fingerlings with the difference that species composition of zoobenthos available in the happa was also analysed apart from Zooplankton (Table 8.2).

Table 8.2: Percent Prey Organisms Available to *C. carpio* Fingerlings, Sub-adults and Adults when Tested for Prey Preference

Food Organisms	*Carp Size–Classes*				
	I	*II*	*III*	*IV*	*V*
Zooplankton					
Protozoa					
Diffugia sp.	8	3	–	4	3
Centropyxis sp.	3	–	–	–	–
Unidentified sp.	1	–	–	–	–
Rotifers					
Rotatoria sp.	4	–	–	–	–
Asplanchna priodonta	4	–	–	–	–
Brachionus calyciflorus	2	9	–	2	3
Chonchilus arboreus	–	–	8	4	–
Cladocera					
Chydorus sphericus	10	31	50	6	7
Alona intermedia	–	4	–	–	–
Ceriodaphnia reticulata	–	10	–	10	16
Daphnia exilis	–	6	–	4	5
Macrothrix rosea	–	–	4	2	2
Simocephalus vetulus	–	14	32	16	12
Copepoda					
Cyclops vicinus	18	3	6	46	36
Mesocyclops sp.	5	10	–	–	6
Paracyclops sp.	5	10	–	–	–
Turbellaria					
Microstomum caudatum	40	–	–	6	8

Contd...

Table 8.2–Contd...

Food Organisms	Carp Size–Classes				
	I	*II*	*III*	*IV*	*V*
Zoobenthos					
Diptera					
Pentaneura sp.	=	=	=	14	16
Chironomus sp.	=	=	=	24	20
Forcipomypia sp.	=	=	=	3	–
Oligochaeta					
Tubifex sp.	=	=	=	24	29
Nais sp.	=	=	=	23	17
Aleosoma sp.	=	=	=	6	9
Stylaria sp.	=	=	=	6	9

=: Availability in the ambient medium (at the bottom) not calculated as *C. carpio* at this stage was found column feeder.

–: Not available in the medium.

The prey type selection in carp was tested by using *Electivity Index E of Ivlev* (1961). The mouth size, when opened at 45° and 90°, was determined by using the formula of Shirota (1970). The categorization of developmental stages of carp is after Titicomb (1910) and Kryzhanovsky (1956).

Results

The study has revealed the following two trends in the foraging behavior of carp fingerling through adult.

1. Carp fingerling (size class I–II) are planktivorous and have a strong preference for cladocerans (*Chydorus*).
2. Sub-adult and adult stages of the carp (size class IV and V) are benthic feeders with preference for chironomids over oligochaetes.

The early carp fingerlings (Size class–I) show preference for cladocerans (*Chydorus*) over copepods (Table 8.3). *Microstomum*, the only turbellarian available in the ambient medium is also selected at this age.

Table 8.3: Prey Selection in *C. carpio* Fingerlingings through Adult

Food Organisms	*Carp Size Classes*									
	I		*II*		*III*		*IV*⁺		*V**	
	Percent Ingested	*Ivlev's Index 'E'*	*Percent Ingested*	*Ivlev's Index 'E'*	*Percent Ingested*	*Ivlev's Index 'E'*	*Percent Ingested*	*Ivelv's Index 'E'*	*Percent Ingested*	*Ivlev's Index 'E'*
Zooplankton										
Cladocera										
Chydorus sp.	14	0.27	70	0.38	85	0.25	–	–	10	0.17
Daphnia sp.	+	+	4	– 0.2	+	+	–	–	–	–
Ceridaphnia sp.	+	+	8	– 0.1	+	+	10	0	–	–
Copepoda										
Cyclops sp.	16	– 0.05	6	0.33	15	0.42	90	0.32	90	0.42
Mesocyclops sp.	–	–	12	0.09	–	–	+	+	–	–
Turbellaria										
Microstomum sp.	70	0.27	–	–	–	–	–	–	–	–
Zoobenthos										
Diptera										
Chironomus sp.	=	=	=	=	=	=	34	0.17	48	0.41
Pentaneura sp.	=	=	=	=	=	=	44	0.51	27	0.37
Oligochaeta										
Nais sp.	=	=	=	=	=	=	18	–0.12	15	– 0.06
Aleosoma sp.	=	=	=	=	=	=	4	– 0.20	–	–
Tubifex sp.	=	=	=	=	=	=	–	–	10	– 0.48

–: Not available in the medium; +: Available but not included in the dietary; *: Selection of Zooplankton & Zoobenthos calculated separately; =: Not tested for their role as food as *C. carpio* is column feeder at this stage.

The fingerlings of size class II are highly selective in respect of their prey within cladocerans, selecting *chydorus* actively (E value +ve; Table 8.3) while including *Daphnia* and *Ceriodaphnia,* non-selectively (E value –ve) in their diet and exhibiting strong disliking for *Simocephalus* (not included in the diet) although available in the medium. This size also exhibits preference for copepodes.

Plant matter, forming 42 per cent of the gut contents in this size group, is also observed for the first time in bulk in the ontogeny of the carp (Table 8.4 and Figure 8.1).

The advanced fingerlings of carp (size class III) show foraging behaviour, more or less similar to that of size class II (Table 8.3).

The sub-adult and adult stages of carp (Size class IV and V) are benthic feeders and show preference for chironomid larvae (*Chironomous* and *Pentaneura*). The oligochaetes (*Nais, Aleosoma* and *Tubifex*) are ingested non-selectively by the carp, (-ve E values; Table 8.3). Besides, carp also feeds on zooplanktonic food at this stage, exhibiting preference for copepods (Cyclops) over cladocerans (*Ceriodaphnia*). However, the zooplankton was found either intact or partly digested in the intestinal bulb, while as chironomids and oligochaetes had suffered maximum digestion in the gut of the carp.

Plant matter forms part of the diet in both sub-adult and adult carp (Figure 8.1).

Besides, the carp of size group IV and V also show detritus in their gut as revealed by Figure 8.1.

Discussion

The present study has revealed that the carp fingerling (size class, I–III) are column feeders as plankton is their main food while as sub-adult and adult carp (size class IV and V) is designated as benthic feeder, for, macrozoobenthos forms their first choice as food.

The continued preference of carp fingerlings for *chydorus* (also preyed upon selectively by advanced carp larvae, Bambroo, 1997) is explained on the basis of phenomenon of 'search images' invoked by fish to seek out the prey to which they have been previously accustomed, the concept proposed by Tinbergen (1960) and later also reported by Ware (1971) in fishes.

The selection of *Microstomum* by early fingerlings appears to be situational as this prey formed 40 per cent of available plankton at

Table 8.4: Plant Food Ingested by *C. carpio* Fingerling through Adult

I	*II*	*III*	*IV*	*V*
Lemna sp.	*Closterium* sp.	*Closterium* sp.	*Lemna* sp.	*Lemna* sp.
Closterium sp.	*Microscystis* sp.	*Melosira* sp.	*Closterium* sp.	*Closterium* sp.
Macrophytic parts	*Synedra* sp.		Leaf fragments	Filamenton algae
	Crucigenia sp.			Macrophytic parts
	Scenedismus sp.			
	Cosmarium sp.			
	Selenastrum sp.			
	Fragilaria sp.			

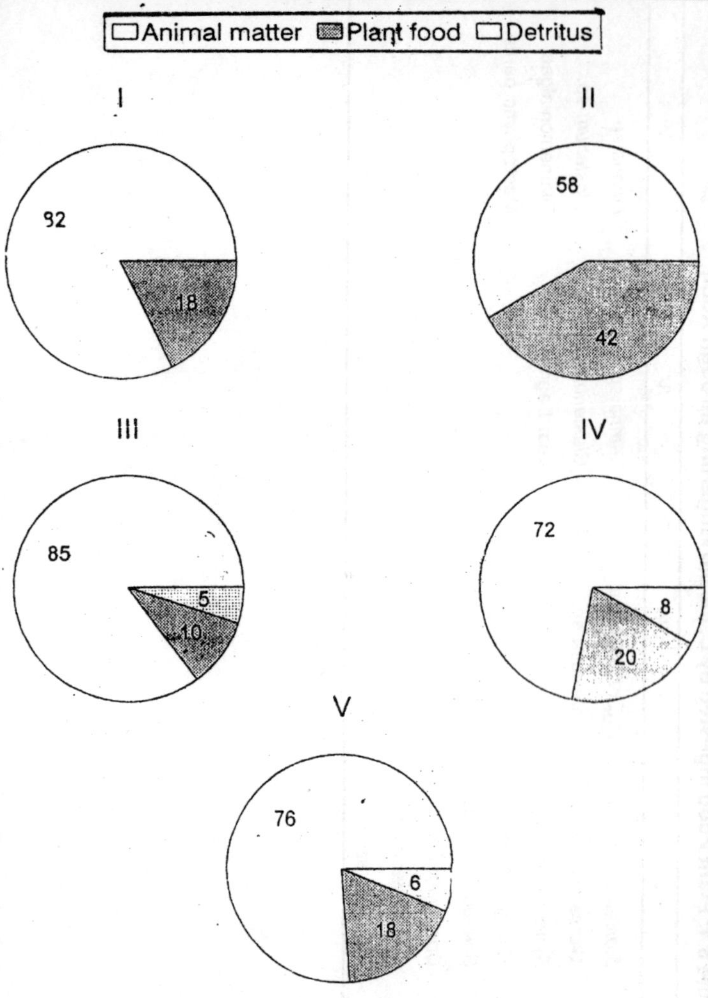

Figure 8.1: Showing Food Spectrum of Carp Fingerling through Adult

this stage and reveals that there is "no all or none law" in the carp for its preferred food.

The selection of evasive copepods by fingerlings (size II–III) over slow moving cladocerans (*Ceriodaphnia*) explains the ability of carp at this stage to *persue* the prey once the attack is initiated.

That phytoplankton forms part of the diet in the fingerlings of carp, is a reflection of their improved ability to digest the plant matter. This is attributed to increased relative length of the gut which exceeds 1 (one) from fingerling onwards (Table 8.1). Earlier Nikolsky (1963) also reported that fish with RLG exceeding 1 are herbivorous. However, delibrate selection of phytoplankton by this fish is ruled out as particulate planktivores (like carp) are not physically equipped to capture small plankton on which filter feeders specialize (Durbin, 1979).

Although adult carp is *omnivorous* (Martyniak 1990), this plasticity permits a certain preference for food items which are easily accessible and are obtained with a lesser effort (Spatru *et al.*, 1980).

The selective feeding by carp (size class IV and V) on chiromomids over oligochaetes appears to be a physiological one; selection facilitated by two pairs of barbels known for their role in food detection, (Nikolsky, 1963) and mastication aided by pharyngeal teeth at this stage.

The degree of digestion, that different groups of food organisms had undergone in the gut of carp sub-adult and adult stage, reveals that carp might have resorted to zooplankton feeding after having consumed all the available preferred benthic food organisms. This points to zooplankton as alternative food of adult carp.

Although benthic feeder yet adult carp does not make use of 'coarse fauna' (water beetles, bugs, larvae of dragon fly) because ratio of mouth size to body size (1 : 16–1 : 17; Table 8.1) does not permit the adult carp to show *macrophagy*. Weatherly and Gill (1987) have also reported that true macrophagists have gape about one-fifth of their standard length as in sculpin *Leptocottus armatus*.

The delibrate intake of detritus is ruled out in the presence of high energy yielding food (zoobenthos), for, the fish is energy maximizer. It is most likely that it (detritus) may have entered passively through the protrusible mouth of carp during its benthic foraging bout.

The *illophagic feeding* behaviour as described by Sunder *et al.* (1984) in respect of carp, cannot be a rule but only 'situational' and in response to the environmental ecological status. That carp is an *opportunistic feeder* is already on record (Spatru *et al.*, 1980).

Acknowledgements

The authors are thankful to Head, P.G. Department of Zoology, University of Jammu for providing the necessary facilities.

References

Bambroo, P., 1997. Studies on early nutritional ecology of *Cyprinus carpio communis* with emphasis on diet substitution and culture of preferred prey from lentic environment of Jammu. *Ph.D. Thesis,* University of Jammu.

Chakrabarti, R. and Jana, B.B, 1990. Food preferences and selection by common carp (*Cyprinus carpio*) fry under different feeding regimes. *J. Appl. Icthyol.*, 6: 223–230.

Durbin, A.G., 1979. Food selection by plankton feeding fishes. In: *Predator Prey Systems in Fishery Institute*, (Ed.) H. Clepper. Washington, pp. 203–218.

Ivlev, V.S., 1961. *Experimental Ecology of the Feeding of Fishes*. Yale University Press, New Haven, CT, USA.

Jhingran, V.G and Pullin, R.S.V., 1985. *A Hatchery Manual for the Common, Chinese and Indian Major Carps*. Asian Development Bank, LCLARM, Manila, Philippines, 191 pp.

Khadka, R.R. and Rao, T.R., 1986. Prey size selection by common carp (*Cyprinus carpio* var. *Communis*) larvae in relation to age and prey density. *Aquaculture,* 54: 89–96.

Kryzhanovsky, S.G., 1956. Development of clupeid fishes. *Tr. Insorfol. Zhivot Serercova* 17: 1–254 (In Russian).

Martyniak, A., 1990. Feeding of carp *Cyprinus carpio* in the artificially aerated lake Mutek. Folia. *Zoologica,* 39(3): 279–284.

Martyshev, F.G., 1983. *Pond Fisheries*. Amerind Publishing Co. Pvt. Ltd., New Delhi, pp. 233.

Mookerji, N. and Rao, T.R., 1993. Patterns of prey selection in rohu (*Labeo rohita*) and singhi (*Heteropreustis fossilis*) larvae under light and dark conditions. *Aquaculture,* 118: 85–104.

Nikolsky, G.V., 1963. *The Ecology of Fishes.* Academic Press, London and New York.

Pillay, T.V.R., 1990. *Aquaculture: Principles and Practices.* Fishing News Books, UK, 575 pp.

Shirota, A., 1970. Studies on the mouth size of fish larvae. *Bull. Jpn. Soc. Sci. Fish.*, 36: 353–368.

Spataru, P., Hepher, B. and Halevy, A., 1980. The effect of the method of supplementary feed application on the feeding habits of carp (*Cyprinus carpio* L.) with regard to natural food in ponds. *Hydrobiologia*, 72: 131–178.

Sunder, S., Kumar, K and Raina, H.S., 1984. Food and feeding habits and length-weight-relationship of *Cyprinus carpio specularis* L. of Dal Lake, Kashmir. *J. Fish.*, 31(1): 90–99.

Tinbergen, N., 1951. *Study of Instinct*. Claredon Press, Oxford.

Titicomb, J.W., 1910. Fish-cultural practices in the United States Bureau of Fisheries. *Bull. U.S. Bur. Fish.*, 28(2): 697–657.

Ware, D.M., 1971. Predation by rainbow trout (*Salmo gairdnerii*): The effect of experience. *J. Fish. Res. Bd. Com.*, 28; 1847–1852.

Weatherley, A.H. and Gill, H.S., 1987. *The Biology of Fish Growth*. Academic Press, London.

Chapter 9

Ectoparasitic Diseases in Freshwater Fishes and their Control

R.K. Gupta, N.K. Yadava, S.K. Gahlawat, K.L. Jain and A.S. Burman

Department of Zoology and Aquaculture, C.C.S. Haryana Agricultural University, Hisar, India

An overview of the global fisheries trends brings out the growing importance of aquafarming in recent years. Indian freshwater aquaculture has evolved from the stage of domestic activity in West Bengal and Orissa to that of an industry in recent years with states like Andhra Pradesh, Punjab, Haryana and Maharashtra taking up fish culture as trade. Fish has long been an important source of food for people all over the world. The importance of fish as a source of high quality, balanced and easily digestible proteins, essential amino acids and nutraceuticals like PUFA is well understood. According to an FAO survey, the number of people involved in fishing and fish farming worldwide has more than doubled since 1970. Most of this growth has taken place in the 1980's and mainly in Asian countries where four fifths of the world fish farmers dwell. In India, the importance of fisheries sector is well known as it provides livelihood to seven million farmers, prime mover of coastal economy and a major foreign exchange earner. Development of several suitable technologies coupled with extension activities during the last five decades has pushed Indian fish production manifold, particularly to become second largest aquaculture fish producer and a thousand fold increase in the value of marine products exports. Presently the contribution of fisheries to the GDP of the country is estimated at 1.4 per cent whereas the export earnings of fisheries products reached Rs. 6,443 crores.

The country is endowed with vast and varied resources possessing river ecological heritage and rich biodiversity. The share of inland fisheries sector, which was 29 per cent in 1950–51 has gone up over 49 per cent in 2001 indicating increasing contribution of the inland sector to the total fish production.

Table 9.1: Inland Fishery Resources

Sl.No.	*Resources*	*Area (hecs)*
1.	Ponds and tanks	2.36 million ha
2.	Oxbow lakes and derolect water bodies	1.3 million ha
3.	Reservoirs	2.05 million ha
4.	Rivers	45,000 km
5.	Canals	1,26,334 km
6.	Estuaries	2.7 million ha
7.	Brackish water	1.42 million ha

Aquaculture has become the fastest growing food production system with a growth rate of over 9 per cent in the last 10 years. Considering the increasing demand of fish to meet the nutritional requirements of the growing population in the country and the potentials of the fisheries sector in general and freshwater aquaculture in particular with regard to the contribution to the country's food basket, several technologies have been developed and evaluated for different agro-ecological situations for increasing fish production in the country.

Dictionary defines disease as the deviation from the state of normal health. Every living organism is prone to several diseases, and fish makes no exception. In aquatic ecosystem there are present so many pathogens that causes various diseases of culturable fishes *viz.*, virus, bacteria, fungi and other parasites. Parasites are those micro or microorganism that live on the surface of the host or embedded inside the body cavity. These are called ectoparasite and endoparasite, respectively. Ectoparasites are very much important. The wound or injury caused by the ectoparasite are the open invitation of other pathogens like fungi, virus, bacteria etc.

Fish under normal condition is in a state of equilibrium with its environment and disease producing organism. Any alteration in the environment disturbs this equilibrium and the fish become more vulnerable to disease. Wedemeyer (1970) has defined the relationship of environment vis-à-vis pathogen vis-à-vis fish and explained that the presence of the fish pathogens will result in the epizootics only if unfavorable environmental condition also exist and the host defence mechanism has been compromised. Intensification of aquaculture, especially in the absence of adequate knowledge about environmental hygiene, balanced nutrition and host-pathogen relationship has resulted in the occurrence and spread of serious diseases in the industry. Disease outbreaks cause significant losses in aquaculture production and are affecting economic development.

Species of Gyrodactylus (monogenea) and *Trichodina* (ciliophora), are common ectoparasites of fish, which multiply under the stressful conditions of aquaculture, both are associated to varying degrees, with chronic dermatitis of the fins and body surface (Lom and Dykova, 1992; Lom, 1995; Sterud *et al.*, 1998). The resulting conditions have traditionally been described as separate diseases, "*gyrodactylosis*" and "*Trichodiniasis*" (Bauer, 1961).

Manna *et al.* (1999) revealed that the females were more susceptible to Lernaea infection than males. They also found that the occurrence of infection was 5 times more in summer season than that in winter and rainy season. The parasite anchored to the vent and at the base of pectoral fin, particularly, to the right pectoral fin of the host body.

In a comparative analysis, it has been suggested that parasite body size could be an important factor determining the aspects of parasite abundance and distribution, including aggregation (Robert, 1999).

Van der Salm *et al.* (2000) reported that the ectoparasities when administered cortisol with feed, induced effects, which were functionally adaptive in the upper pavement cells, while combining the two treatments also induced changes indicative of immunosuppression.

Wang *et al.* (2001) indicated that a 40 days fumagillin treatment was an effective measure against thelohanellosis of allogynogenetic, silver carp. Prevalence and abundance of spores and plasmodia as well as the morbidity and mortality, was drastically reduced in the fumagillin treated fish. They also found that the drug was also very effective against developing spores of *Thelohanellus wuhanensis.*

Nielsen and Buchmann (2003) have given the treatment *Dexamethasone* (immunosuppressant) by bath or injection to Atlantic salmon, *Salmo salar* to study whether that treatment could affect the susceptibility of infections with a Danish strain of *Gyrodactylus derjavini* (Monogenea). Three groups of *S. salar* were immersion treated either with 10, 60 or 240 µg dexamethasone l^{-1} water, respectively. In addition, one was treated intraperitonealy with 200 µg dexamethsone per fish and one control group was kept untreated. A single *G. derjavini* parasite was placed on anal fin of each fish and the infection was subsequently monitored weekly for 6 weeks. An increase in parasite population on the salmon was positively correlated with the amount of immunosuppressant used. Infection levels in the immersion treated with dexamethasone (240 µg l^{-1} water) intraperitoneally infected positive control group were significantly higher compared to untreated control group.

Herms (1983) reported the most important parasitic diseases in carp culture that are caused by *Gyrodactylus sprostonae, Ichthyophthirius, Bothriocephalus acheilogeathis, Trichodina* and *Dactylogyrus.* He also described the diagnosis and therapy of these diseases. Occurrence of ciliates and flukes were recorded in the fry of Indian major carps (Maity *et al.,* 1986) stocked at a density of 2 million/ha. The worms were mainly monogenetic trematodes belonging to the family gyrodactylidae. Helminthic fauna belonging to monogenetic trematode, digenetic treamatode and cestodes on Common carp, Grass carp and Silver carp were reported by Salih *et al.* (1988) in Iraq. Monogenetic species were *Dactylogyrus arguatas, D. extensus, D. solidus, D. vastator, D. inexpectatus, D. camellatus, D. hypophthyolmichthys, D. serjabina, Gyrodactylus buicalensis* and *G. elegans.*

The invasion of parasites (*G. schulmani*) in carp fry reaches their maximum in the warmest summer season and decreases gradually down to their minimum in the autumn Zetnam (1984). The first scientific record of the infection of the carp fingerlings with *Gyrodactylus elegans* in Iraq was reported by Ali *et al.* (1985). He also revealed that fingerlings were infected by *Lernaea* and *Gyrodactylus.*

This article includes the common ectoparasitic diseases of culturable freshwater fishes in India and their management practices.

Table 9.2

Name of the Disease	*Causative Agent*	*Most Affected Species*	*External Symptoms*
1. White spot or itch diseases	*Ichthyophtherics multifilis*	Fry and fingerling of *Catla catla*, *Labeo rohita*, *Cirrhinus mirgala* and other carps	(a) Minute white spots are visible on the surface of fish. (b) Rubbing against hard object may be common (c) Attachment tissue become reddish
2. Trichodiniasis	*Trichodina reticulata*, *T. niger*	Fry and fingerling of *Catla catla*, *Labeo rohita*, *Cirrhinus mirgala* and other carps in nursery and rearing ponds	(a) Colours of the gill turn pale (b) Fish secrete excessive amount mucus (c) Respiratory function will hamper
3. White scale spot disease	*Myxobolus* sp.	All most all culturable fishes can attacked by this parasite	(a) White cysts are found on surface of scale and body of fishes (b) Scale become loose and perforated in many places (c) Ulcer may develop at the attachment site
4. White gill spot disease	*Myxobulus* sp., *Thelohanellus* sp.	All the carps (major, minor Chinese carp) and other freshwater fishes	(a) White cysts are visible on the surface of the gill (b) Excessive mucus secretion occur (c) Gill lamellae will be damaged (d) Surfacing of fishes may be common

Contd...

Table 9.2–Contd...

Name of the Disease	*Causative Agent*	*Most Affected Species*	*External Symptoms*
5. Costiasis	*Costia necatrix*	All the freshwater fishes are prone to attack by this parasite	(a) Colour of the body of fish fade (b) Fish become restless and rubbing against object (c) Fin may damaged (d) Surfacing of fishes are common
6. Argulosis (fish louse)	*Argulus* sp.	*Catla catla, Labeo rohita, Cirrhinus mrigals* and other carps	(a) Adult parasite visible moving freely on the surface of the fish (b) Fishes become restless with erratic swimming (c) Scrap against hard parts (d) Attachment site become ulcerated
7. Lernaeosis (anchor work)	*Lernaea* sp.	All the major carps, common carp and Chinese carps are attacked by the parasite	(a) Fishes become restless and scrap against object (b) Fishes show erratic swimming behaviour (c) Skin redden many places (d) Attachment site show ulceration
8. Ergasilosis	*Ergasilus* sp.	All the freshwater fishes can attach by the parasite	(a) Infection occur in gill, buccal cavity and fin (b) Parasites look like white bodies less than 2 mm long (c) Gill tissue damaged (d) Surfacing of fishes

Contd...

Table 9.2–Contd...

Name of the Disease	*Causative Agent*	*Most Affected Species*	*External Symptoms*
9. Dactylogyrosis (Gill fluke)	*Dactylogyrus* sp.	Fry and fingerling of carps	(a) Fading of normal colour of gill (b) Fishes are restless and scrap against object (c) Gill moves rapidly (d) Surfacing may common
10. Gyrodactylosis (skin fluke)	*Gyrodactylus* sp.	Fry and fingerling of carps	(a) Colour of the body fade (b) Dropping of scale (c) Scrap against object due to irritation (d) Excessive mucus secretion occur
11. Black spot disease	*Diplostomum* sp.	All small and adult carps can attack by the parasite	(a) Black ovoid patches on the surface (b) Cysts vary from few to hundred
12. Leech infection	*Picicola* sp.	All the freshwater fishes are attacked by the parasite	(a) Infection occur on skin, gill and fins (b) Haemmorhage found on the site of attachment (c) Fishes become restless and rubbed against hard part

Table 9.3: Treatment

Name of the Disease	*Treatment Recommended*	*Dose Applied*
White spot disease	1. Formalin bath treatment	200 mg/L for 1 hr.
	2. Malachite green in pond	1.5 mg/L
	3. Methylene blue	1–3 mg/L
	4. $KMNO_4$ bath treatment	4 mg/L for 1 hr.
	5. NaCl bath treatment	3% for 1 hr.
Trichodiniases	1. Maintain the water quality by adding freshwater	–
	2. Decrease the stocking density	–
	3. NaCl bath treatment	2–3% for 1 hr
	4. $KMNO_4$ in pond	4 mg/L
	5. Formalin in pond	25 mg/L
White scale spot disease	1. Decrease the density of fish	–
	2. Mohua oil cake in pond	25 mg/L
	3. NaCl bath	3–5% for 30 min for 1 hr.
Costiasis	1. Acetic acid	2–5%
	2. $CuSO_4$ bath treatment	1–4 mg/L for 3–5 min.
	3. Formalin dip treatment	400 mg/L
	4. Malachite green dip treatment	60 mg/L
	5. $KMNO_4$ bath treatment	4 mg/L for 1 hr.
	6. NaCl dip treatment	3–5 per cent

Contd...

Table 9.3–Contd...

Name of the Disease	Treatment Recommended	Dose Applied
Crustacean disease (Argulosis, Lernaeosis, Argasilosis)	1. NH_4Cl bath treatment	1000 mg/L for 4 hrs.
	2. Benzine hexa chloride (BHC)	0.12 mg/L
	3. Masoten	0.25 mg/L
	4. NaCl bath treatment	3–5% for 1 min
	5. NaCl in pond	0.8–1.1% for 3 days
	6. Diflubenzuron	0.01 m L
Dactylogyrosis and Gyrodactylosis	1. Acetic acid dip	5% for 1 min
	2. Formalin bath treatment	250 mg/L for 1 hr.
	3. Malachitegreen bath	5 mg/L for 1 hr.
	4. Mebendazole bath	0.01 mg/L for 24 hrs.
	5. Methylene blue	2.9 mg/L
	6. Masoton	0.25 mg/L
	7. $KMNO_4$	2.4 mg/L
	8. NaCl bath treatment	3–5 per cent for 1 hr.
	9. Tricholoroform bath	1.6–0.4 mg/L for 24 hrs.

Contd...

Table 9.3–Contd...

Name of the Disease	*Treatment Recommended*	*Dose Applied*
Black spot disease	1. Pond should be free from molluscan population	–
	2. Prevent entering of bird	–
	3. Pond may be dried	–
	4. Picric acid bath	2–7 : 100000 for 1 hr.
Leech infection	1. NaCl bath treatment	3–5% for 10–15 min
	2. Masotin in pond	0.5–1.0 mg/L
	3. Glacial acetic acid dip	1 : 1000 for 1–2 min
	4. Adult parasite can be pull out with the help of a forceps	
	5. Pond may be dried and treated with chlorinated line.	

Followings are the ectoparasites that causes various diseases in freshwater fishes:

Groups of Parasite	*Name of the Parasite*
1. Protozoa	(a) *Ichthyopthirias* spp.
	(b) *Trichodina* spp.
	(c) *Myxobolus* spp.
	(d) *Costia* spp.
2. Crustacea	(a) *Argulus* spp.
	(b) *Ergasilus* spp.
	(c) *Lernea* spp.
3. Helminth	(a) *Dactylogyrus* spp.
	(b) *Gyrodactylus* spp.
	(c) *Diplostomum* spp.

Now different ectoparasitic diseases caused by the above parasite and their common symptoms are discussed in Table 9.2.

References

Ali, N.M., Salih, N.E. and Abdul-Ameer, K.N., 1985. Protozoa and crustacean infesting three species of carp raised in ponds in Iraq. *J. Biol. Sci. Res.*, 19(2): 287–399.

Bauer, O.N., 1961. Parasitic diseases of cultured fishes and methods of their prevention and treatment. In: *Parasitology of Fishes,* (Ed.) V.A. Dogiel, G.K. Petrushevski and Y.I. Polyanski, Translated version by Z. Kabata, pp. 265–298, Oliver and Boyd Ltd., London.

Herms, J., 1983. Experience in handling parasitic disease in carp culture. *Z. binnefish* D.D.R., 30(1): 24–27.

Lom, J., 1995. Trichodinidae and other ciliates (Phylum Ciliophora). In: *Fish Diseases and Disorders, Vol. 1: Protozoan and Metazoan Infection,* (Ed.) P.T. Woo, pp. 229–262. CAB International, Oxon.

Lom, J. and Dykova, I., 1992. (Phylum ciliophora Dofelin, 1901). In: *Protozoan Parasites of Fishes,* (Ed.) J. Lom and Dykova, pp. 237–288. Elsevier, Amsterdam.

Maity, S.S., Roy, R.N. and Haldar, D.P., 1986. Infection by a ciliate and a fluke on the fry of Indian Major carps. *Curr. Sci.*, 56(19): 1023–1030.

Manna, A.K., Paria, T. and Gosh, S., 1999. Incidence and intensity of anchor worm lernea infection on the asiatic rice-fish oryzias metastigma. *Environment and Ecology*, 17(1): 73–75.

Nielsen, C.V. and Buchmann, K., 2003. Increased susceptibility of Atlantic salmon *Salmo solar* to infection with *Gyrodactylus derjavini* induced by dexamethazone bath treatments. *Journal of Helminthology*, 77: 65–68.

Robert, P., 1999. Body size vs. abundance among parasite species: Positive relationships. *Ecography*, 25(5): 246–250.

Salih, N.E., Ali, N.M. and Abdul-Ameer, K.N., 1988. Helminthic fauna of three species of carp raised in Iraq. *J. Biol. Sci. Res.*, 19(3): 369–387.

Sterud, E. Harris, P.D. and Bakke, T.A., 1998. The influence of *Gyurodactylus salaris* M (1957) salmon, *Salmo solar* L. and brook trout, *Salvilinus frotinalis* (Mitchill): Experimental studies. *Journal of Fish Diseases*, 21: 257–263.

Van der Salm, A.L., Nolan, D.T., Spanings, F.A.T., and Wendelaar Banga, S.E., 2000. Effect of infection with the ectoparasitic *Argulus Japonicus* (Thiele) and administration of cortisol on cellular prolifernation and apoptosis in the epidermis of common carp, *Cyprinus carpio* L. (Skin). *J. Fish Diseases*, 23: 173–184.

Wang, G.T., Yao, W.J., Wang, J.G. and Lu, Y.S., 2001. Occurrence of thelohanellosis caused of *Thelohanellus wohenensis* (Myxosporea) in Juvenile allogynogenetic silver crucian carp, *Carassius auratus* Gibelio (Bloch), with on observation on the efficacy of fumagillin as a treatment. *Journal of Fish Diseases*, 24: 57–60.

Wedmeyer, G.A., 1970. The role of stress in the disease resistance of fishes. In: *Symp. Diseases of Fishes and Shellfishes*, (Ed.) S.-F. Snieszko. Amer. Fish. Soc. Publn., (5): 30–35.

Zetnam, R., 1984. New helminthose of the carp in the Carpathian region of Czechoslovakia. *Symp. Biol. Hung.*, 23: 157–171.

Chapter 10

Anthropogenic Impacts and the Distribution of Crocodile Salamander (*Tylotriton verrucosus* Anderson, 1871) in the East Nepal

Kalu Ram Rai, Bharat Raj Subba** and Kul Prasad Limbu***

**Mechi Campus, Bbadrapur, Nepal, E-mail: raikr@ntc.net.np*

***Post Graduate Campus, Biratnagar, Nepal, E-mail: limbukp@yahoo.com*

ABSTRACT

Due to progressive industrial development and encroachment, some of the rare and endangered species of wildlife are on the brink of extinction. The crocodile salamander, a unique species of tailed amphibian in Nepal attracts the attention of conservationists who desperately fight for its survival. The explosion of the human population resulting in considerable problems of land use and pollution is the main factor for habitat loss. During recent survey of East Nepal this species was recorded from only two districts, Ilam and Panchthar. Especially the salamander was found in or near ponds, wetlands and damp or marshy places, but its population was found rapidly declining by anthropogenic activities.

Mainly at Hans Pokhari, Sribung, Dharapani, Mai Pokhari, Suke Pokhari, Jorpokhari, Pokna Pokhari and Panchkanya,

Crocodile salamanders were still observed and were represented in viable populations. Among studied habitats, the impacts were recorded to be especially high at Jor Pokhari and Hans Pokhari due to road construction, introduction of carps and tea cultivation.

Keywords: *Crocodile salamander (Tylotriton verrucosus), Systematics, Environmental condition, Distribution, Human impacts.*

Introduction

East Nepal is very promising for herpetological research, because several rare and endangered species of amphibians and reptiles which were hitherto not recorded for Nepal were discovered one after another. Among them is the Crocodile salamander (*Tylotriton verrucosus*), a natural heritage of the world (Wongratana 1984). The etymological meaning of *Tylotriton* is a newt with a bulge at the ear region. Vernacular Nepalese names are Panigohoro or Thakthake Longlang. Its maximum total length (head-body+tail) is 105+95 mm. Its body is flat and smooth, with two rows of dorso-lateral warts. Its head is strongly flattened with a central ridge and two marginal ones. Its coloration is dark brown, with bright orange confined to the ventral edge of the tail (Figure 10.1). It is distinguished from other members of the genus by an entirely brown dorsum in combination with a strongly developed cranial crest. Reports on breeding behaviour, reproductive mode and distribution pattern in Nepal are still controversial. This species is found in two districts of Nepal *i.e.*, Ilam and Panchthar. Both district lies in the midland between 1000 msl and 3000 msl having a temperate climate and became the ideal place for Crocodile salamander. But now due to the deforestation, urbanization, road construction and tea cultivation, direct impacts on its natural habitats are increasing. Presently 9 localities were selected to study the human impacts on habitats and distribution of the endangered species.

Systematics

The first description of *Tylotriton verrucosus* was made by Anderson (1871), followed by Boulenger (1882, 1890) and Smith (1924), with a few mentionable publications on it. Thereafter, Thorn (1968) made a clear summary of the genus and Nussbaum and Brodie

Figure 10.1: An Adult Crocodile Salamander (*Tylotriton verrucosus*)

(1982) classified the genus into *Tylotriton* and *Echinotriton*. The latter was considered as a subgenus by Zhao and Hu (1988) but later Zhao realized its generic status (Anders *et al.* 1998). Previously its taxonomy was confusing but Nussbaum *et al.* (1995) designated a new species for the orange patterned form (now *Tylotriton shanjing*) and restricted the name *T. verrucosus* to the westemmost brownish form (Anders *et al.*, 1998). The lack of intergrades among specimens from Western Yunnan, where both forms occur close together, also suggests that two species are involved (Probst, 1998). Before the description of *T. shanjing*, the Southeast Asian salamander genus *Tylotriton* Anderson 1871 consisted of three species: *asperrimus*, *kweichowensis* and *verrucosus*.

Descriptions of its reproductive behaviour were given by Annandale (1907), Shrestha (1989, 1994), Scholz (1993) and Anders *et al.* (1998). Parental care was mentioned by Dasgupta (1984) and Shrestha (1989, 1994) and habitats were described by Chaudhari (1966), Shrestha (1984) and Dasgupta (1990).Investigations regarding the status as an endangered form was published by Shrestha (1989, 1996), Dasgupta and Dasgupta (1990), Dasgupta (1990) and Schleich and Kaestle (2002).

Until recently the species was treated also as *Pleurodeles verrucosus* by Dutta (1992), following the publication of Dubois (1987).

In a country report for Nepal, Shrestha (1996) regards *T. verrucosus* as rare and affected by habitat change. The status given by HMG Nepal, IUCN and CITES is unknown for Nepal (Shah, 1995). The existing data are not yet sufficient enough to make definite statements and suggestions strategies (Anders *et al.*, 1998).

Environmental Conditions

Environmental impacts can be categorized into socio-economic, biological and physico-chemical ones. The impacts caused by economic and social changes in a community or its vicinity are called socio-economic ones. Those on vegetation, wildlife, crops and aquatic life refer to biological ones; those on soil erosion, flood and sedimentation are physical impacts, while those on air, water and soil quality are classed as chemical ones.

Climatic Factors

According to the annual records of temperature, rainfall and humidity, the midlands of the Nepalese East are more humid, but March, October and November round arid. The present natural processes together with human activities in the area are especially critical for the mountain environment and the development of infrastructures (Poudel, 1999).

The relative humidity remains almost same all the year around in Ilam and Panchthar districts, creating ideal habitats for amphibians. Hans Pokhari, Panchkanya, Dharanpani, Chanpgairi, Maipokhari, Suke Pokhari and Jor Pokhari areas always seem under a cover of clouds where temperature fluctuates from 15°C to 30°C.

Water Condition

During the survey water analysis considering the main parameters such as temperature, pH, oxidation-reduction potential (ORP), dissolved solids and content of oxygen were performed. The results (Table 10.1) largely agree with Shrestha (1998) and Anders *et al.* (1998) and show that the water in the breeding pools, mostly resulting from spring rain and melting snow, tends to be rather acidic due to the accumulation of humic acids. The pH was recorded as very low, ranging from 5.5-6.6. The higher value of dissolved solid contents from the small pond (Chanpgairi) indicates a considerable contamination by faeces and urine, both from humans and domestic animals. Pollution by pesticides from the tea plantation is also very probable.

Table 10.1: Water Analysis of Sampling Station Temperature (°C)

Locality	Temperature (°C)		pH	ORP (mv)	Dissolved Solids (ppm)	Oxygen	
	Air	Water				%	ppm
Hans Pokhari	23.4	24.8	6.3	61	107	11	0.8
Phikal	18.1	16.6	6.1	138	352	31	2.3
Chanpgairi			6.6	84	39	42	3.6
Mai Pokhari	17.3	20.6	5.5	174	21	24	2.1

Land Use Pattern

The land utilization system in Nepal is mainly farming which includes forestry, agriculture, fishery and animal husbandry which directly affect most animal habitats. In the East Nepal, 26.8 per cent of total land is utilized for agriculture, 28.9 per cent is still in forest cover but very thin. The main crops are paddy, maize, potato, millet, wheat, oats etc. and other cash crops are tea, cardamom, bouquet grass, nettle, ginger etc. The past land use conflict between agriculture and forestry has now a new scenario, agriculture vs. conservation (Gurung, 1999). Road construction is said to initiate a matter of economic development as it provides the basic infrastructure for any kind of investment and for harnessing of economic potential of that area (Agrawal, 2000). Vibrations caused by heavy vehicles on the road also cause the cramping of landscape. The highland is utilized by the process of transhumance, shifting cultivation and sedentary agriculture. The distribution of forest land is 21.6 per cent in mountain, 42.5 per cent in hills, 15.9 per cent in the inner Terai and 20 per cent in Terai (Gurung, 1999) (Tab 12.2).

Materials and Methods

Data on the biology of the Crocodile salamander were collected annually (from 1998 to 2002) during field surveys. The biannual average values are presented in Table 10.3. As far as possible, data on adult salamanders were based on specimen collection whereas notes on larvae or juveniles were based on observation. Human impact on the species was studied from 9 localities, situated at different altitudes (Table 10.3).

Table 10.2: Land Use Type in Eastern Development Region of Nepal

Sl.No.	*Land Use Type*	*Eastern Development Region*		*Whole Nepal*	
		Area ha. '000	*%*	*Area ha. '000*	*%*
1.	Agriculture land	859.2	25.6	3052.7	20.7
2.	Forest/afforested land	923.7	27.5	5512.2	37.4
3.	Pasture/meadow land	174.4	5.2	1746.7	11.8
4.	Shrub land/degraded forest	539.2	16.1	709	4.8
5.	Non-agriculture land	238.6	7.1	997.6	6.8
6.	Other land	460.1	13.7	2729.8	18.5
	Total	**3349.5**	**23**	**14748.4**	**100**

Results

The problems resulting from natural hazards and human exploitation of the natural environment is increasing with unexpected speed. The following habitats were studied regarding the human impact situation on the distribution of this species:

Hans Pokhari

It is located at the northern face of Kuti Danda on the road to Phikal, Ilam at an altitude of 1620 msl. It is a temporary monsoon pond of 10 × 15 m with a depth of 5 cm near the shore and 70 cm in the centre (Anders *et al.*, 1998). It is an ideal habitat for *T. verrucosus* which was regularly observed here during the last five years. The pond vegetation consists of *Polygonum hydropiper* and grasses. It is surrounded by subtropical hill forest, dominated by *Schimi wallichi* and *Albizzia* sp., with a lower vegetation of *Sauraria nepalensis, Rubus ellipticus* and *Eupatorium adenophorum* (Singh, 1996; Anders *et al.*, 1998).

Impacts

A considerable impact on Hans Pokhari was the vibrations caused by heavy vehicles on the road its side. Due to regular vibration by heavy vehicles, it was noticed for early drying up.

Table 10.3: Distribution and Observation from East Nepal

Locality	*District*	*Altitude (m)*	*1998*		*2000*		*2002*	
			Tadpoles	*Adults*	*Tadpoles*	*Adults*	*Tadpoles*	*Adults*
Hans Pokhari	Ilam	1620	Numerous	13	Numerous	10	Very thin	5
Panchakanya	Ilam	1700	Numerous	10	Numerous	11	Numerous	6
Siribung	Ilam	1900	Scanty	6	Scanty	3	Scanty	4
Dharapani	Ilam	1650	Clusters	14	Clusters	5	Clusters	6
Chanpgairi	Ilam	1940	Scanty	2	Scanty	6	Scanty	5
Mai Pokhari	Ilam	2150	Clusters	35	Clusters	4	–	–
Suke Pokhari	Panchthar	2640	Scanty	2	–	1	–	–
Jor Pokhari	Panchthar	1980	Numerous	20	Clusters	4	Clusters	–
Pokna Pokhari	Panchthar	2500	–	–	–	1	Scanty	2

Other impacts were the religious and recreational activities. Each week on Tuesday and Saturday, a large number of people visit the temple at the top of Kuti Danda. After praying and viewing sight scene they come down to the pond to eat and enjoy themselves. Also, many picnickers from the Terai and far away come there to enjoy the scene around the pond. Before leaving, rubbish is thrown into the pond and around its periphery.

Another impact is the deposition of leaves in the lake as it is surrounded by thin bushy deciduous forest. Such facts were found responsible for the filling up of the pond which slowly dries up. Due to slow decomposition of the leaf litter, the water becomes more acidic. These factors heavily threaten the population of the Crocodile salamander. As protective measures a program should be managed to reconstruct the pond and, on another hand, litter bins should be provided to collect the scrap. Road signs to slow down the speed of vehicles could eliminate another source of heavy disturbance.

Pancha Kanya

It is a small, marshy pond located in between Phikal and Tin Ghare. Its south side is occupied by the Pancha Kanya temple. The north side is surrounded by forests and slopes down to Mai Khola. Around this area the vegetation is dominated by cash crops *Thea sinensis* (tea plants), and other main components are *Cryptomeria japonica* (Japanese cedar) and *Elletaria cardamomum* (Cardamom). It lies at 1700m and is occasionally covered with clouds or fog. In that marshy pond, adults and juvenile *T. verrucosus* were found. The adults were also found migrating to the swampy drinking spring area.

Impacts

The villagers called them *paniko mau* (*mother of water*) and in general do not kill them. The local people reported that the newts were seen during rainy season but disappeared during winter. The impact on the distribution of the newts is only indirectly. The heavy vehicles that ply the neighbouring road were producing vibrations that create holes and leaks that made the pond water dry up. The pond was seen in the same condition since many years and no clean up was carried up. This is because as per local people belief it belongs to the goddess, and to move anything from the pond is disagreeable to her. If anybody disobeys her, she will become illic suffers from head-ache, stomach-ache, diarrhoea and dysentery etc.

The aquatic plants growing inside the pond die, the bottom layer of organic material increases and the decomposing leaves and stems decay, forming humus, which generally spoils the habitable environment of tadpoles and juvenile newts. As a consequence they are compelled to move away from the water and become vulnerable to birds, mammals and reptiles which predate on them.

Siribung

It is located at an altitude of 1900 msl on the eastern border of the Ilam district, close to the Mirik Division of Darjeeling, India. Its western side slope up to the river (Siddhi Khola) valley. The east side is connected to the Tea Estate of Mirik. On the Nepal side, there is a small spring from where Mechi River originates. Siribung village situated in the lap of a hilltop, capped by a small forest, looked a very wet area. Several small springs, were used for drinking and watering the tea (*Thea sinensis*) garden.

Impacts

It is north-west facing shadow habitat, which receives sunlight after 8 or 9 AM and lasts in average for 6 hours per day. With the help of local people, several juveniles and adults of the Crocodile salamander were collected from the spring water and other permanent water ditches under the shadow canopy of *Alnus nepalensis* (Nepalese alder, uttish) forest. Asked about the newts, the villagers said that they called them 'thak thake' or 'pani chheparo' or 'paniko mau'. Due to restricted area and isolated occurrence it lacks a common Nepalese name.

To make a comparative study of samples, the researchers went to a famous lake called Mirik Jheel (Darjeeling Region) which was only a few kilometers away. The newts in that lake were of similar size and colour. Anyway the survey to that side was helpful to compare the Nepalese newt with Darjeelings'. The use of pesticides in the tea gardens was found to have negative impacts on the newts. This area has a tremendous potential for increasing tea gardens, and a big processing factory is being established at the top of Siribung Tea Estate. When the factory begins to run with full capacity then its effluents and disposals will surely ruin the habitats of the newts.

Dharapani/Jasbire/Charkhol

This area is located on the way from Ilam bazaar to Mai Pokhari. On the road, at an altitude of 1650 msl, a drinking spring fountain is found, called 'dhara' in Nepali. Therefore the name of this place became 'dharapani' (Nepali "pani" = water). It is a north-east facing mountain village, surrounded by *Alnus nepalensis* (Nepalese alder) trees. The upper part includes the localities of Charkhol and Jasbire, which are used for tea plantation. The lower part slopes down over agricultural land towards the Mai River. The source of water originates from a V-shaped valley surrounded by four tops called Charkhol (the old name of Ilam). Between the Charkhol ridges, there are small streams coming out and flowing into small ditches where several tadpoles as well as juveniles of the Crocodile salamanders were observed.

Impacts

Due to their seasonal migratory habits the Crocodile salamanders of the locality are exposed to predation. In spring they move uphill and reach the Charkhol-Jasbire as the valleys at midsummer. As the winter sets in they move down to Dharapani, with an altitudinal distance of about 1000 msl. Direct human impact on the salamanders was not observed, but after tea plantations, the farmers use fertilizers and pesticides. The springs sources are found either in the centre of the tea garden or near its marginal area. Wherever fertilizers and chemical pesticides are used, they are washed out with rainwater getting into ditches and small streams. In the long turn, they deteriorate the habitats of the newts which vanish without being noticed. If not protected on time, it will be too late.

Chanpgairi

It is also located on the way from Ilam bazaar to Mai Pokhari, not far from Jasbire. The only difference to the preceding habitat is that it is located in a depression at the height of 1940 msl. It is also a temporary summer pond; occupying an area of about 15 × 20 m. It is surrounded by mountain forest with *Schima wallichi* mixed with *Alnus nepalensis* and *Cryptomeria japonica* (Japanese cedar). Aquatic vegetation consists of *Nymphea* sp., *Persicaria hydropiper*, *Persicaria posambu* and grasses while the shore is grown with *Potentilla* sp.,

Eupatorium adenophorum, *Plantago lanceolata* etc. (Singh, 1996; Anders *et al.*, 1998).

Impacts

During midsummer the pond was full of newts with larvae and juveniles. The bottom of the pond is muddy and the water also looked turbid. Only when the newts came to the surface of the water, they could be seen. The pond lies in V-shaped valley and there is no blocking dam to control its outlet. On its east side there are a few shops, the owners of which keep buffaloes. The animals use that pond for wallowing. Thus the water is never clear. Paper wrappers, plastics and other used-up material are thrown into the kitchen garden from where the wastes are carried to the pond and enter the specific food chain. *Cryptomeria japonica* turns both the soil and water somewhat acidic and the eggs and tadpoles are susceptible to such condition (Duellman, 1993). So this small but very important habitat of the Crocodile salamander should be protected by educating the local people and fencing that area.

Mai Pokhari

It is located at the height of 2150 msl on top of ridges, surrounded by mixed forest of the wild chestnuts *Castanopsis indica* and *Castanopsis tribuloides*. The vegetation is dominated by *Sphagum moss* and *Nymphea stellata*. At the east of the pond is a temple and schools, to the west is a Botanical Park under construction, north is afforested with pines (*Pinus* sp.), Juniper (*Juniperus* sp.) and Japanese cedar, and to the south is opening gate, inhabited by local people.

Impacts

The main lake is irregular-sized and it was recently cleaned. Its northern shore is a little elevated and has five ridges and in between each of them there is a small ditch-like pond with a little water. Since time immemorial *Carcassius auratus* (Gold fish) have been kept for ornamental purpose. Now it has overpopulated and hundreds of them are found swimming and moving like a red-flame. The lake was minutely observed for *T. verrucosus*, but not even a single one was found. It may so due to the introduction of exotic *C. auratus*. But there are many small newts living in the small summer ponds on the northern ridges (Anders *et al.*, 1998). Previously it was completely covered with a carpeting of *Sphagnum* moss and *Nymphea stellata*.

Recently herpetologists (Scheich pers. com) have suggested for removal of the gold fish from the lake and the reintroduction of the Crocodile newts if it should be conserved as a last heritage of nature. Other impacts observed at the lake is worshipping by local people on Hindu festival days, where a lot of seeds are thrown into the water, and also other leave-containers, perfumery paper and plastics things. Boating in the lake causes more pollution than revenue. So the anthropogenic impact is rather pronounced in that lake.

Suke Pokhari/Ithung/Ranitar

Suke Pokhari is a lake found at the highest altitude (2640 msl) among other lakes. This was reached through Patenagi, after the survey of Gangate pani. It is surrounded by three small ridges. The northern side is open and slopes towards Phidim (Panchthar). The southern side also opens and slopes down to Deomai Khola. It is surrounded by *Rhododendron barbatum* dominated forests.

Impacts

It is a small natural lake where *Nymphea* sp. covered a considerable part of the surface area of the pond. It was searched minutely for warty newts during the spring season, but not a single sample was found. When enquiries were made with local people, they said that several newts were seen in the lake one decade ago. During the discussion it became clear that introduction of *Nymphea* sp., *Carassius auratus* and freely wallowing buffaloes had affected the condition of the lake resulting in the disappearance of all the Crocodile salamanders. As survey continued in a spring at 2500 msl (Ithung), a warty newt was collected extending the distribution of salamander upto 2500 msl.

Jor Pokhari

Jor Pokhari is situated at the height of 1980 msl. It is along eutrophicated natural lake, located at the north side of Phidim-Taplejung road. At its east side lies Agejung Danda and north is Gudgude Danda. West is sloping to Tamor River, and south steep down to Hewa Khola. A fast growing market, a temple and a Sanskrit school have been established to the east and south side of the lake. About 100 msl north side of the main lake, a small but much eutrophicated another sister lake is there which is called Siddhi Pokhari. The combination of these two ponds is called Jor Pokhari (meaning "paired ponds"). It is surrounded by mixed forest of *Castanopsis indica.* Many Crocodile salamanders were observed there in 1998, but in the recent survey (August, 2001), only small newt larvae at the developing stages were recorded.

Impacts

Recently the lake was found in much polluted condition. Its water was shallow and muddy and mucky. The VDC has given the lake to Jor Pokhari Protection Committee, and the committee has decided to keep carps in the lake. The grown up voracious carps were finding it difficult just to swim around in the water. Therefore at the last observation, no tadpoles and juveniles were found. These might be eaten up by the carps. Jor Pokhari was an original and natural pond four years ago but now it is in a very pitiable condition. Apart from the carps, another noticeable factor is detritus formation by the deposition of *Castanopsis* leaves. The outlet has been dammed at the eastern bazaar side but it has no inlet of feeding source. The water which is collected during the rainy season will last for the whole year. But at the marginal level, it is covered with submerged *Nymphea* sp. and aquatic weeds. In addition *Cryptomeria japonica* was found as the responsible factor for the increasing the acidity of the pond water. Such an increase in acidity is naturally very fatal for the egg and juveniles of the newts. As a result one of the last natural heritages in East Nepal is disappearing due to human impacts.

Pokna Pokhari

It is a small natural marshy lake within dense mixed forest of *Castanopsis* and Rhododendron. Very few tadpoles were observed during survey journey in post monsoon and only one specimen was collected.

Impacts

Due to location at the border of Sikkim, it was non-disturbed area and found still in innate state. Even there, local people come to dig yellow and white lime soil for washing their houses, which can damage the nests of Crocodile salamander.

Discussion

During the survey period, the Crocodile salamanders were observed in different parts of Ilam and Panchthar, but not recorded from other districts of East Nepal. Previously Shrestha (1981) reported that Arun valley was one of the ideal habitats of salamander in east Nepal. Crocodile salamander is also called Himalayan newt, recorded from Kuti Danda, Kolbung VDC of Ilam as the southern extension of its distribution, located at the altitude of 1700 msl. From there, the survey was continued in different seasons since last 5 years. The newts were found in Panch Kanya village, not far from Kuti Danda, Hans Pokhari, which was about at the same height. In this way, the survey was also made to the eastern side of Kuti Danda, at a locality called Siribung near the border with the Darjeeling District of India. They were found in small summer ditches and small springs, at larval, juveniles and adult stages. After that, the survey continued to the north of Ilam district. At Ilam bazaar (1200 msl), they were completely absent but found at Dharapani (1700 msl). They were then observed continuously on the route through Jasbire, Chanpgairi, Mai Pokhari, Deorali and Mai Majhuwa. The survey was continued straight to Panchthar, following the ancient trail from Deorali to Puwa Majhuwa and then Patenagi (2900 msl). From there it descended to Suke Pokhari (2700 msl) which is a small but natural pond, located in the centre of three hill tops. It was continued to Ithung (2400 msl), Ranitar, Bharapa and Jor Pokhari. From there, survey was made up to Taplejung to Surumkhim but not recorded even a single specimen. Thereafter survey was extended towards Pokna Pokhari (2550 msl), where one specimen was collected with the help of local persons. Along the survey route, the highest place recorded was Pokna Pokhari. Sukna Pokhari was also investigated and enquiries made about the newts, but collection was not done. From the above survey it was confirmed that *T. verrucosus* is distributed between 1200 msl to 2700 msl. So far it was called the Himalayan Newts, could not be recorded above the limit of the altitude 2700 msl. But regarding the altitudinal distribution, several authors had given several figures such as the lowest record at 1200 msl and the highest at 3350 msl from Ukhrul/Manipur (India) (Dutta, 1992). But in Nepal, the lowest record was from 1300 msl (Shah, 1995), the highest from 2150 msl (Anders *et al.*, 1998). So far as field surveys are concerned, distribution range of *T. verrucosus* was not directly disturbing the populations, and many people were even not aware of the newt presence. Even though, the present results show that Crocodile salamander were recorded from the east of Tamor River and south from Kabeli River only. Therefore, the river could be the barrier for horizontal distribution.

As observed the use of puddle and ponds by domestic cattle as well as introduction of fishes (Mai Pokhari and Jor Pokhar) are first sign of irreversible habitat destruction. Other threatening impacts observed was also the use of pesticides around Ilam's tea plantation, which could be regarded as steady menace to the species and its habitats. Other impacts observed were use of slope land for growing various cash crops (cardamom, bouquet grass etc.). Horticulture also occupies a large area of wetlands and becomes the factor for habitat loss. In Ilam and Panchthar, there are many dairy farms and milk collecting centre where a large number of people are using horse to transport the heavy load of milk container to the road side. Therefore, keeping many cows for milk, causes overgrazing and loaded horses erode the trail resulting in the landslides.

Additional impacts from developmental projects are hydropower and road network construction. The damming of the Puwa Mai Power Project has established in the north of Ilam bazaar whose ecological effect is neutral at the moment, but the sub-highway construction from Charali to Taplejung lead to habitat destruction of Crocodile salamanders. It destroys the wetland and springs directly by heavy traffic and leads to a slow leakage of the water bodies which ultimately fall dry.

Another impact is the felling of *Alnus nepalensis* for plywood causes direct deforestation and makes slopes susceptible for landslides. Other minor causes for population depletion of Crocodile salamander is due to collection for research and aesthetic purposes.

Conclusions

Every research results in its own set of data and due to some circumstances it could differ from the results of other authors, but on the whole an updating of distribution records would be necessary for a complete summary of environmental impacts. The major anthropogenic threat to *T. verrucosus* is habitat destruction by deforestation and road construction. In some extent use of fertilizers and biocides also caused declination in its population. Natural processes like increasing of acidity by fallen leaves, accumulation of organic waste and finally drying up are major threat to salamander.

Jor Pokhari, Mai Pokhari and Hans Pokhari are in serious condition by introduction of exotic carps which fed on eggs, tadpoles and juveniles of Crocodile salamander. The decomposition of organic matter in pond deteriorates the water quality leading to an increased demand of oxygen for salamander.

The next threat observed was the vehicle along the pond sides producing noise pollution and vibration which increases the leakage of the water through originally impervious layers. Schools, temples and bazaars near Jor Pokhari and Hans Pokhari also disturbed the salamander's behaviour. The situation of Hans Pokhari was the worst last year, but at the moment it is ameliorating by a damming towards the roadside. So it is expected that this habitat of the Crocodile salamanders could recover.

Acknowledgements

The authors express their sincere gratitude to Prof. Dr. Hermann Schleich and Dr. Werner Kaestle for their critical comments and suggestion. The authors would like to thank Prof. Dr. GPS Ghimire, Dean of IOST, T.U., Prof. Dr. T.K. Shrestha, Head of the Central Department of Zoology (CDZ), T.U., Prof. Dr. Umakant Ray Yadav (CDZ), Prof. Dr. I. Shrestha(CDZ), Prof. Dr. P.N. Mishra (CDZ), Prof. Dr. V.K. Thapa (CDZ), Prof. Dr. A.S. Tamrakar (CDZ), Dr. Khadka Basnet (CDZ), Prof. K.B. Shah (Natural History Museum), Dr. T.M. Maskey (DNPWC), Dr. C. Anders (Munich, Germany), Mr. H. Zwartepoorte and Mr. G. Vissar (Royal Zoological Garden, Netherlands), Mr. H. Sastre (University of Wuppertal, Germany) for their encouragement and help.

References

Agrawal, D.K., 2000. Complexities and measurement for environmental management in the Indian Himalayan region. In: *Environmental Impacts Assessment of Water Resources Projects: Concerns, Policy Issues, Perceptions and Scientific Analysis*, (Ed.) Radhey Shyam Goel. Oxford and IBH Publishing Co. Pvt. Ltd., New Delhi, pp. 404–413.

Anders, C.C., Schleich, H.H. and Shah, K.B., 1998. Contributions to the biology of *Tylototriton verrucosus* Anderson, 1871 from East Nepal (Amphibia : Caudata, Salamandridae). In: *Contributions to the Herpetology of South Asia (Nepal, India)*, (Eds.) Schleich, H.H. and Kaestle, W. Fuhlrott-Mus., 4: 1–26. Wuppertal, Fuhlrott–Museum.

Anderson, J., 1871. Description of a new genus of newts from western Yunnan. *Proc. Zool. Soc.*, London, 1871: 423–425.

Annandale, N., 1907. Eggs of *Tylotriton verrucosus. Rec. Ind. Mus.*, 1: 278–279.

Boulenger, G.A., 1882. *Catalogue of the Batrachia Gradientas, Caudata and Batrachia Apoda in the collection of the British Museum, 2nd ed.* Taylor and Francis, London, pp. 127

Choudhary, S.K., 1966. Studies on *Tylotriton verrucosus* (Himalayan Newt) found in Darjeeling. *J. Beng. At. Hist. Soc.*, 35: 32–36.

Dasgupta, R., 1990. Distribution and conservation problems of the Himalayan Newt (*Tylotriton verrucosus*) in the Darjeeling Himalayas. *Hamadryad*, 15(1): 13–15.

Dubois, A., 1987. Miscellanea taxonomica batrachlogica (I). *Alytes*, 5: 7–95.

Duellman, W.E., 1993. *Amphibian Species of the World: Additions and Corrections*. Univ. Kansas Mus. Nat. Hist., spec. publ., 21: 372; Laswrence.

Dutta, S.K., 1992. Amphibians of India: Updated species list with distribution record. *Hamdryad*, 17: 1–13.

Gurung, H.B., 1999. Land use in mountain environment–J. of Geography. *The Geographer's Point*; 9(1&2): 1–13.

Kuzmin, S.L., Dasgupta, R. and Smirina, E.M., 1994. Ecology of the Himalayan newt (*Tylototriton verrucosus*) in Darjeeling Himalayas, India. *Russ. J. Herpetol.*, 1(1): 69–76.

Nussbaum, R.A. and Brodie, E.D. (Jr), 1982. Partitioning of the salamandrid genus *Tylotriton* Anderson (Amphibia: Caudata) with a description of a new genus. *Herpetologica*, 38(2): 320–322.

Poudel, L.P., 1999. Role of geology in the morphological development and environmental degradation in the western hill of Nepal. J. of geography: Geographer's point; Centre for Nepalese Geography, Kathmandu, 9(1&2): 14–20.

Schleich, H.H., 1993. Contribution to the systematics and to a bibliography of the amphibians and reptiles from Nepal. *J. Nepal Res. Centre*, 9: 141–168.

Schleich, H.H. and Kaestle, W. eds, 2002. *Amphibians and Reptiles of Nepal*. Koenigstein, Germany; Koeltz Scientific Books, 1200 pp.

Scholz, K.P., 1993. Undersuchungen zur Phylogenie der Familie Salamandridae–Diplomarbeit, Phillips–Universitaet Marburg.

Shah, K.B., 1995. Biodiversity Profile Project, Enumeration of the Amphibians and Reptiles of Nepal. BPP Tech. Publ. No 2, DNPWC, MOFSC, HMG, Kathmandu.

Shrestha, T.K., 1981. Distribution of Amphibians in Nepal. In: *Wild is Beautiful*, (Ed.) Majpuriya, T.C., p. 142–146.

Shrestha, T.K., 1984. On the distribution and habitat of the Himalayan Newt (*Tylotriton verrucosus*) in eastern Nepal. *J. Bombay Nat. Hist. Mus.*, 81: 485–487.

Shrestha, T.K., 1989. Ecological Aspects of the life history of the Himalayan newt *Tylototriton verrucosus* (Anderson) with reference to the conservation and management. *J. Bombay Nat. Hist. Soc.*, 86: 333–338.

Shrestha, T.K., 1994. The Himalayan Newt: A cryptological enterprise. In: *Encounter with the Himalayan Wildlife*, Kathmandu, 147 pp.

Shrestha, T.K., 1996. Country report for Nepal. In: Abstracts: *International Conference on the Biology and Conservation of the South Asian Amphibians and Reptiles*, 1–5 August, 1996, Sri Lanka.

Shrestha, T.K., 1998. Conservation strategy for amphibians and reptiles of Nepal. *Zoos' Print, Amphibian Spec. Issue*, Coimbatore, 11(5): 19–21

Singh, S.C., 1996. A Rapid Observation on Vegetation enroute Jhapa-Ilam-Phidim sector in East Nepal. Natural History Museum, Kathmandu, T.U.

Smith, M.A., 1924. The tadpole of *Tylotriton verrucosus* Anderson. *Rec. Indian Mus.*, 26: 309–310.

Soman, P.W., 1966. An addition to the amphibians of Nepal and extension of the range of the Indian Newt *Tylotriton verrucosus*. *Cuv. Sci. Cult.*, 32: 427–428.

Wongratana, T., 1984. Range extension of the Crocodile salamander, *Tylotriton verrucosus*, Phu Luang. Thailand, *Nat. Hist. Bull. Siam. Soc.*, 32: 107–110.

Zhao, E. and Hu, Q., 1988. Studies on Chinese tailed amphibians. In: *Studies on Chinese Salamanders. Contr. Herpetol.*, 4: 1–44.

Chapter 11

Population Trends of Indian Mugger Crocodiles *Crocodylus palustris* at Some Selected Sanctuaries in Andhra Pradesh, South India

V. Vijaya Kumar*, B.C. Choudhury and V.C. Soni*****

**Gujarat Institute of Desert Ecology, Post Box No. 83, Mundra Road, Bhuj-Kachchh – 370 001*

***Wildlife Institute of India, P.O. No. 18. Dehra Dun – 248 001*

****Department of Biosciences, Saurashtra University, Rajkot – 360 005*

ABSTRACT

Monitoring studies on the restocked mugger crocodiles *Crocodylus palustris* in Andhra Pradesh between January 1987 and April 1990 showed a decreasing trend in population of *C. palustris* in Manjira and Siwaram wildlife sanctuaries, and increasing in Ethipothalla falls. The fluctuation was recorded only among non-breeders in which very low percentage was recorded. The characteristics of the habitats and the interaction of restocked crocodiles with existing population structure of adult *C. palustris* were found to be the two major factors, which have effected on the non-breeding population structure in the study areas. The biotic interferences, which might also play an important role in the dispersal of the non-breeders, were found to be high in Manjira and Siwaram, but minimum in Ethipothalla falls.

Introduction

This article provides information on the population trend of *Crocodylus palustris* from three different study areas; Manjira wildlife sanctuary (reservoir), Ethipothalla falls (water fall) and Siwaram wildlife sanctuary (open river) in Andhra Pradesh (A.P) based on long-term census data collected during the study period from January 1987 to April 1990.

Crocodile Population and Conservation Project in A.P. State

The A.P state had populations of *Crocodylus palustris* and *Crocodylus porosus* while the latter became extinct in late 1940's (Vijaya Kumar, 1993). Even *C. palustris* population had also reduced by the mid 1970's. However, the state, with two major rivers, Krishna and Godavari and their numerous tributaries, and the deltaic mangroves, had many suitable *C. palustris* (Figure 11.1) habitats as well as limited habitat for the *C. porosus.*

In A.P. the systematic survey of crocodile habitat began in late 1976 after the Crocodile Conservation Programme was initiated. Prior to 1976, there was no published record on the status of *C. palustris* in A.P. However, since 1976 the State Forest Department officials carried out surveys on *C. palustris* (Figure 11.1) in two major river systems of A.P. (Krishna and Godavari and its numerous tributaries) and estimated population to be around 50 individuals (Choudhury 1979), with a record of six breeding locations in the wild. In view of improving the status of wild *C. palustris* population, in 1977 and subsequent years, 264 *C. palustris* were reintroduced in six different sanctuaries of A.P. (Figure 11.2). The monitoring on the restocked crocodiles by the Forest Department officials was intensive only before and after breeding season.

The present study on *C. palustris* population was conducted mainly to assess the population trend in the three study areas based on year round intensive surveys.

Study Area

C. palustris is known to inhabit all kinds of freshwater habitats such as large and small rivers, man-made reservoirs, lakes, village ponds and even below water falls. In A.P. the *C. palustris* are distributed in such areas in the wild. The six protected areas where

Figure 11.1

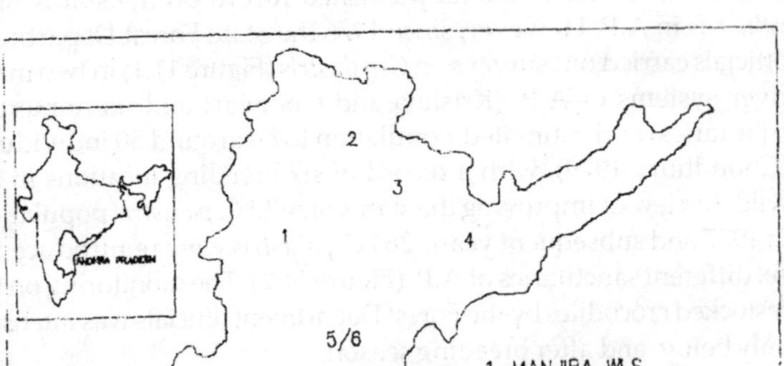

Figure 11.2: *Crocodylus palustris* Reintroduction Sites in Andhra Pradesh

C. palustris were reintroduced in A.P. represent open river (Siwaram wildlife sanctuary), lakes (Pakhal wildlife sanctuary), reservoirs (Manjira, and Kinnarsani wildlife sanctuary) and waterfall (Ethipothalla–Nagarjuna Sagar Srisailum Tiger Reserve). The areas chosen for the intensive study, therefore, took into account three representative habitat types.

1. Reservoir (Manjira wildlife sanctuary)
2. Waterfall (Ethipothalla falls)
3. Open river (Siwaram wildlife sanctuary)

Manjira Wildlife Sanctuary

Manjira wildlife sanctuary, a man-made reservoir on Manjira river (Figure 11.3) is located in Medak district, 50 km northwest of Hyderabad (A.P. state headquarter). The Manjira River is a tributary of the Godavari river system. The Manjira reservoir formed due to construction of barrage across Manjira river at Kalabgoor village in the year 1969, spreads over 20 km^2. It was declared a sanctuary in 1978 due to the presence of a stable *C. palustris* population identified during a state wide crocodile surveys in 1976–77 (Choudhury Unpubl. Report). Since the reservoir is used for the drinking water supply to Hyderabad City, water is available in the reservoir round the year. It supports a varied submerged and shoreline vegetation. A thin margin of *Typha* sp. *Ipomoea* sp. and *Acacia* sp. along the fringes of water line, while agricultural land (Paddy *Oryza sativa*, Maize *Zea mays*, Jowar *Sorgham vulgare* and groundnut *Arachis hypogia*) constitutes the next line of vegetation towards the land. The river does not flow through forested area, but in the dry undulated savanna type land, where the predominant species was *Acacia* sp., *Azadiracta* sp., *Annona* sp. and *Prosopis* sp. The aquatic plants include *Ipomea aquatica*, *Nymphoides* sp., *Hydrilla* sp., *Potamogeton* sp., *Polygonum* sp., *Vallisneria* sp., *Najas* sp., *Chara* sp., and many algae. The reservoir also has several islands with extensive marsh along its fringes.

Even though, the A.P. crocodile conservation project's main objectives and activities from 1976–1984 were to locate and collect crocodile eggs from wild crocodile habitats for hatching and rearing in captive for reintroduction in the wild. The A.P. Forest Department had left Manjira Wildlife sanctuary population undisturbed. No eggs were collected from this protected area but every year monitoring

Figure 11.3

of nests and hatchlings was done during the nesting season (Varaprasad and Choudhury, 1980). The Manjira barrage at Kalabgoor provided very little opportunity for the reintroduced crocodile to migrate downstream at the time of reintroduction in 1985. Further, at the initiation of this study in 1987 one more dam was constructed across the Manjira river, 32 km upstream (at Singoor) making the study area into a closed water body. The sanctuary receives water every year during the south-west monsoon, between June and September. Water depth in the reservoir varies from 1 to 10 m. During summer, temperature shoots up to 42°C and falls to 15°C in winter.

Ethipothalla Falls

Ethipothalla falls (Figure 11.4) is located 180 km south of Hyderabad, in the eastern most limits of the Nagarjuna Sagar–Srisailam Tiger Reserve covers an area of 3600 km^2 including much excellent *C. palustris* and tiger habitat in which, a part of this Reserve covering the waterfalls covering an area of about 3 km^2 was declared reserved and gazeted through the crocodile project in June 1978. The rationale of selecting this habitat as one of the sites for study was, the impossibility of reintroduced *C. palustris* going upstream

Figure 11.4

due to the blockade by the 25 m high waterfall. An additional reason for considering this as a major study site was the availability of continuous monitoring data from the sanctuary staff right from the day of the first release in February, 1977.

Siwaram Wildlife Sanctuary

Siwaram Wildlife Sanctuary located in Karimanagar district encompasses about 3 km of the Godavari River stretch that provides a natural habitat for *C. palustris.* Following the 1976–77 statewide crocodile survey, the area was declared as a crocodile sanctuary in 1978. The river at this site is 300 m wide and 12 m deep even during the dry season. For the reintroduced *C. palustris,* the open riverine habitat provided both upstream and downstream movement opportunities. The north bank is bushy boundary (*Salix* sp.) with patches of tree line of *Terminalia arjuna.* The southern bank is sandy and open while the northern bank is silty with regular deposition from the floods. The south bank had mostly fallow-lands where occasional dry land farming and cultivation of crops such as chilly *Capsicum annum* and grams *Cicer arietium* are being done. Siwaram experiences extreme climatic conditions. The annual temperature varies from 11°C (December–January) to 47°C (May–June). Most of the precipitation is from the southwest monsoon during June to September.

Methodology

In all the study areas the river side was sub-divided into grids which were marked using riverside florescent kilometer markers to facilitate recording of locations of the crocodiles sighted along with other habitat parameters.

Surveys were carried out both during day and night. Day time surveys were carried out by boat and on foot walking along the banks. During day time boat surveys, the disturbance factors along the river being greater, indirect evidences such as spoor marks, fecal droppings and disturbance factors like human interference, cattle grazing and fishing were also collected.

Night surveys were carried out by boat, using a heavy-duty 12-volt Q-beam spotlight and a torch. The spotlight was held at the eye level and the river banks were scanned for any signs of crocodiles. The tapetum layer of the crocodilian eye reflects characteristic ruby red or yellow. Once such reflection was noticed the boat was moved closer to the crocodile to estimate its size and to make note on the habitat parameters. For each sighting, the location and activity of the spotted *C. palustris* and the environmental parameters like water temperature, air temperature and water depth in the location were also recorded.

Results

In total 211 surveys were carried out during day and night between January 1987 and April 1990 to monitor the *C. palustris* population trend in the three study areas. The surveys were grouped into three seasons; winter (November–February), Summer (March–June) and monsoon (July–October) (Table 11.1).

Status of *C. palustris* in the Study Areas–Before January 1987

Manjira Wildlife Sanctuary

The estimated wild *C. palustris* population during 1978 at Manjira was seven adults and five sub-adults (Varaprasad and Choudhury, 1980) (Table 11.2). In addition, ten juveniles (mean length 81.15 cm) were introduced in January 1985. Therefore at the

Table 11.1: Number of Crocodile Surveys Carried Out in Different Seasons in the Three Study Areas

Study Area	*Day Surveys*	*Night Surveys*	*Total Surveys*	*Seasons*								
				Winter			*Summer*			*Monsoon*		
				Total Day Counts	*Total Night Counts*	*Total*	*Total Day Counts*	*Total Night Counts*	*Total*	*Total Day Counts*	*Total Night Counts*	*Total*
Manjira	62	52	114	28	17	45	19	26	45	15	9	24
Ethipothalla	34	21	55	10	6	16	15	9	24	9	6	15
Siwararn	22	20	42	3	3	6	14	12	26	5	5	10

Table 11.2: Mugger Population Status at the Start of the Study

Study Area	*Manjira WLS*			*Ethipothalla Falls*			*Siwaram WLS*		
Size Class	*Adult*	*Sub Adult*	*Juvenile*	*Adult*	*Sub Adult*	*Juvenile*	*Adult*	*Sub Adult*	*Juvenile*
Known Natural Population	7	5	–	–	–	–	7	5	–
Details of Released Stock									
No of Mugger Released			10		8			7	8
Year of Release			1985		1978			1985	1985
Expected Minimum Population in January 1987	12	10	–	8	–	–	12	15	–
Total Expected Population		**22**			**8**			**27**	

beginning of this study in January, 1987, total population of *C. palustris* were estimated to be 12 adults (7 adults and 5 sub-adults of 1978 which would have become adults in 1987), 10 sub-adults (reintroduced juvenile stock may enter sub-adult class) and an unknown number of juveniles breed in the wild. Thus, this location represented a natural age / size class distribution.

Ethipothalla Falls

There were no wild *C. palustris* reported at the Ethipothalla falls in 1977. Between 1977 and 1980 the Forest Department introduced eight sub-adults (mean length 139.13 cm) in this area in phases (four in 1977, two in 1978 and two in 1980). The first breeding of these introduced stocks in the wild was reported in 1981 (Choudhury and Bustard, 1982). Nests of the females were monitored by the A.P. Forest Department since 1981 and jointly by the Crocodile Research Center of the Wildlife Institute of India since 1985. Though nests were recorded, no hatchlings or juveniles were seen till 1986. Further, no *C. palustris* eggs were collected from this location for captive hatching and rearing purposes. Therefore, it was thought that this location would have an admixture of adults (reintroduced stocks), sub-adults, juveniles and hatchlings (from the nests recorded since 1981).

Siwaram Wildlife Sanctuary

In Siwaram the crocodile population reported in 1977 included, seven adults and five sub-adults (Choudhury, Unpubl. data). In January 1985, seven sub-adults (mean length 126.7 cm) and eight juveniles (mean length 84.93 cm) were released. Assuming that no mortality or dispersal had occurred, the population would have been twelve adults (seven resident adults and five sub-adult of 1977 may become adults in 1986) and fifteen sub-adults [average growth rate per year 15 to 21 cm–Acharjyo and Mahapatra (1977), Choudhury and Bustard (1982)]. A few 1985 released sub-adults were presumed to have joined the adult size class in January 1987, at the beginning of this study. In addition, between 1979 and 1986 a maximum of three nests per year were reported from this location. The eggs laid wild were collected by the Forest Department up to 1983 after which collection was stopped. Therefore, it is presumed to have the hatchlings and yearling of 1984, 1985 and 1986 in the population. Thus, this site was also expected to have a representation of all size class.

Status of *C. palustris* in the Study Areas after 1987

Population Density

The number of crocodiles sighted per kilometer of river is taken as an estimate of minimum density. Table 11.4 gives the comparison between the population density of the expected minimum population at the beginning of the study in January 1987 and the survey results during 1987. Maximum number of crocodiles sighted during a single survey in the three study areas is given in Table 11.3. From the expected population only 49.3 per cent were sighted in Manjira and 33.3 per cent were sighted at Siwaram. This indicates that, there is a decrease of *C. palustris* population at Manjira and Siwaram Sanctuary limits whereas it increased at Ethipothalla falls (125.4 per cent). Such densities to study population trend were calculated for all seasons.

Population density trends from January 1987 to April 1990 showed that Manjira had a low density (mean density 0.41) of crocodiles. Siwaram had a high density (mean density 3.5) and the population was fluctuating but not increasing during 1987 to 1990. Ethipothalla falls had a stable population during the study period; however, the 1990 data shows an increase in the population (Table 11.4). Increasing trend in the population of *C. palustris* of each study population is revealed after the hatchling data was excluded. Hatchling numbers were highly variable in contrast to the relatively stable trends in non-hatchling numbers.

Table 11.3: Observed Population Density (Non-hatchlings: Based on Maximum Number of Sightings during a Single Survey)

Study Area	*Season*	*Day/Night*	*Year*	*No of Mugger/ Distance (km)*	*Mean Density*
Manjira	Summer	Night	1987	11/32	0.34
	Summer	Night	1988	10/32	0.031
	Summer	Night	1989	18/32	0.56
	Summer	Night	1990	14/32	0.43
Ethipothalla	Winter	Day	1987	10/3.5	2.86
	Summer	Night	1988	10/3.5	2.86
	Summer	Night	1989	10/3.5	2.86
	Summer	Night	1990	11/3.5	3.14
Siwaram	Summer	Night	1987	9/2.5	3.6
	Summer	Night	1988	7/2.5	2.8
	Summer	Night	1989	10/2.5	4.0
	Summer	Night	1990	9/2.5	3.6

Table 11.4: Comparison of Population Density Between Expected Population and 1987 Survey Results

Study Area	Expected Population Density in January 1987		Observed Population Density During January 1987		Percentage of Mugger Sighted from the Expected Population	Population Trend
	Tot. No. Distance in km	Density	Tot. No. Distance in km	Density/km		
Manjira	22/32	0.69	11/32	0.34	49.28	Decreased
Ethipothalla	8/3.5	2.28	10/3.5	2.86	125.44	Increased
Siwaram	27/2.5	10.8	9/2.5	3.60	33.33	Decreased

Table 11.5 gives more details of two major size-classes within the non-hatchling; 1–1.5 m and greater than 2 m size-class. The former is referred to as juvenile or sub-adults (non-breeders) and the later as adults (breeders). In Manjira and Ethipothalla falls the adults were stable during the study period. But the non-breeders population in Manjira and Siwaram were fluctuating and showed a decrease in 1990. In Ethipothalla it was stable from 1987 to 1989 but showed an increase in 1990.

Table 11.5: Population Density between Breeders and Non-breeders

Study Area	*Year*	*Breeders*		*Non-breeders*	
Manjira	1987	7	0.22	4	0.13
	1988	7	0.22	3	0.09
	1989	7	0.22	11	0.34
	1990	7	0.22	7	0.32
Ethipothalla	1987	6	1.71	4	1.14
	1988	6	1.71	4	1.14
	1989	6	1.71	4	1.14
	1990	6	1.71	5	1.42
Siwaram	1987	7	2.8	2	0.8
	1988	6	2.4	l	0.4
	1989	6	2.4	4	1.6
	1990	7	2.8	2	0.8

Factors Affecting the Sightings of *C. palustris*

A total of 13 major factors were identified to limit the activities of the *C. palustris*. Table 11.6 highlights the important abiotic and biotic factors identified and recorded in the three study areas. The abiotic factors were maximum in Manjira (20 per cent), minimum in Ethipothalla falls (11.3 per cent) as well as in Siwaram (11.3 per cent). However, the biotic factors were of importance, of which disturbance due to 'religious function' was recorded maximum in Ethipothalla falls. Even though it was for a short duration (about 3–5 days in a year) hundreds of people stay on the banks and used the potential crocodile habitat in addition to creating noise pollution by drums that causes heavy disturbance to the *C. palustris* in that area.

Table 11.6: Factors Affecting Population of Mugger in the Study Areas (A Ten Point Rating System)

Study Area	*Manjira*				*Ethipothalla*				*Siwaram*			
Study Period	*'87*	*'88*	*'89*	*'90*	*'87*	*'88*	*'89*	*'90*	*'87*	*'88*	*'89*	*'90*
Abiotic												
Drought			10									
Low Water Level					2	2	2	2		1	1	
Fire												
Flood	6	6	10		1	1	8		6	4	6	
Sub Total–I	**32/160 (20%)**				**18/160 (11.30%)**				**18/160 (11.30%)**			
Biotic												
1) Fishing												
Set Net	10	10	10	10					10	6	10	10
Dynamite												
Other Technique		10	10			2			1	1		
2) Cattle Grazing												
3) Removal of Vegetation												
Shoreline Vegetation	8	6	4	4	2	1	1	1	3	2	2	1
Reed beds	10	10	8	6								
Submerged Vegetation	2	2	1	1								

Contd...

Table 11.6–Contd...

Study Area	*Manjira*				*Ethipothalla*				*Siwaram*			
Study Period	*'87*	*'88*	*'89*	*'90*	*'87*	*'88*	*'89*	*'90*	*'87*	*'88*	*'89*	*'90*
4) Chemical Pollution												
Effluent									2	2	2	2
Pesticides						2	2		2	2	2	2
Film Shooting Colours							5					
5) Lift Irrigation	6	6	6	6	1	1	1	1				
6) Domestic Use	10	10	10	10	3	3	2	3				
7) Tourism												
8) Irrigation Dept.	8	6	10	10	3	3	2	3				
Sub Total-II	**274/560 (48.90%)**				**79/560 (14.10%)**				**106/560 (18.90%)**			
9) Religious Function					10	10	10	10	1	1	1	1

Further Table 11.6 also highlighted that maximum biotic factors were recorded in Manjira (48.9 per cent) followed by Siwaram (18.9 per cent) with the minimum in Ethipothalla falls (14.1 per cent).

Day and Night Surveys, Season and Water Temperature

Apart from the biotic and abiotic factors there are some other factors that also governs the activities and thereby the sightings of *C. palustris*. Standard survey techniques of day and night counts were utilized to monitor the crocodile populations. In all the three study areas, of the total sightings, maximum number in the size class categorizes (Juveniles, Sub Adults and Adults) were recorded in summer than in other seasons; hatchlings which were highly unstable were not taken into account. In summer more *C. palustris* were observed on the shoreline or floating on open water (Tables 11.7 and 11.8). Thus summer months proved to be suitable season for estimating the *C. palustris* population.

Table 11.7: Number of Sightings of Adult *C. palustris* in Different Seasons During Night Surveys in the Three Study Areas

Study Area	*Total No. of Sub Adult C. Palustris Sighted*	*Seasons*		
		Winter	*Summer*	*Monsoon*
Manjira WLS	161	56 (35%)	88 (55%)	17 (10%)
Ethipothalla	117	35 (30%)	56 (48%)	26 (55%)
Siwaram WLS	93	21 (23%)	62 (67%)	10 (55%)

Table 11.8: Number of Sightings of Sub Adult *C. palustris* in Different Seasons During Night Surveys in the Three Study Areas

Study Area	*Total No. of Sub Adult C. Palustris Sighted*	*Seasons*		
		Winter	*Summer*	*Monsoon*
Manjira WLS	99	11 (11%)	72 (73%)	16 (16%)
Ethipothalla	28	5 (18%)	18 (64%)	5 (18%)
Siwaram WLS	13	2 (15%)	11 (85%)	0 (0%)

The Table 11.9 highlights the fact that night surveys are more fruitful than the day surveys. Sightings were more successful during night in the three study areas with 94.23 per cent sightings in Manjira WLS, 95.24 per cent in Ethipothalla falls and 95 per cent in Siwaram WLS.

Table 11.9: Successful Surveys during Day and Night Surveys in the Three Study Areas

Study Area	*Total Surveys*	*Day Survey*		*Night Survey*	
		Total Survey	*% of Sighting Success*	*Total Survey*	*% of Sighting Success*
Manjira	114	62	41.93	52	94.23
Ethipothalla	55	34	67.64	21	95.24
Siwaram	42	22	63.63	20	95.00

Mean number of *C. palustris*/survey/km (same distance surveyed during day and night was taken into account) was higher in summer night (Tables 11.8–11.10) in all the three study areas indicating the suitability of summer night for population counts.

Table 11.10: Sightings of *C. palustris* During Day and Night Surveys in the Three Study Areas

Study Area	*Seasons*	*Night Surveys*			*Day Surveys*		
		'n'	*Mean*	*SD*	*'n'*	*Mean*	*SD*
Manjira	Winter	16	3.13	1.74	13	1.30	0.75
	Summer	25	6.32	3.11	9	1.22	0.66
	Monsoon	8	2.50	1.41	4	3.25	1.25
Ethipothalla	Winter	6	3.16	0.98	7	2.14	1.06
	Summer	9	3.22	0.97	10	2.00	1.05
	Monsoon	5	2.60	1.14	6	1.83	0.75
Siwaram	Winter	3	3.00	0	3	5.00	1.73
	Summer	12	5.72	2.19	8	2.00	1.41
	Monsoon	4	1.75	0.50	3	1.33	0.57

An experiment conducted in semi natural condition in captivity also confirmed the same. As an ectothermic animal their daily and seasonal activity patterns were mainly related to ambient temperature. Therefore, as the water and air temperature changes, so does the activity and visibility of *C. palustris*. The correlation of air

Table 11.11: Correlation Between Water Temperature and Number of Sightings of Different Size Class *C. palustris* in During Night Surveys

Water Temp. (°C)	*Manjira*				*Ethipothalla*				*Siwaram*			
	AD	*SA*	*JU*	*Tot. No.*	*AD*	*SA*	*JU*	*Tot. No.*	*AD*	*SA*	*JU*	*Tot. No.*
20–22	7	0	0	7	0	0	0	0	0	0	0	0
22–24	10	0	0	10	0	0	0	0	3	0	0	3
24–26	26	10	1	37	14	7	3	24	9	0	0	9
26–28	56	64	8	128	25	9	4	38	45	9	4	58
28–30	31	15	0	46	11	1	2	14	9	0	0	9
Total no. of *C. palustris*	130	89	9	228	50	17	9	76	66	9	4	79
% of *C. palustris* sighted, between 24–28°C Water temperature	63%	83%	89%	73%	78%	94%	78%	81%	99%	100%	100%	85%

temperature with the number of sightings was not significant but the water temperature and water level with the number of *C. palustris* sighted was siginificant. The counts were positively correlated with water temperature up to an optimum temperature (28°C).

Maximum number of sightings of all size classes were recorded between 24°C and 28°C in the three study areas (Table 11.11). The small size classes (< 1.5 m–juvenile and sub adults) were not sighted up to 24°C water temperature. Out of the total number of *C. palustris* recorded during the study period, about 73 per cent in Manjira, 81 per cent in Ethipothella and 85 per cent in Siwaram were sighted only in water temperature ranging between 24 to 28°C. Such an ideal water temperature was available only during night period of summer months (Table 11.12).

Table 11.12: Mean Water Temperature in Three Seasons at Manjira WLS during Night Surveys

Season	*Mean Water Temperature (°C)*	*Mean Air Temperature (°C)*
Winter	24.01	21.34
Summer	27.45	26.30
Monsoon	25.37	24.95

Discussion

The monitoring of *C. palustris* population during the study period had shown retrogressive results in Manjira and Siwaram whereas it was progressive in Ethipothalla falls. Analysis of number and size class of *C. palustris* sighted during the study period in the three study areas indicates that habitat and existing wild adult *C. palustris* population in that habitat were the two important factors that influenced the populations of *C. palustris*. Tables 11.2 and 11.3 revealed presence of low per cent of *C. palustris* population than expected population in Manjira and Siwaram while high in Ethipothalla falls. In the present study, data on only non-hatchlings were taken as Bayliss and Messel (1988) had stated that, while assessing rates of population increase, greater stability will result if the hatchling size class is deleted and only non-hatchlings were considered. Many vertebrate species exhibit variable recruitment in relation to environmental variability (Nichols 1976), and this may also apply to crocodilians (Fagarty, 1974; Joanen and Mc Nease,

1975; Nichols, 1987). This was also confirmed in the present study as the hatchling population showed a high fluctuation (Vijaya Kumar, 1993). Hence, data on hatchlings were not taken in to account. The density of non-hatchlings was high at Siwaram and it was low at Manjira. The density was fluctuating between the years but in Ethipothalla falls it was stable from 1987 to 1989 with increase shown only in 1990. The non-hatchlings include yearlings, juveniles, sub-adults and adults. Therefore, it is crucial to find out in which age-class the fluctuation occurs. For this reason, following Bayliss and Messel (1988), the non-hatchlings were further classified into non-breeders and breeders. Table 11.5 revealed that, except for Siwaram where the breeders showed a decrease during 1988 arid 1989, in the other two study areas the number of breeders remained stable from 1987 to 1990. The reasons for the decreased sightings of breeders at Siwaram were studied during Godavari survey trip conducted in April, 1990, the local villagers and fishermen confirmed that, an adult *C. palustris* was seen during the year 1988 and 1989 at Sudarsal Madugu, a small deep pool, about 30 km downstream from the Siwaram Sanctuary. In an open fast flowing river system (Godavari river) it is possible that, the *C. palustris* could have migrated in 1987 monsoon floods, where it had found suitable habitat and settled there for some time. Later, may be due to anthropogenic pressure or due to tendency for homing (Webb *et al.* 1983), it might have returned to its original area (Siwaram) as reflected by the 1990 survey results. In the population, fluctuation occurs only in non-breeder and very low percentage was sighted than the expected population. Except in Ethipothalla falls where the population remained stable from 1987 to 1989 and showed an increase in 1990. In other two study areas it was highly fluctuating. From three and a half years of field study it is very clear that, the habitat characteristics and the existing *C. palustris* population have a major effect on the non-breeders in the three study areas. Manjira reservoir having 32 km long river with considerable depth and water spread with mid-river islands facilitate the adults to move longer distances. In Siwaram 2.5 km stretch of the Godavari River has a deep pool which harboured all *C. palustris* whereas in Ethipothalla falls a series of 11 deep pools provide haunts for the *C. palustrts*, each adult occupying individual pools. It was observed that, there was no increase in number either in the breeders or in non-breeders (except in Ethipothalla falls) during the study period. In general, in the wild population, the natural recruitment into non-breeder is very low due to low survival of hatchlings (Vijaya Kumar, 1993). Whitaker and Whitaker (1977) observed that, the hatchling survival rate was almost zero in Tamil Nadu. In the wild, high mortality in crocodilian between egg and hatchling stages was reported by Gorzula (1978), Whitaker and Whitaker (1997), Choudhury and Bustard (1979), Medem (1981, 1983), Hutton and Child (1989), Thorbjarnarsan (1989) and Vijaya Kumar (1993). Messel *et al.* (1984) accounted that the loss of major proportions of sub-adults *C. porosus* may be due to predation by the adults. During breeding season adults exhibited territorial behaviour and readily chased any intruders. On 27 occasions five adult female alligators were observed attacking sub-adults (Hunt, 1988). Territoriality and dominance of adult *C. palustris* has been reported by Whitaker and Whitaker (1984, 1989). The existing adult population in Manjira and Siwaram might have chased away the non-breeders from the study area or may be due to the extensive movements of the smaller size classes. De Vas (1984) reported on the ability of *C. palustris* to move long distance over land. Hence, they found new areas for their survival. This was confirmed during the upstream surveys from Manjira during 1987 and 1989 and down stream survey in Godavari River during 1990. Some of the stocks released in 1985 at Manjira, were located 30–40 km upstream from the release site and about 60 km downstream in Siwaram Sanctuaries (Vijaya Kumar and Choudhury, 1990 and Vijaya Kumar, 1993). In Ethipothalla falls there were no wild *C. palustris* at the time of 1978–80 release, while the sub-adults released had become adults. Out of eight reintroduced *C. palustris* (1977–78) six were recorded in this area after 12 years (1990). Table 11.6 highlights that, apart from these reasons, the anthropogenic pressures like use of river water for bathing, washing the cloths, cleaning and watering the cattle, fishing and land use practices along the river banks were maximum in Manjira and Siwaram. These might also play an important role in the dispersal of non-breeders. But, Ethipothalla falls, being located in the Nagarjunasager–Srisailam Tiger Reserve limits, such pressures were either very meager or absent.

The study also revealed that night counts were more suitable than day counts as reported by Whitaker (1977), Missel and Vorlicek (1980), Montaque (1983) and Woodward and Marion (1978) for different species of crocodiles. Standardization of survey time and season is of primary importance. The representation of all size classes in summer night counts proved to be better than the other season counts. However the fluctuation of number of crocodiles in summer night counts prompted us to analyze other parameters too. Missel and Vorlicek (1980) (*C. porosus*) and Woodward and Marion (1978) studied on the effect of environmental parameters on the visibility of crocodiles (*Alligator mississippiensis*) during night counts. The results focused that the sightings of *C. palustris* were positively correlated with water temperature up to 28°C and is negatively associated with water depth as in case with *Alligator mississippiensis* (Murphy, 1977; Woodward and Marion, 1978). Present study reveals that 24–28°C is favorable water temperature that gives reasonably high counts and the juveniles were not recorded in temperature below 24°C.

Acknowledgement

We are thankful to the Wildlife Institute of India, Dehra Dun, for funding the study. We also take this opportunity to thank Mr. Pusph Kumar (Retd. Principal Chief Conservator of Forest, A.P.) and staffs of Manjira, Siwaram and Ethipothalla sanctuaries for constant support during the field study.

References

Acharjyo, L.N. and S. Mahapatra, 1977. Growth rate of mugger (*Crocodylus palustris*). *Hamadryad*. 2: 9.

Bayliss, P.G. and H. Messel, 1988. The population dynamics of Estuarine crocodiles: An assessment of long-term census data. In: 9th *Working Meeting CSG of the IUCN*. October, 1988, Lae, New Guinea.

Choudhury, B.C. and H.R. Bustard, 1979. The status of mugger in Tamil Nadu. In: *Proc. First Indian Crocodilian Researcher's Symposium*, Katerniyaghat, U.P., India 21–25 January, 1979.

Choudhury, B.C. and H.R. Bustard, 1982. Restocking mugger crocodile (*Crocodylus palustris*) in Andhra Pradesh. Evaluation of a pilot release. *J. Bombay Nat. Hist. Soc.*, 79(2): 275–279.

De Vas. A., 1984. Crocodile conservation in India. *Biological Conservation*, 29:183–189.

Fagarty, M.J., 1974. The ecology of the Everglades alligators. In: *Environments of South Florida: Present and Past*, (Ed.) P.J. Gleason, pp. 364–374. Memoir 2: Miami Geological Society, 452 pp.

Gorzula, S.J. 1978. An ecological study of *Caimen crocodilus crocodilus* inhabiting savanna lagoons in the Venezuelan Guayana. *Oecologia (Berl.)*, 35: 21–34.

Hunt, R.H., 1988. Aggressive behavior of adult Alligators *Alligator mississippiensis* towards sub-adults in Okefenokee swamp. In: 9th *Working Meeting IUCN–CSG*, Lae, Papua New Guinea, pp. 360–369.

Hutton, J.M. and G.T.F. Child, 1989. Crocodile management in Zimbabwe. In: *Crocodiles*, IUCN Publ. (NS). Gland, pp. 62–79.

Joanen, T. and L. McNease, 1975. Notes on the reproductive biology and captive propagation of the American alligators. In: *Proc. Ani. Conf. S.E. Assoc. Game and Fish Comm.*, 32: 182–189.

Medem, F., 1981. *Los, Crocodylia de sur America. Vol. I: Los crocodilia de colombia*. Colciensias. Bogota, Colombia.

Medem, F., 1983. *Los Crocodylia de sur America. Vol. II Colciencias.* Bogota, Colombia.

Messel, H., G.C. Vorlicek, 1980. Surveys of tidal river systems in the northern territory of Australia and their crocodile populations. Mono G. 12 Limmen Bright Towns. Pergamen Press. Sydney.

Messel, H., G.C. Vorlicek, W.J. Green and I.C. Onley, 1984. The continuing and mysterious disappearance of a major fraction of sub-adult *Crocodylus porosus* from tidal water ways in Northern Australia. In: *Crocodile. Proc. of the 6th IUCN/SSC/CSG Meeting.*

Montaque, J.J., 1983. Influence of water level, hunting pressure and habitat type on crocodile abundance in fly river drainage, Papua New Guinea. *Bio. Cons.*, 26: 309–339.

Murphy, T.M., 1977. Distribution, movement and population dynamics of the American alligator in a thermally altered reservoir. *M.Sc. Thesis*, University of Georgia, Athens.

Nichols, J.D., 1976. Simulation of a commercially harvested alligator population in Louisiana. *Louisiana State University Bull.* No. 691.

Nichols, J.D., 1987. Population models and crocodile management. In: *Wildlife Management of Crocodiles and Alligators*, Chap. 17: 177–187.

Thorbjarnarsan, J.B., 1989. Ecology of American crocodile *Crocodylus acutus.* In: *Crocodiles*, pp. 228–258. IUCN Publ. (NS). Gland.

Varaprasad, S. and B.C. Choudhury, 1980. Crocodile survey in Manjira Crocodile Sanctuary. *Makara*, 2(1): 9–10.

Vijaya Kumar, V. and B.C. Choudhury, 1990. Evaluation of mugger crocodiles restocking by monitoring studies and its use in long-term management of the species: Interim study report. Wildlife Institute of India, Dehra Dun, pp. 35.

Vijaya Kumar, V., 1993. Studies on the Indian mugger crocodiles *Crocodylus palustris* (Lesson) in Andhra Pradesh, India. *Ph.D Thesis*, Saurashtra University, Rajkot, India.

Webb, G.J. *et al.*, 1983. *Crocodylus johnstoni* in the Mc Kinlay river area N.T.I.V: A demonstration of homing. *Aust. Wildl. Res.*, p. 403–406.

Whitaker, R. and Z. Whitaker, 1977. Note on the natural history of *Crocodylus palustris. J. Bombay Nat. Hist. Soc.*, 74(2): 358–360.

Whitaker, R. and Z. Whitaker, 1984. Reproductive biology of mugger (*Crocodylus palustris*). *J. Bombay Nat. Hist. Soc.*, 81(2): 297–316.

Whitaker, R. and Z. Whitaker, 1989. Ecology of the mugger crocodile. In *Crocodiles*. IUCN Publ. (NS), pp.276–296.

Woodward, A.R. and W.R. Marion, 1978. An evaluation of factors affecting night-light counts of Alligators. In: *Proc. S.E. Assoc. Fish and Wildl. Agencies*, 32: 291–302.

Chapter 12

Management, Conservation and Farming of Saltwater Crocodiles: An Australian Case Study of Sustainable Commercial Use

Clem Tisdell and Hemanath Swarna Nantha***

School of Economics, The University of Queensland, Brisbane 4072 QLD, Australia

ABSTRACT

Opinions differ about what types of policies are likely to be most effective in conserving wildlife species. For example, the Convention on International Trade in Endangered Species of Wild Flora and Fauna (CITES) is based on the premise that curbing the commercial use of endangered species favours their conservation, whereas the Convention on Biological Diversity envisages the possibility that such use may contribute to the conservation of species. In Australia, as illustrated in the case of the saltwater crocodile, the governments of the Northern Territory and Western Australia have favoured the latter policy in recent years whereas Queensland has favoured the former approach. The saltwater crocodile management plan of the Northern Territory provides an instructive case study of the consequences of adopting a commercial use strategy to promote

*Tel: 61-7-33656570, Fax: 61-7-33657299, E-mail: c.tisdell@economics.uq.edu.au

**Tel: 61-7-33656645, Fax: 61-7-33657299, E-mail: h.swarnanantha@uq.edu.au

wildlife conservation. The methodology used in this study, which involves a survey of crocodile farm managers and managers of cattle properties in the Northern Territory as well as secondary data, is outlined, after providing background on the conservation status of saltwater crocodiles in Australia and the saltwater crocodile management plan of the Northern Territory. In the results section, after outlining the nature and structure of the Northern Territory crocodile farming industry, evidence is presented on whether or not the crocodile management plan of the Northern Territory encourages pastoralists to conserve crocodiles on their properties. This study concludes with a discussion of the overall conservation effectiveness of the crocodile management scheme of the Northern Territory and considers its possible implications for saltwater crocodile management in areas of Asia where the species occurs.

Keywords: *Australia; Conservation economics; Convention on Biological Diversity; Convention on international trade in endangered species; Crocodylus porosus; Property rights; Saltwater crocodiles; Sustainable use; Wildlife conservation.*

Introduction

There are several different approaches to conserving wildlife species. One that has obtained growing support in recent years is the use of economic incentives to encourage the sustainable use of wildlife. This policy has, for example, been espoused in *Caring for the Earth* (IUCN-UNEP-WWF, 1991) and has been reflected in the framing of the Convention of Biological Diversity. This approach is motivated by the view that in the absence of economic incentives to conserve wildlife species on their land, landholders (which can include private, state and communal landholders) will obtain no economic benefits from wildlife on their properties. They will, therefore, have little or no economic reason to conserve wildlife because of their inability to appropriate economic returns from it. With some qualifications, this point of view has been endorsed by economists such as Swanson (1997). The ban on trade in ivory which was once adopted under CITES (Convention on International Trade in Endangered Species of Wild Flora and Fauna) as a strategy to prevent the decline in African elephant populations has been used as an example (Barbier *et al.*, 1990). This strategy was claimed to be counterproductive because it reduced funds available to those African governments which were genuinely pursuing the conservation of elephants. It also denied local communities legal economic benefits from elephants thereby taking away any motivation they might have had to help conserve them.

However, this sustainable use strategy is in conflict with the conservation policy underlying CITES. When a species is endangered, the main strategy adopted by CITES is to ban trade in the species, thereby removing market-based economic incentives for its commercial exploitation. If the species is an open-access resource (freely available to all to exploit), this can reduce harvesting pressure on it and help conserve it. This would be so, for instance, for species in the open seas (cf. Hardin, 1968). In the case of a species exploited by open-access, the higher the market gains from its exploitation, the higher is the intensity of exploitation. In this case, an increase in commercial economic gains raises the likelihood of extinction of the species (Tisdell, Ch. 8, 2002; 2005). By contrast, if populations of the species are the exclusive and effective property of landholders (or sea-space holders), the opposite result follows: as the commercial or marketable economic benefits from the species rises, landholders will become more eager to manage the species so that its population is sustained. However, the market-based gains must be enough to give a positive return to the landholder from husbanding the species for the landholder to have an incentive to conserve it (cf. Tisdell, 2004a, b).

The saltwater crocodile *Crocodylus porosus* (illustrated by Figure 12.1) is a species with considerable commercial use value. Therefore, it is possible that it might be conserved by giving landholders exclusive property rights to crocodiles on their properties and by allowing the marketing of crocodiles and products derived from them. This sustainable use approach to conserving crocodiles has been adopted by the governments of the Northern Territory (of Australia) (NT) and of Western Australia (WA) whereas the Queensland government continues to ban the commercial sale of crocodiles (including their eggs) from the wild. The result has been that ranching of saltwater crocodiles occurs in NT and WA but not in Queensland, the only Australian states in which crocodiles occur in the wild (see for example, Tisdell, *et al.*, 2005). However, commercial crocodile farms exist in all three Australian states. Farms in Queensland rely mostly on the incubation of eggs laid by crocodile

Figure 12.1: A Photograph of an Adult Saltwater Crocodile in the Northern Territory of Australia

brood stock on the farms whereas those in NT and WA rely heavily on crocodile eggs and adults collected from the wild and for which they pay landholders a fee.

The purpose of this article is to consider and assess the policy of the NT for managing and conserving saltwater crocodiles using information provided to the authors by crocodile farmers and by managers of pastoral properties primarily producing beef cattle. The presentation is developed as follows: After providing some brief historical background on the status of saltwater crocodiles in Australia and the crocodile management plan of the Northern Territory, the methodology used for surveying crocodile farmers and pastoralists in NT is outlined. The results are then reported and discussed. In this study, we give particular attention to whether crocodile farming is likely to remain commercially viable in the NT. In the absence of the continuing economic viability of this sector, the NT's commercial use strategy for conservation of crocodiles will be jeopardised. Secondly, we enquire whether managers of NT pastoral properties from which crocodile eggs are collected by crocodile farmers and for which the landholders receive a payment do in fact adopt practices that help to conserve saltwater crocodiles as a result of the payments they receive.

Brief Background on the Conservation Status of Saltwater Crocodiles in Australia and the Crocodile Management Plan of NT

Crocodiles were effectively open-access resources in Australia until 1969. From the mid-1940s onwards, they were increasingly hunted for recreational and commercial purposes. The commercial harvesting of saltwater crocodiles was mainly for the skin trade (Webb *et al.*, 1987; PWCNT, undated). By around 1965, over harvesting had greatly reduced Australia's population of saltwater crocodiles. In the NT, for example, the estimated initial population of saltwater crocodiles of about 100,000 individuals (Webb *et al.*, 1984) had been reduced by around 95 per cent to around 5,000 individuals (Webb *et al.*, 2000). The species had become endangered in Australia. This resulted in both WA and NT fully protecting crocodiles in 1969 and 1971, respectively, and Queensland followed suit in 1974 (Letts, 1987). Recreational harvesting and hunting of crocodiles for commercial purposes were banned. However, Australian Aborigines still retained the right to hunt crocodiles for their own subsistence.

Research on the possibility of farming saltwater crocodile possibly began in 1969 in Queensland when a crocodile farm was set up under a State government scheme. It aimed at conserving crocodiles while at the same time providing employment to local indigenous people (Onions, 1987). No commercial crocodile farming operations were in existence then.

As a result of protection given to saltwater crocodiles in the NT, their population rose and it is estimated that by the early 1990s their population had returned to pre-exploitation levels (Webb *et al.*, 1994). Even by the beginning of the 1980s, the number of crocodiles had increased and there were calls for culling them, especially after crocodile attacks were reported (Webb, 2002, p. 17). Australia's management of saltwater crocodiles responded to include a public education programme about the recovering crocodile population and dangers that they may pose, a programme to remove problem crocodiles so as to minimise hazards to livestock and humans, and crocodile farming and ranching (the collecting of crocodile eggs from the landholders in exchange for money) (Webb, 2002, p. 17). Egg harvesting from the wild started in NT in 1979/1980 on a small scale with collections numbering in the hundreds. In the mid-1980s, when Australia's proposal to move its saltwater crocodile population to Appendix II of CITES succeeded, collections were scaled up, and since 1989/1990, between ten to twenty thousand eggs were harvested annually. In the mid-1990s, the NT government started to allow limited harvesting of hatchlings, juveniles and adults by commercial crocodile farms. The number of adult crocodiles harvested in 1997 was 17 individuals and recently, it was as high as 600 adult individuals. Increasing commercial harvesting since the 1980s has been accompanied by gradually increasing saltwater crocodile populations in the wild (Webb *et al.*, 2000; PWCNT, undated).

The saltwater crocodile was considered to be an endangered and vulnerable species in the past, but now is considered at 'lower risk' or of 'least concern' according to the IUCN (World Conservation Union) Red List (Crocodile Specialist Group, 1996). The species is classified under Appendix II in CITES and controlled trade is permitted for Australia, Indonesia and Papua New Guinea. For all other countries including India (UNEP-WCMC, 2005), where its populations have been extirpated from most of its range and are low, the saltwater crocodile remains listed under Appendix I, prohibiting harvest for trade. India's remaining populations are found mainly in nature reserves in Orissa and West Bengal and in the Nicobar and Andaman Islands (Whitaker, 1987).

Saltwater crocodile skins produce the highest quality leather of all species of crocodiles (Peucker, 1997; Department of Primary Industries and Fisheries, 2005), and these are the primary products of Australia's crocodile farms. Australia exports crocodile skins for the manufacture of leather goods and, on a smaller scale, produces its own finished products such as shoes, belts, watchstraps and handbags (Ashley and David, 1987; Brazaitis, 1987). Crocodile meat, a byproduct, is mainly consumed domestically but it is also exported (Peucker, 1997; Department of Primary Industries and Fisheries, 2005). The fascination with crocodiles as powerful and dangerous beasts in the wild has spawned a vibrant tourism industry based on crocodile viewing in the Northern Territory, Western Australia and Queensland (Ryan, 1998). The crocodile skin, meat and tourism businesses generate millions of dollars in revenue annually (Stubbs, 1998; Australian Senate, 1998a).

In summary, the NT scheme for sustainable commercial use of saltwater crocodiles has the following characteristics:

1. Quotas for crocodile (mostly eggs and some hatchlings) are allocated by government permit to landholders;
2. Registered crocodile harvesters, almost invariably crocodile farm operators, negotiate with landholders to harvest crocodile eggs, hatchlings and other crocodiles covered by the landholders' quota. They agree to pay a fee to landholders for permission to harvest, for example, an agreed price per egg collected.

Harvesting quotas are determined by the Parks and Wildlife Commission of the Northern Territory (referred to as PWCNT hereafter) in a way that sustains the population of saltwater crocodiles in the NT. Quota setting is informed by annual or biannual spotlight monitoring of saltwater crocodiles in several important rivers in the NT and by survey of data obtained from harvesting (PWCNT, undated). Thus the approach of PWCNT to managing saltwater crocodile populations is a regulated market-based approach in which it assigns limited property rights in crocodiles to landholders. Consider now is the methodology adopted to study this management strategy.

Study Methodology

Some crocodile farm managers and pastoral landholders in the Northern Territory were interviewed during a field trip in 2004, and a more extensive survey of crocodile farms and landholders in the Northern Territory was conducted using structured questionnaires in 2005. Aboriginal landowners in native title land also participate in crocodile harvesting, but they are not part of this study because of difficulty in procuring information and cost constraints. However, an officer from the Northern Land Council, an organization representing indigenous (Aboriginal) communities of the Northern Territory, was interviewed about the crocodile harvesting activities of Aboriginal communities and the possible benefits these communities may derive from these activities. Also interviewed about crocodile harvesting was a government scientist from the Parks and Wildlife Commission of the Northern Territory.

Two sets of questionnaires were prepared for the two groups, and accompanying each questionnaire was a self-addressed postage-paid envelope for their convenient return. The questionnaires for crocodile farmers were mailed directly to each of the then six crocodile farms in the Northern Territory. Out of these, three responded whereas two did not. The sixth crocodile farm is no longer operational. One of the crocodile farmers who responded and one who did not respond (owners of two of the largest crocodile farms in the Northern Territory) were previously interviewed. Therefore, feedback received through the questionnaires account for more than half of all operating commercial crocodile farms in the Northern Territory, and if the interviews were also considered, 80 per cent (all but one) of the operating crocodile farms were surveyed for the study.

Forty questionnaires for pastoral landholders were sent to the Cattlemen's Association of the Northern Territory, the umbrella body that represents Northern Territory pastoralists. The association then distributed the questionnaires to its members. We have obtained responses from seven pastoralists so far. This is in addition to two other pastoralists who were interviewed on an earlier occasion. The surveyed pastoral properties lie 100 to 850 kilometres away from Darwin, the capital city of the Northern Territory. The sizes of these pastoral properties range between 200 to 4,500 square kilometres. In total, the properties of the surveyed pastoralists cover an area of at least 16,420 square kilometres, more than twice the size of the Indian state of Sikkim or more than a third the size of the state of Haryana.

Crocodile farm managers and landholders were asked to provide background data about their crocodile farms or cattle properties. Both groups were asked to comment on the crocodile management plan of the Northern Territory government. Views of crocodile farmers were sought about the features of their farming operations, trends in the crocodile farming industry, harvesting practices, and whether the attitudes of landholders towards the conservation of crocodiles had altered as a result of the NT crocodile management plan. Pastoral landholders were questioned about the crocodiles and crocodile habitat on their property in general, their thoughts about crocodile harvesting and sustainable use, and their attitude towards the conservation of crocodiles and their habitat.

The preliminary findings presented in the results section are based on the interviews and questionnaires as well as information obtained from relevant secondary sources, such as management plan reports and scientific literature.

Results

The Structure of the NT Crocodile Farming Industry

Figure 12.2 provides a schematic overview of the crocodile farming sector in NT. The crocodile farming industry is centred on crocodile farms and there are currently five of them operating in NT. A sixth has been closed due to injuries sustained by the owner-manager while collecting crocodile eggs. It is not clear if it will re-open and when. All five farms produce physical crocodile products–meats and skins. For two farms, tourism (see Figure 12.3) is also a significant source of revenue, and in the case of a third, there is some income from tourism (*e.g.*, crocodile viewing). One farm, Crocodylus Park, engages heavily in research and tourism. Research into production technology is viewed as an important component in crocodile farming as farms seed to improve production and reduce production costs to compete internationally.

Farm stocks of crocodiles are replenished from two sources: (1) eggs laid by brood stock on the farm which are hatched and raised to 2 or 3 years old of age before processing, and (2) eggs collected from landholdings for which landholders have been allocated a harvest quota by the PWCNT, and to some extent hatchlings, juveniles and occasionally adults harvested from the same source. The major source of eggs for restocking farms is ranching rather than eggs laid on crocodile farms. The quantity of crocodile eggs and adults harvested by some of the farms surveyed are shown in Table 12.1. Present annual quotas for wild harvests are 25,000 eggs and 575 adult crocodiles (see PWCNT, undated).

The largest proportion of eggs collected from landholders is from communal lands held by Australian Aborigines under Native Land Title; around 70 per cent and 80 per cent of the crocodile eggs collected by two of the largest crocodile farms came from this source (Table 12.1). On average, about 65 per cent of eggs are collected by all farms from Aboriginal land. The remainder are obtained from pastoral landholdings. Almost 50 per cent of NT's area is under pastoral lease, and these beef cattle properties are very large in size (the largest single pastoral property in NT is more than 12,000 sq. km). Because of the ease of establishing communication with the latter group, due to assistance from NT Cattlemen's Association, we have concentrated on feedback from this latter group.

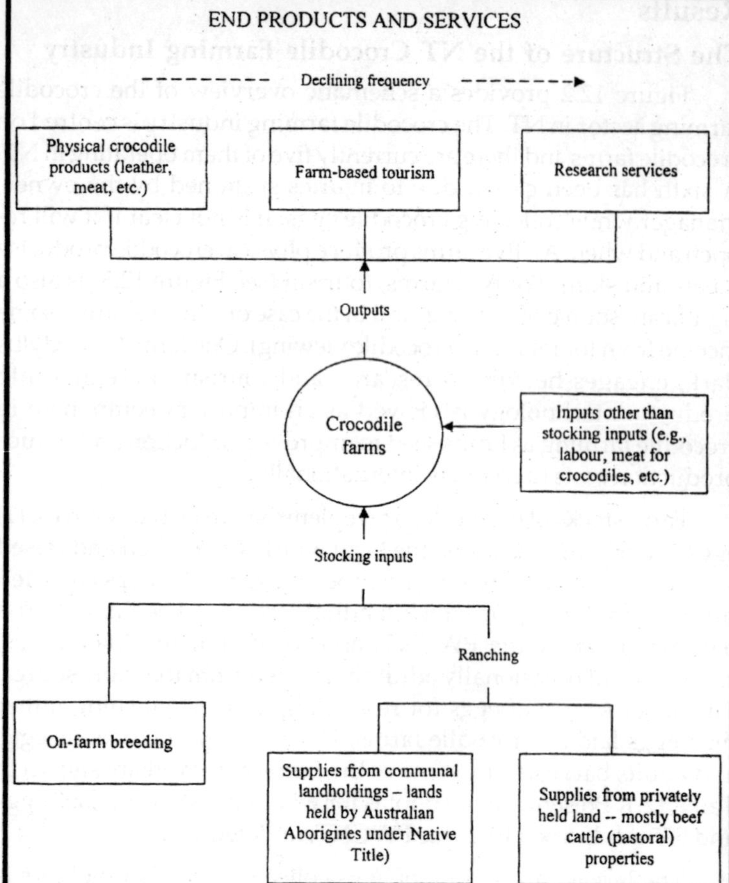

Figure 12.2: The Product Chain in the NT Crocodile Farming Industry

The Crocodile Farming Sector: Nature, Economic Viability, Future Trends

The five commercially operating crocodile farms in NT at present process between 3,000 and 9,000 head of crocodiles annually (refer Table 12.1), with one farm planning to increase its output to process 10,000 crocodiles annually in the near future. Crocodile farms in NT produce 60 per cent of Australia's exported saltwater crocodile skins,

Figure 12.3 (a): Keepers Feeding a Fowl to a Crocodile at Darwin Crocodile Farm in NT to entertain Tourists. The tourists are watching behind an iron-mesh fence but cannot be seen in this photograph.

Figure 12.3 (b): These Wild (feral) Asian Buffaloes (*Bubalus bubalis*) were seen near the Mary River in the Northern Territory in an area where Saltwater Crocodiles are Abundant. There are plenty of nesting materials for crocodiles in this vicinity. However, the manager of the 'Melaleuca' cattle property located downstream reported that while crocodiles occur on this property, they do not nest because of lack of nesting materials.

Table 12.1: Size of the Crocodile Harvests and Production Statistics of Crocodile Farms in NT, 2003/2004 (information obtained from interviews and survey questionnaires except where stated otherwise)

Crocodile Farm	*Number of Eggs Harvested Overall in the Last Year (2003/2004)*	*Number of Adult Crocodiles Harvested in the Last Year*[a]	*Percentage of Collected Eggs from Aboriginal Land*	*Head of Crocodiles Processed Annually (as of 2004, unless stated otherwise)*
A	8,000	200	70	9,000
B	N/a	N/a	80	5,000[b]
C	4,000	20	40	3,000
D	2,428	0	0	0
E	N/a	N/a	N/a	4,800[c]

[a] Numbers may be in addition to problem crocodiles caught separately and sold to farmers by the Parks and Wildlife Commission of the Northern Territory.

[b] Estimate (possibly for 2004) obtained from data available on the crocodile farm's Internet Website (Porosus Pty. Ltd., 2004). The estimate was obtained as follows. The farm plans to increase output to 10,000 animals a year from 320 female breeders and a stock of 30,000 individuals. It currently has 148 female breeders (half of the planned 320 breeders). Therefore, 10,000 × ½ = 5,000 processed animals a year at present. The number of one- and two-year old animals in this farm is about 5,000 individuals. This value may also serve as a surrogate measure of animals skinned a year, and is consistent with the value calculated above.

[c] Estimate (possibly for 2000) from the Australian Government Rural Industries Research and Development Corporation publication (Hyde, 2000). The book features the farm concerned. In the relevant section, the farm's production is revealed to be 400 skins a month. Multiplying this by 12 months gives the tabulated value.

and their main export destinations are France, Italy, Japan and Singapore (Isberg *et al.*, 2004).

Each viable egg can cost between Aus$5 to Aus$20 (Aus$1 = US$0.75; 18 July 2005), according to an interviewed crocodile farmer. Other crocodile farmers stated that the average price they paid for each crocodile egg was Aus$8 (one of them), and Aus$20 (two of them). The crocodile farmers state that the main factors that result in the variation in crocodile egg prices between landholdings are the cost of collection (ease of access, logistics) and competition. One crocodile farmer stated that he paid the same price for all eggs. First grade crocodile skins that are exported by crocodile farms to overseas markets fetch prices that range between Aus$4.50 per cm for skins that are 18 to 24 cm wide and up to Aus$10.00 per cm for those with widths of 40 to 45 cm (Isberg *et al.*, 2004).

Crocodile farmers stated that the advantages of crocodile farming in the Northern Territory are the climate (the heat suits crocodile growth) and the easily available crocodile and crocodile egg stocks. Among the disadvantages cited are the costs of acquiring crocodile feed, the cost of labour and capital costs.

The number of crocodile farms operating in NT rose from its level in 1984 to reach a peak in 1997 and has declined since then. Crocodile farms were established in NT in the early and especially mid-1980s when the CITES listing of Australian saltwater crocodile populations was changed from Appendix I to II. In 1984, there were three crocodile farms in NT (Onions, 1987). Between 1987 and 1990, there were four crocodile farms in NT (PWCNT, undated). By 1991, this number had risen to six (Australian Senate, 1998b; PWCNT, undated) and by 1997, to eight (PWCNT, undated). It fell by 2001 to six farms (PWCNT, undated), and at present, there are only five commercially operating crocodile farms in NT.

The crocodile farmers interviewed and those surveyed by post gave their reasons for why there has been a rise and decline in the number of crocodile farms in NT. Competition from overseas, such as from Nile crocodile (*Crocodylus niloticus*) farms in southern Africa, could have, in their opinion, led to the closure of some local farms. Economic competition as well as the importance of economies of scale resulted in greater consolidation of farms, according to another crocodile farmer. Differences in business skills and experience in crocodile farming of entrants may also have influenced survival in the industry. Crocodile farming is difficult because there is no established history of raising crocodiles as with domesticated animals such as cattle.

Three out of the four crocodile farmers surveyed believe that the number of crocodile farms in NT will remain unchanged. They claim that all, or virtually all, egg collection areas are being exploited and that the cost of entering the industry is high. Therefore, new farms are unlikely. However, one farmer stated that the process of specialisation in specific parts of production could bring in new players. All the crocodile farmers nevertheless expect the total production of NT crocodile industry to increase in the future. This is anticipated as demand rises and as learning and experimenting in the production process leads to possible falls in production costs. Most farms put crocodile production ahead of tourism as a contributor to the success of their businesses. Only one crocodile farms placed a higher priority on tourism as a contributor to its business success. The continuing economic viability of saltwater crocodile farming in NT is crucial for the maintenance of the NT's crocodile management programme. If this sector should become uneconomic, then there will be no demand for harvesting of crocodile eggs or crocodiles in other development stages from landholdings. The strategy of crocodile use via sustainable use would become unworkable.

Crocodile Farmers' and Pastoral Landholders' Assessment of the Northern Territory Government's Crocodile Management Plan

The crocodile farm managers/owners stated that they had very good or good knowledge of the current NT crocodile management plan. None stated that their knowledge of it was poor. The crocodile farmers surveyed expressed their satisfaction with it, and do not think that any significant changes to the current plan are required. However, two out of the three respondents to the posted questionnaire stated that they would like to see changes in the Australian Government's policies that affect their industry. This concerns CITES. One respondent stated that he would like speedier application and processing of CITES export permits and would prefer permits be issued by State governments rather than by the Federal government. Another stated that live commercial exports should be allowed. Such comments may come as no surprise to some considering that the requirements of the Australian Federal government for trading in crocodiles are more stringent than those of CITES itself (Onions, 1987).

Pastoralists had varying degrees of knowledge about the crocodile management plan. One did not answer this question about knowledge of the management plan (this landholder stated that he did not have crocodiles on his property) and one stated no knowledge of the management plan (he stated that he does not know if he has crocodiles on his property). Of the landholders who said they have crocodiles on their property, two stated that they are aware of the management plan but their knowledge of it is poor, three mentioned that their knowledge of it is moderately good, and one maintained that it was excellent (one of the interviewed pastoral landholders).

Pastoral landholders gave their opinions about the NT crocodile management plan. Most stated that management is good and is better than if there were no management at all (*e.g.*, the PWCNT helps remove problem crocodiles), but there were others who stated that (*i*) crocodiles are thriving on their property and the present quota for the harvest of adult saltwater crocodiles of 600 individuals is too low, and (*ii*) that management appears to be more concentrated on populated areas and pastoralists are 'ignored'–not listened to. They suggest that more open communication is required between pastoralists and those administering the crocodile management plan.

Crocodiles on Pastoral Properties and Crocodile Conservation Impacts on Pastoralists of the NT Crocodile Management Scheme

The Prevalence of Crocodiles on the Properties of the Pastoralists Surveyed

Most of the pastoralists said that they have crocodiles on their property. Of the two pastoralists who did not answer 'yes', one said that he does not have crocodiles on his property and the other said he was unsure of whether he had crocodiles on his property. At least 10 per cent of the land area of the properties of each pastoralist surveyed constituted crocodile territory if they had crocodiles on their land. One pastoralist even stated that 100 per cent of his land is crocodile territory.

Table 12.2: The Prevalence of Crocodiles on the Properties of the Pastoral Landholders Surveyed

No. of Pastoralists	*No. Who have Saltwater Crocodiles on their Property*	*No. Who have Problem Crocodiles on their Property*	*No. Who have Saltwater Crocodiles Nesting on their Property*	*No. Stating the Percentage of their Property Considered Crocodile Territory (Applies to those who said 'yes' to having crocodiles on their properties)*		
				Less than 10%	*10% to 25%*	*More than 50%*
Surveyed with Questionnaires (n = 7)	5	4	3	–	2	3
Interviewed (n = 2)	2	2	1	–	–	–

Attitudes Towards Crocodile Number on Pastoral Properties

Virtually all the pastoralists surveyed stated that the number of crocodiles on their property has increased over the past decade, and most would like numbers to decrease (Table 12.3). Almost all of the pastoralists responding thought that the present annual quota for crocodile harvesting was too low and are in favour of increasing its size. They unanimously consider crocodiles as both pests (because attack cattle stock and endanger humans) and assets (because they have financial value).

The main reason given by all pastoralists for obtaining permits to harvesting saltwater crocodiles and crocodile eggs on their properties is to curb the loss of cattle resulting from crocodile attacks. Their following comments are illustrative:

> *Mainly to reduce crocodile numbers; income too, but what is received is not much; the amount received is not worth worrying about (Pastoralist No. 1)*
>
> *To reduce crocodile numbers; [for income] to neutralise or recoup losses incurred from crocodile kills of cattle; if you don't manage, in five years plus you will be overrun by cow killers (Pastoralist No. 3)*
>
> *To reduce crocodile numbers; to support local business (Pastoralist No. 6)*
>
> *To decrease crocodile numbers (Pastoralist No. 8)*
>
> *Done just to reduce crocodile numbers (Pastoralist No. 9) [For] income (Pastoralist No.2)*

Do Pastoralists find Crocodile Harvesting Economically Worthwhile to them?

Despite acknowledging the financial value of saltwater crocodiles, almost all of the pastoralists surveyed consider their earnings from crocodile harvesting to be insignificant (Table 12.4). The interviewed pastoralists stressed that their main business and concern is cattle-raising. Only one landholder (Pastoralist No. 3) thought that earnings from crocodile harvesting are significant. Nevertheless, income from having permits to harvest crocodiles constitutes a very small portion of the net income obtained on all pastoralists' properties. At most, it is 5 per cent of total earnings (in the case of Pastoralist No. 3 who claimed that income from crocodile

Table 12.3: Views of Pastoralists on Crocodiles on their Properties and their Attitudes Towards them

Pastoralist	*Has Number of Crocodiles on Property Increased in the Last 10 Years?*	*Are Saltwater Crocodiles on Your Property Pests, Assets or Both?*	*Would you Like to See Saltwater Crocodile Numbers on Your Property Increase/Decrease/Remain about the Same?*	*Size of the Annual Quota for Harvesting Crocodiles: Favour Increase, Decrease or Prefer it Left Unchanged?*
Pastoralists Surveyed with Questionnaires				
1	Yes	Both	Decrease	Yes, increase considerably
2	Yes	Both	Decrease	Yes, increase a little
3	Yes	Both	Decrease	Yes, increase considerably
4	Yes	N/r	N/r	Yes, increase a little
5	Yes	Both	Remain about the same	Yes, increase a little
6	Yes	Both	Decrease	Yes, increase a little
7 (no croc. on property)	–	–	–	–
Interviewed Pastoralists				
8	Yes	Both	Decrease	–
9	–	–	Remain about the same	Yes

harvesting is significant). Some landholders (particularly those with smaller-scale cattle operations–less than 4,000 head of cattle) said that earnings from crocodile harvesting may compensate for some (but not all) of their cattle losses due to crocodile attacks (see last column, Table 12.4). In contrast, Pastoralist No. 6, who runs an operation of a more typical size and has had fairly large losses of cattle from crocodile attacks, considers coverage of these losses from gains in crocodile harvesting to be nil.

Views on Whether Land-use Practices of Managers of Beef Properties are Altered to Conserve Crocodiles and their Habitats

Three out of four crocodile farmers are of the opinion that paying landholders for rights to harvest crocodiles and their eggs will encourage landholders to conserve crocodiles on their property (Table 12.5). Their argument is that payments would make landholders more tolerant of crocodiles on their property. The fourth crocodile farmer was unsure whether crocodiles will be conserved by all landholders as a result of payments received from harvesting. He added that only smaller graziers, rather than larger landholders, may find it a worthwhile engagement.

Crocodile farmers were also asked whether holders of cattle property are likely to change land use practices to conserve more crocodile habitat as a result of crocodile harvesting on their land. Three out of four crocodile farmers were unsure of this, and the only farmer who gave a positive response indicated that it is conditional upon farming becoming a more profitable business and so that crocodile farmers could pay landholders more money for their adult crocodiles and crocodile eggs (see Table 12.5).

The views of the crocodile farmers mirror those of the pastoralists themselves: virtually all pastoralists stated that income from allowing crocodile harvesting has not induced them to alter their land-use practices or pastoral management in any way, or to conserve any habitat or area suitable for saltwater crocodiles or crocodile nests (Table 12.6). Only Pastoralist No. 3 stated that income from crocodile harvesting provided an incentive to alter land-use practices or to conserve crocodile habitat. This landholder stated that she maintained existing crocodile nesting sites and tried to increase available nesting areas by removing a harmful introduced weed (*Mimosa pigra*). Another landholder (Pastoralist No.9) would like to encourage crocodiles to nest on his property so he can harvest

Table 12.4: Economic Aspects of Crocodile Harvesting, Including Estimates of Losses Resulting from Crocodile Predation on Cattle as Reported by Pastoralists

Pastoralists	*Pastoral Property Size in sq. km. (and no. of cattle held)*	*Ever Applied for a Permit Allowing the Harvesting of Saltwater Crocodiles on Property?*	*Do You Regard Income Earned from Crocodile Harvesting as Significant?*	*Income from having Permits for Crocodile Harvesting as a Percentage of Net Income from Property*	*No. of Cattle Lost Annually*	*Total Value in Australian Dollars of Cattle Stock Lost Annually to Crocodile Attacks*	*Percentage Dollar Amount of Stock Losses Covered by Earnings from Crocodile Harvesting*
Pastoralists Surveyed with Questionnaires							
1	200 (4,000)	Yes	No	0.2	50	33,000	2
2	320 (2,500)	No*	No	N/r	10	8,000	6
3	400 (1,100)	Yes	Yes	5	25	30,000	90
4	3,800 (15,000)	No	N/r	N/r	0	N/r	N/r
5	4,200 (18,000)	No**	N/r	N/r	0	0	0
6	4,500 (30,000)	Yes	No	N/r	300	180,000	0
7 (no croc. on property)	3,000 (27,000)	–	–	–	–	–	–

Contd...

Table 12.4–Contd...

Pastoralists	Pastoral Property Size in sq. km. (and no. of cattle held)	Ever Applied for a Permit Allowing the Harvesting of Saltwater Crocodiles on Property?	Do You Regard Income Earned from Crocodile Harvesting as Significant?	Income from having Permits for Crocodile Harvesting as a Percentage of Net Income from Property	No. of Cattle Lost Annually	Total Value in Australian Dollars of Cattle Stock Lost Annually to Crocodile Attacks	Percentage Dollar Amount of Stock Losses Covered by Earnings from Crocodile Harvesting
Interviewed Pastoralists							
8	–		No	0.01	12 to 20	N/r	N/r
9	– (5,000)		No	N/r	150	75,000	N/r

*: This pastoralist stated that he does not apply for a permit to harvest but has 'professionals' (farmers/harvesters) come over to do the harvesting for him (which would still require him, the landholder, to provide a written consent for permit applications).

**: This pastoralist may not have applied for permits to harvest crocodiles but may have either had farmers/harvesters do the harvesting for him (with him giving a written consent for permit applications, as in the case of Pastoralist No. 2) or may have had problem crocodiles removed by the PWCNT.

Table 12.5: Views of Crocodile Farmers on Whether Pastoralists will Conserve Crocodiles and Crocodile Habitat as a Result of Crocodile Harvesting on their Lands

Crocodile Farmer	*Do you Think Paying Landholders for Rights to Collect Crocodile Eggs and Crocodiles on their Land Encourages them to Conserve Crocodiles?*	*Further Comments*	*Do you Think Pastoralists are Likely to Conserve more Crocodile Habitat as a Result of Receiving Payments for Crocodile Harvesting?*	*Further Comments*
		Farmers Surveyed with Questionnaires		
A	Yes	Pastoralists are tolerant of and more interested in crocodiles since some money is received	Yes	Not all pastoralists alter land management to take advantage of crocodile business. As farming/harvesting becomes more professional, efficient, and lucrative, some stations are seeing the logic of this [conserving crocodile habitat]. However, many stations have absentee owners relying on managers who may not be professional or interested
C	Yes	–	Unsure	Cow feed is at a premium and as cattle prices increase, this can compromise (crocodile habitat)
D	Yes	Positive attitude towards crocodiles and greater likelihood to tolerate crocodiles on their land	Unsure	–

Contd...

Table 12.5–Contd...

Crocodile Farmer	*Do you Think Paying Landholders for Rights to Collect Crocodile Eggs and Crocodiles on their Land Encourages them to Conserve Crocodiles?*	*Further Comments*	*Do you Think Pastoralists are Likely to Conserve more Crocodile Habitat as a Result of Receiving Payments for Crocodile Harvesting?*	*Further Comments*
		Interviewed Farmer		
B	Unsure	Large landholders might not care, probably want to get rid of them, but smaller graziers may earn something substantial	Unsure	They would not necessarily change land-use practices to suit crocodile harvesting

Table 12.6: Whether Pastoralists Change their Land-use Practices because of Income from Crocodile Harvesting

Pastoralist	*Has your Income from Allowing Crocodile Harvesting Altered your Land-use practices or Farm Management in Any way?*	*Have you Conserved (not altered) any Habitat or Area that Suits Saltwater Crocodiles or their Eggs because you have Obtained Income from their Harvesting on your Property?*
	Pastoralists Surveyed with Questionnaires	
1	No	No
2	No	No
3	Yes	Yes
4	–	–
5	No	No
6	No	No
7 (no croc. on property)	–	–
	Interviewed Pastoralists	
8	No	N/r
9	N/r	May do so (wants to have nesting area for crocodiles)

the eggs. Pastoralist No. 8 said that while he would be happy to sell farmers/harvesters crocodile existing eggs from his property, but he would not change his land practices to suit crocodile harvesting. He stated that the returns from wildlife must be worth its while or significant enough for cattle station owners to seriously consider taking action to conserve wildlife.

Pastoralists View on the Trophy Hunting of Crocodiles

Most pastoralists are in support of trophy hunting (six out of nine), because it could add more value to crocodiles. But some (two) commented that the big crocodiles that are the proposed targets of trophy hunting (animals over 3.5 m in length) (PWCNT, undated) are "not the problem", and that the smaller animals (juveniles) should be hunted too. According to one interviewed pastoralist (Pastoralist No. 9), the big crocodiles keep many smaller crocodiles at bay and hence reduce the number of cattle taken.

Discussion

The NT crocodile management plan allows the regulated commercial use of saltwater crocodiles. There are claims that this plan encourages landholders to conserve saltwater crocodiles (Australian Senate, 1998b). However, although they are favour of sustainable commercial harvesting of crocodiles, our surveys indicate that NT cattlemen have not altered their land management significantly as a result of being able to earn some income from crocodiles on their properties. This is because their returns from sales of crocodile eggs and from crocodiles in the later stages of their development are too low to make it profitable to 'husband' crocodiles. On all cattle properties surveyed, revenues from sales of the rights to crocodile farmers to collect crocodile eggs and crocodiles are insufficient to compensate landholders for their estimated annual loss of cattle due to predation by crocodiles. Thus for all holders of cattle properties surveyed having crocodiles, saltwater crocodiles remain a net pest. It is, therefore, not surprising that virtually all holders of cattle properties surveyed would like to see larger harvests of crocodile populations to reduce their numbers.

Only one cattle property indicated that it had changed its land management as a result of the NT crocodile management plan. It had used the money paid by crocodile farmers for rights to collect crocodile eggs and crocodiles on its property to remove the weedy invasive shrub (*Mimosa pigra*). This removal may have benefited the nesting of saltwater crocodiles as well as provided more grass for cattle.

Much higher prices could need to be paid for crocodile eggs and crocodiles harvested from cattle properties before pastoralists would be likely to husband crocodiles or significantly alter their land use practices so as to conserve crocodiles. Our results accord with a general proposition of Ross (2001) that "the relative returns of sustainable levels of crocodilian use are much lower than alternative uses". However, the present NT scheme does provide some, but far from complete, financial compensation to pastoralists for stock losses caused by crocodiles and does allow cattlemen to maintain limited population control of saltwater crocodiles on their properties. This increases the social acceptability of the current management scheme to NT cattlemen. Naturally, those who operate crocodile farms support the scheme and the general community can see economic benefits from the scheme. This support is reinforced by economic benefits that have been realised from the substantial tourism industry that has developed in NT based on the viewing of crocodiles in the wild.

The sustainability of the NT crocodile management scheme depends on the continuing economic viability of crocodile farming in NT. Crocodile farmers are confident of its continuing viability and foresee the possibility of a limited expansion in production. At the same time, they are aware of considerable economic competition, particularly internationally from production in Africa based on utilisation of the Nile crocodile.

There may be some pressures from cattlemen in NT to increase quotas for harvesting crocodiles, including juveniles as well as adults. Just how much scope there is for increasing harvesting quotas without reducing stocks of wild saltwater crocodiles significantly in NT is not known. Now that populations of saltwater crocodiles in NT have recovered to pre-1945–1965 harvest levels, they may be able to withstand considerable harvests before registering a large decline in their populations because they seem to exhibit a high degree of intraspecific competition (Tisdell *et al.*, 2005).

The present (crocodile harvesting) scheme of NT is weighted in favour of egg collection and crocodiles in their early stages of development. This has much to recommend it because less than 25 per cent of saltwater crocodile eggs laid in the wild hatch in the NT, and of those hatching, only about 50 per cent appear to survive until about, one year of age (Webb and Manolis, 1989, pp. 82–83). Therefore, the annual collection of 25,000 crocodile eggs from the wild (the NT quota) could be expected to reduce the population of one-year old crocodiles by about 3,000, or only by about three per cent of the estimated current population. The hatching rate of crocodile eggs on crocodile farms using artificial incubation is high as is the survival rate of hatchlings.

Thus current crocodile harvesting policies pose no danger to wild population of crocodiles in the NT. However, if harvesting policies should become heavily biased in favour of harvesting adult crocodiles, the impacts on wild populations would become more pronounced. So, caution is required in altering the weighting.

Whether similar sustainable use policies to those of NT could be adopted successfully in eastern India, Myanmar and Southeast Asian countries to conserve their saltwater crocodiles is unclear. Given the large size of landholdings in NT and that only extensive land use is economic there, it is doubtful if NT experience would transfer easily to Asian countries where landholdings are small and intensive land-use is the rule rather than the exception.

Acknowledgement

This study was supported by an Australian Research Council *Discovery* Grant awarded for the 'Economics of Conserving Australia's Tropical Wildlife' project. We wish to thank all those who have helped us with our surveys and the provision of secondary materials.

References

Ashley, D. and David, D.N., 1987. Marketing crocodilian skins. In: *Wildlife Management: Crocodile and Alligators,* (Eds.) G.J.W. Webb, S.C. Manolis and P.J. Whitehead, pp. 397–403. Chipping Norton, Australia: Surrey Beatty and Sons in association with the Conservation Commission of the Northern Territory.

Australian Senate, 1998a. Report of the Rural and Regional Affairs and Transport References Committee: Commercial utilisation of Australian native wildlife, Chapter 5: Economic Viability of Commercial Activities. Canberra: Parliament of Australia.

Available from: http://www.aph.gov.au/senate/committee/rrat_ctte/completed_inquiries/1996-99/wild/report/c05.htm [accessed 4 July 2005].

Australian Senate, 1998b. Report of the Rural and Regional Affairs and Transport–References Committee: Commercial utilisation of Australian native wildlife, Chapter 11: Crocodiles. Canberra: Parliament of Australia. Available from: http://www.aph.gov.au/senate/committee/rrat_ctte/completed_inquiries/1996-99/wild/report/c11.htm [accessed 4 July 2005].

Barbier, E., Burgess, J., Swanson, T. and Pearce, D., 1990. *Elephants, Economics and Ivory*. Earthscan Publications, London.

Brazaitis, P., 1987. The identification of crocodilian skins and products. In: *Wildlife Management: Crocodile and Alligators,* (Eds.) G.J.W. Webb, S.C. Manolis, P.J. Whitehead, pp. 373–386. Chipping Norton, Australia: Surrey Beatty and Sons in association with the Conservation Commission of the Northern Territory.

Crocodile Specialist Group, 1996. *Crocodylus porosus*. In: IUCN (2004), *2004 IUCN Red List of Threatened Species*. Available from: www.redlist.org [accessed 3 July 2005].

Department of Primary Industries and Fisheries (Government of Queensland), 2005. Crocodile farming: A general overview. Available from: http://www.dpi.qld.gov.au/crocodiles/12600.html [accessed 4 July].

Hardin, G., 1968. The tragedy of the commons. *Science,* 162: 1243–1248.

Hyde, K. (ed.), 2000. *Thirty Australian Champions: Shaping the Future for Rural Australia*. Barton, A.C.T.: Rural Industries Research and Development Corporation.

Isberg, S., Thomson, P., Nicholas, F., Barker, S. and Moran, C., 2004. *Farmed Saltwater Crocodiles: A Genetic Improvement Program*. Barton, A.C.T.: Rural Industries Research and Development Corporation.

IUCN-UNEP-WWF, 1991. *Caring for the Earth: A Strategy for Sustainable Living*. Gland, Switzerland: World Conservation Union.

Letts, G.A., 1987. The management of crocodilians in Australia: Introductory Comments. In: *Wildlife Management: Crocodile and Alligators,* (Eds.) G.J.W. Webb, S.C. Manolis and P.J. Whitehead, pp. 103–105. Chipping Norton, Australia: Surrey Beatty and Sons in association with the Conservation Commission of the Northern Territory.

Onions, V., 1987. Crocodile farming and ranching in Australia. In *Wildlife Management: Crocodile and Alligators,* (Eds.) G.J.W. Webb, S.C. Manolis and P.J. Whitehead, pp. 345–348. Chipping Norton, Australia: Surrey Beatty and Sons in association with the Conservation Commission of the Northern Territory.

Peucker, S., 1997. The crocodile industry. In: *The New Rural Industries: A Handbook for Farmers and Investors,* (Ed.) K. Hyde. ACT, Australia: RIRDC (The Australian Government Rural Industries Research and Development Corporation). Available from: http://www.rirdc.gov.au/pub/handbook/croco.html [accessed 7 July 2005].

Porosus Pty. Ltd., 2004. About us. Available from: http://www.4porosus.com/aboutus.asp [accessed 18 July 2005].

PWCNT (Parks and Wildlife Commission of the Northern Territory) (Undated). *Draft Management Plan for Crocodylus porosus in the Northern Territory* (For 2004–2008). Darwin: Parks and Wildlife Commission of the Northern Territory.

Ross, J.P., 2001. Commercial captive breeding of crocodilians. In *Selected Briefing Notes of the IUCN/SSC Conservation Impacts of Commercial Captive Breeding Workshop,* December 7–9, 2001, White Oak Foundation, Jacksonville, Florida, USA. Gland, Switzerland: IUCN. Available from: http://www.iucn.org/webfiles/doc/SSC/CCP_/selected_briefing_papers_fin_secure.pdf [accessed 21 July 2005].

Ryan, C., 1998. Saltwater crocodiles as tourist attractions. *Journal of Sustainable Tourism,* 6: 314–327.

Stubbs, A., 1998. *Information Systems for New Animal Industries.* RIRDC (Rural Industries Research and Development Corporation)–Publication No. 98/139. Canberra: RIRDC. Available from: http://www.rirdc.gov.au/reports/NAP/PTP-1A.doc [accessed 4 July 2005].

Swanson, T., 1997. *Global Action for Biodiversity.* Earthscan Publication, London.

Tisdell, C.A., 2002. *The Economics of Conserving Wildlife and Natural Areas.* Edward Elgar, Cheltenham, UK and Northampton, MA, USA.

Tisdell, C.A., 2004a. Economic incentives to conserve wildlife on private lands: analysis and policy. *The Environmentalist,* 24: 153–163.

Tisdell, C.A., 2004b. Property rights in non-captive wildlife and biodiversity conservation. *International Journal of Global Environmental Issues,* 4: 195–206.

Tisdell, C.A., 2005. *Economics of Environmental Conservation,* 2nd Edition. Edward Elgar, Cheltenham, UK and Northampton, MA, USA.

Tisdell, C., Swarna Nantha, H. and Wilson, C., 2005. Australian tropical reptile species: Ecological status, public valuation, attitudes to their conservation and commercial use. In: *Trends in Biodiversity Research,* (Ed.) A.R. Burk. Nova Science Publishers (In Press), New York.

UNEP-WCMC (United Nations Environment Programme World Conservation Monitoring Centre), 2005. *UNEP-WCMC Species Database: CITES–Listed Species.* Available from: http://sea.unep-wcmc.org/isdb/CITES/Taxonomy/tax-species-result.cfm?displaylanguage=eng&source=animal&species=%25porosus%25&Genus=Crocodylus&Country=&tabname=legal [accessed 4 July 2005].

Webb, G.J.W., 2002. Conservation and sustainable use of wildlife: An evolving concept. *Pacific Conservation Biology,* 8: 12–26.

Webb, G.J.W., Bayliss, P.G. and Manolis, S.C., 1986. Population research on crocodiles in the Northern Territory, 1984–1996. In: *Crocodiles. Proceedings of the 8th Working Meeting of the IUCN–SSC Crocodile Specialist Group,* Quito, Ecuador, October 1986, pp. 22–59. IUCN, Gland, Switzerland

Webb, G.J.W., Britton, A.R.C., Manolis, S.C., Ottley, B. and Stirrat, S., 2000. The recovery of *Crocodylus porosus* in the Northern Territory of Australia: 1971–1998. In: *The Proceedings of the 15th Meeting of the IUCN–SSC Crocodile Specialist Group*, Varadero, Cuba, 15–20 January, 2000, pp. 195–234. IUCN, Gland, Switzerland.

Webb, G.J.W. and Manolis, S.C., 1989. *Crocodiles of Australia.* Frenchs Forest, NSW: Reed Books.

Webb, G.J.W., Manolis, S.C. and Ottley, B., 1994. Crocodile management and research in the Northern Territory: 1992–1994. In: *Crocodiles. Proceedings of the 12th Working Meeting of the Crocodile Specialist Group*, pp. 167–180. IUCN, Gland, Switzerland.

Webb, G.J.W., Manolis, S.C., Whitehead, P.J. and Letts, G.A., 1984. A proposal for the transfer of the Australian population of *Crocodylus porosus* Schneider (1801), from Appendix I to Appendix II of CITES. Report No. 21. Darwin: Conservation Commission of the Northern Territory.

Webb, G.J.W., Whitehead P.J. and Manolis, C.S., 1987. Crocodile management in the Northern Territory of Australia. In: *Wildlife Management: Crocodile and Alligators*, (Eds.) G.J.W. Webb, S.C. Manolis and P.J. Whitehead, pp. 107–124. Chipping Norton, Australia: Surrey Beatty and Sons in association with the Conservation Commission of the Northern Territory.

Whitaker, R., 1987. The management of crocodilians in India. In: *Wildlife Management: Crocodile and Alligators*, (Eds.) G.J.W. Webb, S.C. Manolis and P.I. Whitehead, pp. 63–72. Chipping Norton, Australia: Surrey Beatty and Sons in association with the Conservation Commission of the Northern Territory.

Chapter 13

Avifauna in Agri-horticultural Ecosystems of Maidan Area of Davangere District, Karnataka, India

S. Basavarajappa

Department of Studies in Zoology, University of Mysore,
Manasagangotri, Mysore – 570 006, Karnataka, India,
E-mail: braj@rediff.com

ABSTRACT

The irrigated rice-fields accompanied with Coconut, Arecanut gardens and wet land ecosystems of Maidan area of Davangere district have provided a congenial set up for the survival of 85 species of birds, which belongs to 29 families. The relative abundance of different species varied considerably. All the existing bird species were broadly classified into resident, resident migrant and migratory birds and which were further grouped based on their feeding habit and habitats. Furthermore, brief information on the foraging, roosting, resting and nesting grounds are collected so as to understand the prevailing constraints if any at their preferred habitats.

Keywords: *Avifauna, Agri-horticultural ecosystem, Preferred habitats, Maidan area.*

Introduction

Birds are widely distributed in an array of diversified ecosystems in the Indian subcontinent. Since prehistoric times, birds have always been valuable creatures for man (Ali, 1996 and Futehally, 2005), their material advantages which human beings received are enormous. The feathered species are considered as potential species because of their ecological, economical, ethical, medicinal and scientific values (Ali and Ripley, 1983).

In Karnataka State, the Davangere district posses diversified agro-horticultural conditions at its Maidan (*i.e.*, Plains) region, which favours the survival of different bird species (Basavarajappa, 2001 a and b, and 2005). The area lies in between 13°27′–14°39′ N. latitude and 74°38′–76°40′ E. longitude with an altitude of 671 meters above msl (Kamath, 1991). The area receives more than 800 mm annual rainfall. The temperature ranges between 17 to 34°C and the relative humidity ranges in between 45 and 75 per cent. In general, the climate is dry with more than 225 days of sunshine. The landscape consists of vast stretches of paddy fields with scattered gardens. The region includes two irrigation systems: Bhadra irrigation system and Santhisagar–locally called "Sulekeri (second biggest lake in South India) irrigation system. These two water systems becomes lifeline of this region. Because, water from Bhadra reservoir and Sulekeri lake drain a large region of Maidan area of Davangere district. Concurrent to the continuous availability of water flow through many canal systems along with microhabitats in canals, rivers and stream course harbour a rich bird fauna (Basavarajappa, 2002). The horticultural gardens, cereals, pulses and millets as rainfed crops at the valley of small and uneven hillocks and paddy, jowar and ragi as irrigated crops on the plains at this region extend the rich food sources and (Basavarajappa, 2005), become favourable habitat for birds. Further, the natural vegetation is confined to canals, streams and tanks, which are bordered with different tree species. Moreover, the scattered scrub-bushy vegetation represented by sedges and wild grass species particularly on the bank of canals and streams of Maidan area become an ideal habitat for different bird species (Basavarajappa, 2005). Over the last few years, efforts were made to enlist the existing bird species in this area. Besides this, some species of game birds (Ex. Peafowl, *Pavo cristatus*) and water bird population status was also carried out. However, the birds, which share varied habitat conditions other than aquatic birds were not considered critically. To have precise and up-do-date information on the status, habit and habitats of bird species at this local conditions and also to initiate conservation activities (Mundkur, 2005) present investigation has been initiated.

Methodology

The study was conducted during the year 2003–2004. The study area Belalagere village is located nearby Sulekeri Lake, Davangere district, Karnataka. The area is dominated with paddy fields with scattered Coconut and Arecanut gardens. About 10 sq. km area was selected at random by covering paddy fields, horticultural gardens and natural vegetation on the bank of various canals, streams and small water bodies. The variable width line transect, 2 km in length was selected. The transect covered all the sites *i.e.*, man-made ecosystems (paddy fields and horticultural gardens) and natural ecosystems (natural vegetation and water canals and small water bodies). The observations began in May 2003 and continued till June 2004. Permanent observation sites were selected at 5 places and observations made from morning to evening at every fifteen days. Altogether, 24 samples of line transect were collected and birds were identified using a binocular (10 × 30) and with the help of field guides and reference books (Ali, 1996 and Ali and Ripley, 1983). The collected data was analyzed and compared by following standard methods as per Saha (1992).

Results and Discussion

The results of the study are presented in Tables 13.1–13.5 and are discussed as follows. Table 13.1 shows the species of birds recorded in the agri-horticultural ecosystem of Maidan area. A total of 85 species of birds were recorded. All these species belongs to 29 families, and from each family, minimum one to maximum five species have been counted during the observations Maximum numbers of bird species were recorded from the family Muscicapidae alone (Table 13.1). The families Accipitridae, Cuculidae, Sturnidae and Ploceidae, whose members were more in the ecosystem, followed this. However, only one each species was recorded from the families Dromadidae, Psittacidae, Upupidae, Laniidae, Dicaeidae and Ciconiidae (Table 12.1). Thus, the members of Muscicapidae have dominated this ecosystem by its different species composition.

Table 13.1: Avifauna and their Status in the Study Area

Sl.No.	Common Name	Scientific Name	Family	Status
1.	Brahminy Kite	*Haliastur Indus*	Accipitridae	R, C
2.	Sparrow Hawk	*Accipiter nisus*	Accipitridae	R, UC
3.	Black Eagle	*Ictinaetus malayensis*	Accipitridae	Rm, C
4.	Short toed eagle	*Circaetus gallicus*	Accipitridae	R, C
5.	Largger falcon	*Falco biramicus jugger*	Falconidae	R, VC
6.	Peregrine falcon	*Falco peregrinus*	Falconidae	R, UC
7.	Grey Partridge	*Francolinus pondicerianus*	Phasianidae	R, C
8.	Grey Jungle fowl	*Gallus sanneratii*	Phasianidae	R, C
9.	Red Jungle fowl	*Gallus gallus*	Phasianidae	R, VC
10.	Common Peafowl	*Pavo cristatus*	Phasianidae	Rm, C
11.	Crab Plover	*Glareola pratincala*	Dromadidae	R, VC
12.	Grey Plover	*Pluvialis squatarola*	Charadriidae	R, C
13.	Red-wattled lapwing	*Vanellus indicus*	Charadriidae	R, VC
14.	Little Ringed Plover	*Charadrius dubius*	Charadriidae	R, C
15.	Blue rock pigeon	*Columba livia*	Columbidae	R, C
16.	Spotted dove	*Streptopelia chinensis*	Columbidae	R, VC
17.	Indian Ring dove	*Streptopelia decaocta*	Columbidae	Rm, C
18.	Roseringed parakeet	*Psittacula krameri*	Psittacidae	R, C
19.	Indian Cuckoo	*Cuculus micropterus*	Cuculidae	R, C
20.	Crow-Phesant	*Centropus sinensis*	Cuculidae	R, C
21.	Indian Drongo Cuckoo	*Surniculus lugubris*	Cuculidae	R, C
22.	Koel	*Eudynamys scolopacea*	Cuculidae	Rm, C
23.	Sirkeer cuckoo	*Taccocua leschenaultii*	Cuculidae	Rm, UC
24.	Barn Owl	*Tyto alba*	Strigidae	R, UC
25.	Brown Wood Owl	*Strix* sp.	Strigidae	R, UC
26.	Barred Jungle owlet	*Glaucidium radiatum*	Strigidae	R, C
27.	Indian edible-nest Swiftlet	*Collocalia unicolar*	Apodidae	Rm, C
28.	House Swift	*Apus affinis*	Apodidae	R, C

Contd...

Table 131.1–Contd...

Sl.No.	Common Name	Scientific Name	Family	Status
29.	Palm Swift	*Cypsiurus parvus*	Apodidae	R, C
30.	Small blue King fisher	*Alcedo atthis*	Alcedinidae	R, C
31.	Pied King fisher	*Ceryle rudis*	Alcedinidae	Rm, C
32.	White breasted King fisher	*Halcyon smyrnensis*	Alcedinidae	R, C
33.	White collored King fisher	*Halcyon chloris*	Alcedinidae	R, UC
34.	Hoope	*Upupa epops*	Upupidae	R, C
35.	Broadbilled Roller	*Eurystomus orientalis*	Coraciidae	R, UC
36.	Roller or Blue Jay	*Coracias benghalensis*	Coraciidae	R, UC
37.	Blue cheeked Bee-eater	*Merops suprcillosus*	Meropidae	R, C
38.	Small Green Bee-eater	*Merops orientalis*	Meropidae	R, C
39.	Common Grey Hornbill	*Tockus birostris*	Buceritidae	R, C
40.	Lessergoldenbacked woodpecker	*Chrysocolaptes festivus*	Picidae	R, C
41.	Indian goldenbacked woodpecker	*Chrysocolaptes lucidus*	Picidae	R, C
42.	Swallow	*Hirundo rustica*	Hirundinidae	R, VC
43.	Indian Cliff Swallow	*Hirundo fluvicola*	Hirundinidae	Rm, C
44.	House Swallow	*Hirundo tahitica*	Hirundinidae	R, VC
45.	Brown Shrike	*Lanius cristatus*	Laniidae	R, C
46.	Black Drango	*Dicrurus adsimilis*	Dicruridae	R, C
47.	Grey Drango	*Dicrurus leucophaeus*	Dicruridae	R, C
48.	Brahminy Myna	*Sturnus pagodarum*	Sturnidae	R, VC
49.	Indian Myna	*Acridotheres tristis*	Sturnidae	R, VC
50.	Pied Myna	*Sturnus contra*	Sturnidae	Rm, UC
51.	Jungle Myna	*Acridotheres fuscus*	Sturnidae	R, C
52.	House crow	*Corvus splendens*	Corvidae	R, VC
53.	Jungle crow	*Corvus macrovhynchos*	Corvidae	R, C
54.	Whitebelled Minivet	*Pericrocotus erythropygius*	Campephagidae	Rm, C

Contd...

Table 131.1–Contd...

Sl.No.	Common Name	Scientific Name	Family	Status
55.	Blackheaded Cuckoo-shrike	*Coracina melanoptera*	Campephagidae	Rm, UC
56.	Common Wood Shrike	*Tephrodornis pondicerianus*	Campephagidae	R, C
57.	Black headed yellow Bulbul	*Pycnonotus melanicterus*	Pycnonotidae	R, C
58.	Redvented bulbul	*Pycnonotus cafer*	Pycnonotidae	R, C
59.	Red whiskered bulbul	*Pycnonotus jocosus*	Pycnonotidae	R, C
60.	Slatyheaded scimitar Babbler	*Pomatorhinus horsfieldi*	Muscicapidae	Rm, UC
61.	Large Grey babbler	*Turdoides malcolmi*	Muscicapidae	R, C
62.	Tailor Bird	*Orthotomus sutorius*	Muscicapidae	R, C
63.	Paddy field Warbler	*Acrocephalus agricola*	Muscicapidae	R, C
64.	Pied Bush Chat	*Saxicola caprata*	Muscicapidae	R, UC
65.	Indian Robin	*Saxicoloides fulicata*	Muscicapidae	R, C
66.	White Wag tail	*Motacilla alba*	Motacillidae	R, C
67.	Paddy field pipit	*Anthus* sp.	Motacillidae	R, C
68.	Thickbilled Flower pecker	*Dicaeum agile*	Dicaeidae	R, UC
69.	Purple Rumped sunbird	*Nectarinia zeylonica*	Nectariniidae	R, C
70.	Purple sun bird	*Nectarinia asiatica*	Nectariniidae	R, C
71.	Small sun bird	*Nectarinia minima*	Nectariniidae	R, C
72.	Loten's sun bird	*Nectarinia lotenia*	Nectariniidae	Rm, UC
73.	Yellow throated sparrow	*Petroniax xanthocollis*	Ploceidae	R, VC
74.	Baya weaver bird	*Ploceus philippinus*	Ploceidae	R, C
75.	House sparrow	*Passer domesticus*	Ploceidae	R, VC
76.	Black headed munia	*Lonchura Malacca*	Ploceidae	Rm, C
77.	Black headed weaver bird	*Ploceus benghalensis*	Ploceidae	Rm, C
78.	Painted Stork	*Mycteria leuiocephala*	Ciconiidae	M, UC

Note: Birds identified as per Ali, S. (1996) and Ali and Ripley (1983).

R: Resident; Rm: Resident migrant; M: Migratory; C: Common; VC: Very Common; UC: Uncommon.

Table 13.2: Avifauna Belongs to Different Families and their Status

Sl.No.	Family	No. oi Species	Relative Abundance	Relative Dominance
1.	Accipitridae	05	5.88	3.30
2.	Alcedinidae	04	4.70	1.95
3.	Apodidae	03	3.52	10.89
4.	Bucerotidae	02	2.35	0.36
5.	Campehagidae	03	3.52	2.57
6.	Charadriidae	04	4.70	5.26
7.	Ciconiidae	01	1.17	0.61
8.	Columbidae	04	4.70	3.18
9.	Coraciidae	02	2.35	0.24
10.	Corvidae	02	2.35	5.50
11.	Cuculidae	05	5.88	4.16
12.	Dicaeidae	01	1.17	0.97
13.	Dicruridae	02	2.35	1.46
14.	Dromadidae	01	1.17	2.44
15.	Falconidae	02	2.35	1.22
16.	Hirundinidae	03	3.52	4.28
17.	Lanilidae	01	1.17	0.24
18.	Meropidae	02	2.35	2.20
19.	Motacillidae	02	2.35	1.46
20.	Muscicapidae	07	8.23	5.99
21.	Nectariniidae	04	4.70	3.67
22.	Phasianidae	04	4.70	4.65
23.	Picidae	02	2.35	0.97
24.	Ploceidae	05	5.88	19.82
25.	Psittacidae	01	1.17	2.44
26.	Pycnonotidae	03	3.52	0.73
27.	Strigidae	04	4.70	1.10
28.	Sturnidae	05	5.88	7.71
29.	Upupidae	01	1.17	0.48

Note: Each value is a mean of 24 observations.

The relative abundance and relative dominance of various species belonging to different families are presented in Table 13.2. The avifauna of the study area included Resident (R), Resident migrant (Rm) and Migratory birds (M). Table 13.3 shows the details on different bird groups, their relative abundance and the relative dominance in this ecosystem. Of all, bird species resident were more (74.1 per cent), followed by Resident migrant (24.7 per cent) species. However, the migratory bird species composition was very low (1.2 per cent). The resident, resident migrant and migratory birds have had 85.43, 13.83 and 0.73 relative dominance in the ecosystem respectively (Table 13.3).

Table 13.3: Different Group of Birds and their Composition

Sl.No.	*Group*	*No. of Species*	*Relative Abundance*	*Relative Dominance*
1.	Resident	63	74.10	85.43
2.	Resident migrant	21	24.70	13.83
3.	Migratory	01	1.2	0.73

Note: Each value is a mean of 24 observations.

Table 13.4: Occurrence of Different Group of Birds in the Study Area

Sl.No.	*Occurrence*	*No. of Species*	*Relative Abundance*	*Relative Dominance*
1.	Common	54	63.5	60.22
2.	Very Common	12	14.1	31.70
3.	Uncommon	19	22.4	8.07

Note: Each value is a mean of 24 observations.

Further, fifty-four bird species have occurred during most of the seasons (Table 13.4). They have appeared regularly at the different places of the ecosystem throughout the year and they were referred as '*Commonly occurring birds*'. However, their density varied throughout the year. About 19 species were not common, and their per cent occurrence was low during some parts of the year. Their relative abundance was 22.4 per cent and the relative dominance was 8.07 (Table 13.4). Hence, they were referred as '*Uncommon birds*'.

Table 13.5: Conditions Prevailed for Activities of Birds

Sl.No.	Foraging	Roosting/Resting	Nesting
1.	Intensively cultivable low land area dominated with paddy *Oryza sativa* cultivation.	1. Different species of trees with moderate to very good ramifyed canopy. 2. Coconut, Aracanut and Banana Gardens with sparsely distributed Mango, Acacia, Pongamia, Syzygium, Terminalia, Tectona species. Rose wood, Ficus, Sandal wood, Neem tree species.	Coconut, Tamarind, Acacia, Ficus, Neem, Jack fruit, Mango, Pongamia tree species used for the construction of nest.
2.	Unevenly distributed small patches of Jowar (*Sorghum vulgure*), Cotton (*Gossypium* sp.), Ragi (*Elusina corana*) fields.	3. Bamboo species, small shrubby bushes.	Wild grassy vegetation with small to medium sized bushes at valleys of waste lands, on the bank of nalas, canal systems and small to medium sized water bodies provided an ideal conditions for nesting.
3.	Horticultural gardens *viz.*, Coconut (*Cocus nucifera*), Areca (*Areca catucha*), Mango (*Mangifera indica*), Banana (*Musa Paradisica*) plantation along with sparse vegetation of *Syzygium* sp. Tamarind, Ficus, Sapota, Chiku and other fruit yielding plants.		
4.	Pest and non-pest insects, crabs, frogs, baby chicks, pullets and etc.		

However, twelve bird species were found '*Very common*' in their occurrence with 14.1 per cent relative abundance. They have had good (31.7) relative dominance (Table 13.4). Their density was almost same throughout the year. It is evident from the data that, the bird species exhibited seasonal abundance.

However, the migratory birds were very scanty, and the Bucerotidae and Ciconiidae family members *viz.*, Painted Stork (*Mycteria leuiocephala*) were the only two species having visited his ecosystem during winter season. The Painted Strok, *Mycteria leuiocephala* are apparently uncommon winter visitors (Table 13.1) to this area, but their density was very less.

While observing resident and resident migrant bird species, few Painted Storks were seen amidst a flock of the aquatic birds *viz.*, White Ibis, Black Ibis and Egrets. The results of the aquatic avifauna has been described elsewhere (Basavarajappa, 2005). The Painted Stork, *Mycteria leucocephala* a resident bird of Indian sub-continent, inhabits freshwater marshes, lakes, ponds, reservoirs, flooded fields and river banks (Grimmett *et al.*, 1998). It was recorded in different localities (*viz.*, Kengalu, Neeravari, and Doddathota) of agri-horticultural ecosystem (Basavarajappa, 2005) of Maidan area. It was also recorded at Sulekeri (second biggest lake in South India) situated at about 60 km away from Sahyadri hills of Western Ghat (Basavarajappa, 2005), where it was seen amidst a flock of Ibis. The bird remained in the area for few days, always in association with the group of White Ibis and Black Ibis species. Surprisingly, there are no records of this species from this region, and it is not known whether this species stay here for breeding. Perhaps, Painted Stork passes through this region on its annual migration within flyways that connect the southern part of Western Ghat. As the area is enriched with wetlands dominated with paddy plantation and innumerable number of small water bodies, water canals and streams, found congenial for Painted Stork to spend few days with other resident migrant bird species. Perhaps, this may be the reason, while flyways to southern part of Indian subcontinent, the Painted Stork might have took a chance to spend some times in this ecosystem. This observation corroborates the findings of Grimmett *et al.* (1998).

Table 13.5 shows the prevailed conditions for the various activities of birds at the agri-horticultural ecosystem of Maidan area.

As birds require ideal foraging, roosting and nesting sites for their survival, the prevailed conditions became congenial both for resident and resident migrant species.

Paddy cultivation is the most important agricultural activities in this area. The paddy production depends on irrigation and proper maintenance of plants. The paddy plantation begins in July–August for rainy crop and January–February for summer crop. The paddy harvesting starts in November–December and May–June respectively for rainy and summer crops. In these ten to eleven months, farmers at every level try very hard to manage water levels 3 to 4 inches constantly in the paddy fields to maximize the yield. The health of paddy plants therefore, becomes a primary concern. As paddy fields and water canals contains innumerable holes and passages of all sizes, in which dwell a great cryptic fauna of Crabs, Snails, Clams, Worms, insect larvae, pupae, fishes and their larvae. Thus, to varying degrees, the paddy fields as well as water canals, streams and small water bodies constituted an energy and nutrient trap. Further, many cryptic animals dependent organic compounds and nutrients retained in the paddy fields, becomes food for all these cryptic animals. It is evident that, the conditions which prevailed at paddy fields have attracted innumerable number of insect pest species *viz.*, sap suckers, shoot-, stem- and root-borers, defoliators etc., which become food for bird species. Perhaps, all these conditions might have helped variety of bird species to forage, and to conduct breeding activities too. Thus, both resident and resident migrant bird species have enjoyed feeding on various insect pest species, molluscans, crabs etc. which might be one of the reasons to see different species of birds in this ecosystem.

The Pycnonotidae members *viz., Pycnonotus atriceps, P. cafer, P. jocosus* (Tables 13.1 and 13.2) are gregarious birds, feeding on fruits of various tree species (Table 13.5). They also feed on nectar from the flowers of different flora. The Ploceidae members *viz., sparrows* (*Petronia axanthocollis* and *Passer domesticus*); *weaver birds* (*Ploceus philippinus* and *P. benghalensis*) and *Munias* (*Lonchura malacca*) were found to feed chiefly on grains, grass seeds and fruit berries of cultivated and non-cultivated flora (Table 13.5). However, the Hirundinidae members were gregarious, but they were found to become insectivorous. The *Hirudo rustica, H. fluvicola* and *H. tahitica* becoming insectivorous to avoid much pressure on the food, which became prime importance to Pycnonotidae and Ploceidae members (Table 13.5).

The Corvidae members such as *Corvus splendens* and *C. macrorhynchos* (Tables 13.1 and 13.2) were largest passerine birds and are called *social birds*. By feeding on plant and animal matter, they become major scavengers of this ecosystem.

About four species which belongs to family Columbidae *viz.*, *Columba livia, Streptopelia chinensis* and *S. decaocta* (Table 13.1 and 13.2) are grainivorous having the habit of feeding on grains of cereals, pulses and millet crops (Table 13.5). The Phasianidae members such as *Pavo cristatus, Francolinus pondicerianus, Gallus sannerati* and *C. gallus* are *'game birds'* having the habit of feeding on grains like Columbidae members. However, they were also feeding on flesh and other insect species at the vicinity of human dwelling places. The Nectariniidae member's *viz., Nectarinia zeylonica, N. minima, N. asiatica* and *N. lotenia* (Tables 13.1 and 13.2) found as *'nectar feeders'*, depended mainly on the flower nectar. Thus, the available food both from plants and animal origin was well shared by all these species so as to avoid competition within and between the species of avifauna. It is environmentally necessary to maintain viable populations of bird species for the health of biodiversity.

Quiet interestingly, about 50 per cent of the bird species were found as *'carnivorous'*. Of all the species nine were raptors and diurnal birds, which belong to the family Accipitridae (*Haliastur indus, Accipiter* sp., *Ictinaetus malayensis, Hieraetus kienerii, Circaetus gallicus*); Falconidae (*Falco biramicus* and *Falco* sp.) and Ciconiidae (*Ciconia* sp. and *Mycteria leucocephala*) (Tables 13.1 and 13.2). Only four species *viz., Tyto alba, Strix* sp. and *Glaucidium cuculoides* were nocturnal birds which belong to family Strigidae (Tables 13.1 and 13.2).

The *wading birds* of the family Charadriidae (*Pluvialis squatarola, Vanellus indicus, Charadrius alexandrinus* and *Dromas ardeola*) and the members of Motacillidae (*Motacilla alba* and *Anthus* sp.); Upupidae (*Upupa epops*) and Laniidae (*Lanius cristatus*) (Tables 13.1 and 13.2) were exclusively carnivorous birds of this area.

The *non-social birds* such as *Alcedo atthis, Ceryle rudis, Halcycon smyrnensis* and *H. chloris* of the family Alcedinidae (Tables 13.1 and 13.2) were also carnivorous in their feeding habits. However, about ten species were identified as '*insectivorous birds*'. They feed chiefly on insects in the paddy fields. They belong to the family Campephagidae (*Pericrocotus erythropygius, Coracina melanoptera, Tephrodornis pondicerianus*); Meropidae (*Merops superciliosus, M. orientalis*); Apodidae (*Collocalia unicolor, Aapus affinis, Cypsiurus parvus*) and Coraciidae (*Eurystomus orientalis* and *Coracias bengalensis*) (Tables 13.1 and 13.2). Thus, these species were considered as insectivorous.

Interestingly, considerable numbers of birds were found non-specific in their feeding habitats. They become *heterogenous in their feeding habit*. They were feeding on fruits, berries, flower nectar, vegetable matter, and grains and by feeding partly on insects they became carnivorous also. The species belonging to family Picidae (*Dinopium benghalensis, Chrysocolaptes lucidus*); Cuculidae (*Cucculus micropterus, Eudynamys scolopacea, Surniculus lugubris, Centropus sinensis, Taccocua* sp.); Bucerotidae (*Tockus birostris* and *T. griseus*); Dicruridae (*Dicrurus adsimilis, D. leucophaeus*); Dicaeidae (*Dicaeum agile*); Saturnidae (*Sturnus pagodarum, Acridotheres tristis, A. ginginianus, A. fusacus* and *Sturnus contra*) (Tables 13.1 and 13.2) were referred as heterogenous birds of this ecosystem.

The heterogenous passerine birds such as *Babblers* (*Pomatroshinus horsfieldii, Turdoides malcolmi*); *Warblers* (*Orthotomus* sp., *O. sutorius* and *Acrocephalus agricola*) and *Chats* (*Saxicola torquata* and *Saxicoloides fulicata*) were also found to become unspecific in their feeding habits. All these species belongs to the family Muscicapidae and were common in this area. Thus, the heterogenious feeding habit of these species might have supported the prosperity of a broad spectrum of birds and it led to a rich avifaunal diversity in this ecosystem. Hence, the man-made ecosystem favors the survival of bird species (Daniel *et al.*, 1990).

Further, to conduct activities such as roosting, resting and nesting there are huge sized tree species in the amidst of this ecosystem, which could become an ideal habitat for many bird species. Sunbirds are observed nesting on the Acacia plants. Many old Coconut and Mango trees (Table 13.5) inside the gardens serve as micro-nesting places for Mayna, Eagles, Hawk, Drango and Crows. The owls are observed breeding in the hollows of the old trees of Mango, Tamarind, Neem and Jamun tree species. The parakeets are observed residing and breeding in several holes made on the older Coconut trees in the gardens. Further, the Acacia, Tamarind, Neem Ficus, and other trees have provided an ideal nesting ground for the rest of bird species, which nest every year. Moreover, along some Bhadra water canals, Sulekeri water canals, streams, ponds and pools, King fishers are seen breeding in the holes. Perhaps, all these conditions might have collectively supported the avifauna of this ecosystem.

Conclusion

Birds are unique creatures, living in diversified agro-ecosystems to conduct normal foraging, roosting, resting and nesting activities. They prefer different types of food from plant and animal matter. Moreover, their roosting and resting places are diversified, bird's nest architecture is diversified and species specific. They use different types of nesting material to construct nests on different objects (*viz.*, trees, bushes, on ground and old houses) to avoid competition among different species. Quiet often birds act as biological controlling agents. Majority of the bird species feed on insect pests, which infest economically important Paddy plants and other garden tree species. Thus, they become friends of farmers in agri-horticultural ecosystem. The ubiquity of species-specific birds life style and the available varied features with respect to foraging, roosting, resting and nesting may indeed be a powerful reason for the avifaunal diversity in this ecosystem. Thus, birds are highly adapted to their particular way of life. They are very widespread and important in agri-horticultural ecosystems.

However, management of habitats of different species of bird's requires coordinated multi-sectoral activities, which include local people. To realize the needs of biodiversity conservation, proper awareness programmes on birds should be conducted to make public aware about these creatures and their role in the ecosystem. Many times, the conservationists, ecologists, researchers, oversight knowingly or unknowingly the locals and are unable to involve local farmers in conservation movements. In addition, changes in cropping pattern and encroachment of uncultivable lands and other barriers have not enabled to co-operate in conservation activities. As the birds visit several wetlands of different ecosystems and the advantages which people received for the well being and conservation of species for the sustainable use of their habitats rests with local people. Therefore, it is high time to survey on existing local conditions, which extend congenial habitat for variety of bird species, to conserve them at their preferred habitats. To pursue success in such activities, base line information on the existing bird fauna and its status may provide a ready rockner to conduct in depth research to plan and implement conservation activities at the agri-horticultural ecosystems of India.

Acknowledgements

I am grateful to Prof. G. Sashidharprasad, Hon'ble Vice-Chancellor, University of Mysore and Shri N.D. Tiwari, Registrar, University of Mysore, Mysore, for their encouragement. I am thankful to Prof. P.R. Naik, Chairman, BOS in Zoology, Manasagangotri, Mysore for facilities and cooperation. I also thankful to Prof. H.A. Ranganath, Senior faculty member, BOS in Zoology, MGM for his support and constant encouragement.

References

Ali, S. and Ripley, S.D., 1983. *Handbook of the Birds of India and Pakistan*. Oxford University Press, Delhi.

Ali, S., 1996. *The Book of Indian Birds*. Bombay Natural History Society, Oxford University Press, Mumbai.

Basavarajappa, S. 2001b. Habitat degradation of Peafowls. In: *Glimpse of Biodiversity*, (Ed.) B.B. Hosetti. Daya Publishers, Delhi.

Basavarajappa, S., 2001a. Studies on habitatic features of Indan Peafowl, *Pavo cristatus* under diversified conditions of Karnataka, India. In: *Proc. 88th ISC*, New Delhi, p. 29.

Basavarajappa, S., 2002. A preliminary report on the Indian crested Peafowl in Channageri area, Karnataka. *Newsletter for Birdwatchers*, 42(2): 20–21.

Basavarajappa, S., 2005. Habitatic features, relative abundance, sex-ratio and population decline of peafowl *Pavo cristatus* in Channageri area of Karnataka, India. In: *Perspectives in Animal Ecology and Reproduction*, (Eds.) V.K. Gupta, Vol. II. Daya Publishing House, Delhi, (In Press).

Daniels, R.J.R., Hegde, M. and Gadgil, M., 1990. Birds of man-made ecosystems: the plantations. In: *Proc. Indian Acad. Sci (Anim. Sci)*, 99(1): 79–89.

Futehally, Z., 2005. Recoveries from Newsletter for Birdwatchers-6. *Indian Birds*, 1(2): 39–40.

Grimmett, R., Inskipp, C. and Inskipp, T., 1998. *Birds of the Indian Sub-continent*. Christopher Helm, London.

Kamath, U.R., 1991. *Davangere District Gazetteer*, Karnataka.

Mundkur, T., 2005. Building bridges for Migratory water bird conservation in the Central Asian Flyway. *Indian Birds*, 1(1): 46–47.

Saha, T.K., 1992. *Biostatistics in Theory and Practice*. Emkay Publications, Delhi.

Chapter 14

Computers in Wildlife: A Tool to Create a Communication with Birds

Manu Malhotra and D.N. Sahi

Department of Zoology, University of Jammu, Jammu – 180 006

Introduction

Computers are being used in every walk of life these days. These machines are involved in the manufacturing from pin to spacecrafts, Medicines, Education, Defence, Training and Research works. Those days are gone when the knowledge of this machine was confined to few people only and working with the help of a computer was known as novelty of the research. Instead, it is one of the most common and important tools for the research activities these days if properly used. Computers are useful to ecologists because they can handle large amount of complex relationships rapidly, Roger Jeffcoate (2002).

Among various types of computers hybrid PCs (Personnel Computers) were the handiest machines of the early days. Earlier we were working on 386 machines, which were quite slow as far as data processing is concerned, and there was hardly any facility for Audio-visual attachments, moreover, the storage capacity of the data was also very meager. Then came the era of 486 machines which were quite hardy machines and were fixed with separate Audio-Visual cards on their motherboards and were capable for processing the Audio-Visual data but due to their low RAM (Read Access

Memory) which used to be between 8–16 MB (Mega byte) only, it was not possible for them to hold huge data at the same time. The storage capacity of the Hard disk was also not upto the mark.

Then came Pentium series with its first machine with the name P-I which brought revolution in the world of data processing. The Intel Company's processor was quite fast, with 32 MB RAM, Improved storage capacity, CD-ROM attachment (Compact Disk–Read Only Memory). The motherboard, as upgradable with onboard Sound and Video cards. This was the time when multimedia became a facility for a common man and was no longer confined to the highly technical people only. With the arrival of P-II, P-III and P-IV machines multimedia became WINDOW based and also became more sophisticated. It was not only the hardware, which grew at enormous speed, but also the software, which grew at a better pace than the hardware.

Nowadays we are working on class of art machines which are not only capable of handling large amount of data but also have the capability of data sharing with other machines through LAN (Local Area Networking), WAN (Wide Area Networking), INTERNET and INTERANET.

As far as computers in the field of Zoology in general and Wildlife in particular is concerned this machine can help in achieving new horizons in research and that day is not far when the researcher will quietly sit in the Lab. and will perform very high tech experiments without disturbing the flora and fauna.

Presently we are using computers for Word Processing, Data Processing, Data Analysis, GIS (Geographical Information System), GPS (Global Positioning System), RS (Remote Sensing) and IPHS (Image Processing and Enhancing System) with the help of their respective software.

Earlier when ever we used to observe the behaviour of a bird or an animal either we used to record it in black and white or capture the particular behaviour of the bird/animal in still camera with different types of lenses fixed in it. Following are the steps that we have used to take care while recording the behaviour of a bird/animal at macro level:

1. Mating behaviour in case of some birds only.
2. Feeding behaviour by visual method in the field only.

3. Frightening behaviour due to human or predator interference.
4 Dominance behaviour or supremacy behaviour to defend the territory.

These behaviours are not sufficient for creating an exact data for developing a language pattern of a bird/animal as it is quite difficult for human to stay in the fields for long and continuously watch their behaviour in order to make a behaviour data bank.

The above mentioned points are no doubt the main backbone of the behaviour recording but it is quite difficult for a person to express the body language of a bird/animal on paper or tape recorder. It is possible but with the help of an On-line Audio-visual aid (Marker, 1995).

During 2002 an attempt was made by Anil Kumar and Dinesh Bhatt, Gurukul Kangri University, Haridwar and they have conducted an experiment on the song in female *Orinetalis Magpie-Robin* with the help of a microphone, tape recorder and a scientific 25MHz digital storage Oscilloscope. As far as the software is concerned they used SIGNAL to analyze the song and then spectrogram of the song of the bird was created showing different characteristics like frequency of max-min, range of frequency, duration of phases and interval between phases. These workers have worked only on sound and have ignored videography.

In the present paper it is suggested that in order to create a language pattern of birds/animals video and audio should simultaneously reach the labs where teachers, scholars, students and programmers can jointly observe the behaviours and sounds of the birds/animals under observation (Thupe, 1956). Birds of selected species are required to be caged in initial stages in different cages located at a distance from each other and their behaviour in the cages is to be recorded with the help of a computer with not less than 256 RAM and 80 GB Hard disk, a professional capture card and a supersensitive microphone is also required. The computer should be aided with the Adobe's 'Premier' software.

Capturing Video

Capturing video is the most important tool and for that 2 to 3 Bita-Cams/Latest Digital handy cams with tripot stands fixed with BNC/Fiber optic/Wireless connectivity to the Lab. on Local Area

Network. Bita-Cams are professional cameras and can give more pixels and better frame rates in their out put and these are the cameras which are used for professional cinematography these days for their better results, moreover, the film is not flickering whenever down loaded on computer as it does in case of normal VHS or cheap digital cameras.

As we have to record the behaviour of more than three caged birds (2 singles and 2 pairs) of the same species in order to make a comparative relationship analysis we have to place equal number of cameras and microphones in their respective cages and then with the help of the above mentioned equipment we will be able to record the following behaviours at micro level in different category files in the computer to create a proper data bank:

Mating Behaviour

1. Artificial Sound disturbances
2. Mating calls of males (before)
3. Mating calls of females (before)
4. Pleasantness (PR)

Reproductive

1. Mating Calls of males (at mating)
2. Mating Call of females (at mating)

Post Reproductive

1. Resting
2. Mating Call of males (after)
3. Mating Call of females (after)
4. Maternity Calls (egg laying)
5. Paternity Calls
6. Removing of a partner

Parental Care

1. Offspring care
2. Neighbourhood stranger
3. Stranger Discrimination
4. Thirst

5. Starvation
6. Acceptance of Wanted food
7. Fear
8. Offspring recognition
9. Predator Alarm
10. Feeding
11. Hunger
12. Flight Learning of off springs

Abiotic

1. Rain
2. Humidity
3. Thunder
4. Extreme Hotness
5. Dust storm
6. Hailstorm
7. Extreme Cold

Biotic

1. Human Interference

Miscellaneous

1. Stranger discrimination
2. Rejection of unwanted food
3. Introducing a new partner
4. Introducing a bird from another species.

The reference bank will be created by sharing all the computers attached with the cameras and than frequency graph of calls of each call along with the behaviour will be made in computer with Win XP based M-Player Oscilloscope.

Sound

It is the sound only with which we can make a language pattern of the bird for future conversation with them. Therefore, there will be a requirement of a proper sound recorder which can make graphical presentation also. All the sounds and their respective videos will be stored in their respective category Bank branches for references with human remarks in those.

Sound Recording

There should be no human interference in this case because human interference during the study can bring abnormal characteristics which we are suppose to record in their micro level behaviour also. In order to avoid human interference we have to make automated or automatic feeder for feed and water in their cages. The cages should be airy, big and natural from inside so that the bird should not feel out of place.

Future Planning

When all the required information of a particular species of birds is gathered then the work of the analysists will began and with the help of computer and its accessories they will create a bank of the particular species which will be called Call Bank with its video. Then a fresh batch of the same species of the birds will be caged and a countercheck of the previous bank will be conducted. This type of analysis requires at least five sets of birds for the experiment. When all the call frequencies will start matching (in terms of graphs) then the work of the programmers will start to create a simulated call for creating a response for the birds and then the actual language pattern will be developed.

Why this Study

As birds/animals can understand our language and are acting according to our will why can't we understand the language of the birds. The day we will start conversing with them the problems of herbal medicines, aviation, migratory routes, thermal dynamics and weather forecasting system etc. will be solved and we will be able to help each other in many respects.

References

Adams, Lowll, 1984.

Kumar, Anil and Dinesh Bhatt, 2002. Song of female *Orientalis Magpie. Robin.*

Marker, P., 1995. Characteristics of some animal call. *Nature.*

Roger Jeffcoate (2002). Multimedia in Practice. pp. 95.

Thupe, W.H., 1956. *Language of Birds.*

Chapter 15

Status and Conservation of Bird Fauna in the Vembanad-kole Ramsar Site, Kerala, India

C. *Sivaperuman* and E.A. *Jayson**

Division of Forest Ecology and Biodiversity Conservation,
Kerala Forest Research Institute, Peechi,
Thrissur, Kerala – 680 653, India

ABSTRACT

The status and conservation of bird fauna in the Vembanad-Kole Ramsar site, Kerala, India was studied from 1998 through 2003. The Study was mainly based on direct observational methods. Total count and point count methods were employed to census the birds. Two hundred and forty seven taxa of birds were recorded from the Vembanad-Kole wetlands. These belong to 60 Families under 16 Orders. Of the recorded bird species 150 were residents, 28 local migrants, 68 trans-continental migrants and one species was a straggler. The main threats are uncontrolled tourism, extensive habitat alteration, poaching, fire, unregulated fishing and extensive use of pesticides. The Vembanad-Kole Ramsar site is having the third highest position in terms of species richness among the Ramsar sites in India. This Ramsar site supported wetland birds similar to the well-known wetland habitats *viz.* Chilka Lake, Pulicat Lake, Great Vedaranyam Swamp, Pichavaram Mangroves. It

* Email: jayson@kfri.org.

supported good habitat for one vulnerable and seven near threatened bird species and the Vembanad-Kole Ramsar site is an ideal habitat of Raptors and Passerines. As this wetland is coming under 'Central Asian–Indian flyway', protection of the migratory species and the habitat is of highest priority in this Ramsar site.

Keywords: *Bird fauna, Conservation, Vembanad-Kole, Ramsar site, Kerala, India.*

Introduction

India is presently having 19 wetlands designated as Ramsar sites distributed in twelve States. In the year November 2002, three wetlands were declared as Ramsar sites from the State of Kerala. Among them the Vembanad-Kole Ramsar site is largest in extent. Other two Ramsar sites in the State are Ashtamudi-Lake and Sasthamkota-Lake. The information on the avifauna of newly declared Ramsar sites is scanty.

Wetlands in Kerala are distributed all along the coast and in the inlands. Prominent coastal wetlands in Kerala are Vellayani Kayal, Aakkulam–Veli backwater stretch, Kayamkulam Pozhi, Kumarakam, Mangalavanam, Kole wetlands, Purathur estuary, Manaloor Kayal, Chervarpur Kayal, Kadalundy estuary, Azhinijilam, Dharmadom estuary, Kattampalli, Ezhimala, Chempallikundu and Mangrove areas. Wetlands in Kerala are under extreme pressure due to the high population density of the State. According to Gopalan (1991), as much as two-third area of Vembanad-Lake has been either reclaimed as land or converted into fields for agricultural and fishery activities. Wetlands in Kerala are mainly used for agriculture, aquaculture, reclamation for housing and industrial purposes, disposing the waste materials, discharging the industrial effluents and municipal waste water, wood seasoning, feeding waters for ducks, dumping dredged soil, coir retting and for fishing (Balachandran *et al.*, 2002).

Apart from few occurrence reports of species no detailed information on the avifauna of this area is available. Jaira and Kumar (1990) recorded the Eurasian Spoonbill (*Platalea leucoradia*) from the Thrissur Kole wetlands. Similarly, Ray (1992) reported the Water Cock (*Gallicrek cinerea*) from Kuttanad. Nature Education Society Thrissur (NEST) reported the list of birds of Kole wetlands of Thrissur, during the winter months of 1992 and 1993.

Many new sighting reports were published on the birds of Kole wetlands and some of them are listed here. Jayson (2000) reported Black Stork (*Ciconia nigra*), Ravindran (2001) reported White-winged Black Tern, Sivaperuman and Jayson (2002) recorded Northern Shoveller (*Anas clypeata*) and Lesser Frigate Bird *Fregata ariel* from Thrissur. Similarly, Sivaperuman and Jayson (2000) published the inventory of the birds of the Kole wetlands and Jayson and Easa (2000) studied the birds of Mangalavanam mangrove area. Inventory of birds from Vembanad Lake were also reported during the period (Anon., 2003). Structure and species composition of birds of Kole wetlands have been reported by Sivaperuman and Jayson (2001). Detailed inventory of avifauna of Vembanad- Kole wetland Ramsar site is not available. An attempt is made to compile the avifauna of Vembanad-Kole wetlands based on the field surveys and published information.

Study Area

The Vembanad-Kole wetland Ramsar site is situated in the humid tropical region on the south west coast of India and is uniquely endowed with all attributes of a tropical wetland *viz.* mangroves, waterfowl habitat, agriculture and fisheries. This area lies between 9° and 10°40′N latitudes and 76°–77°30′E longitudes. The Vembanad-Kole wetland with an extent of 1,51,250 ha is spread over Alappuzha, Thrissur and Malappuram Districts in Kerala State. The altitude of the Vembanad-Kole Ramsar site varies from 3 m below MSL to 1m above MSL and it remains submerged for about six months in a year.These lands were formerly shallow lagoons, which gradually silted up. The Vembanad-Kole Ramsar site is fed by 10 rivers and these rivers originate from the Western Ghats, flows westwards through the wetland system and empty in the Arabian Sea. The mean temperature varies from 29.2°–32.8 °C and the average pH is 6.86 and salinity was 1.2 in the year 1997.

Methodology

The study was mainly based on direct observational methods. Total count and point count methods were employed to census the birds. Data was collected from two areas, namely the Thrissur Kole wetlands and the Vembanad Lake during the years 1998 to 2003. The avifauna of Thrissur Kole wetlands had been prepared as a part of major ecological study during the year 1998 to 2001 *viz.*, Jayson

(2000), Sivaperuman and Jayson (2000), Jayson and Sivaperuman (2003). The information on the birds of Mangalavanam Mangroves (Jayson and Easa, 2000) and census reports of Vembanad-Kole wetlands (Anon., 2003 and 2004) were also consulted. In the month of September 2000 and November 2003, a bird survey was carried out in the Vembanad-Kole wetlands, by boat.

Results

Two hundred and forty seven taxa of birds were recorded from the Vembanad-Kole wetlands. These belong to 60 Families under 16 Orders. Of the recorded bird species 150 were residents, 28 local migrants, 68 trans-continental migrants and one species a straggler (Figure 15.1).

Thrissur Kole Wetlands

Two hundred and three taxa of birds were reported from Thrissur Kole wetland, which belong to 54 families, under 16 orders (Figure 15.1). Out of these, 118 species were residents and 85 species migrants. Among these, 62 were trans-continental migrants, 22 local migrants and one species is a straggler. Out of the 13 species of egrets and herons, Little Egret was abundant followed by Cattle Egret. Species like Black-crowned Night-Heron, Purple Heron, Western Reef Egret, Yellow Bittern, Black Bittern, Cinnamon Bittern and Little Green Heron were recorded in low numbers. Nine species of Storks and Ibis were recorded from the Kole wetlands. Out of these, highest dominance was for Asian Open-bill Stork followed by Black-headed Ibis. Sixty-four species of waders were recorded and these constitute an important group of wetland species. These birds depend heavily on shallow waters and mud flats and normally recorded from September onwards. Ducks recorded from the area were Lesser Whistling Duck, Common Teal, Northern Pintail, Spot-billed Duck, Gadwall, Garganey, Northern Shoveller, Ferruginous Poachard and Cotton Teal. Apart from the above ten species of raptors namely Black-shouldered Kite, Black Kite, Brahminy Kite, Shikra, Eurasian Sparrowhawk. Oriental Honey-Buzzard, Pallid Harried, Pied Harrier, Western Marsh-Harrier and Osprey were also recorded.

Vembanad Lake

One hundred and ninety one species of birds were reported from the Vembanad Lake, which is southern most part of the Ramsar

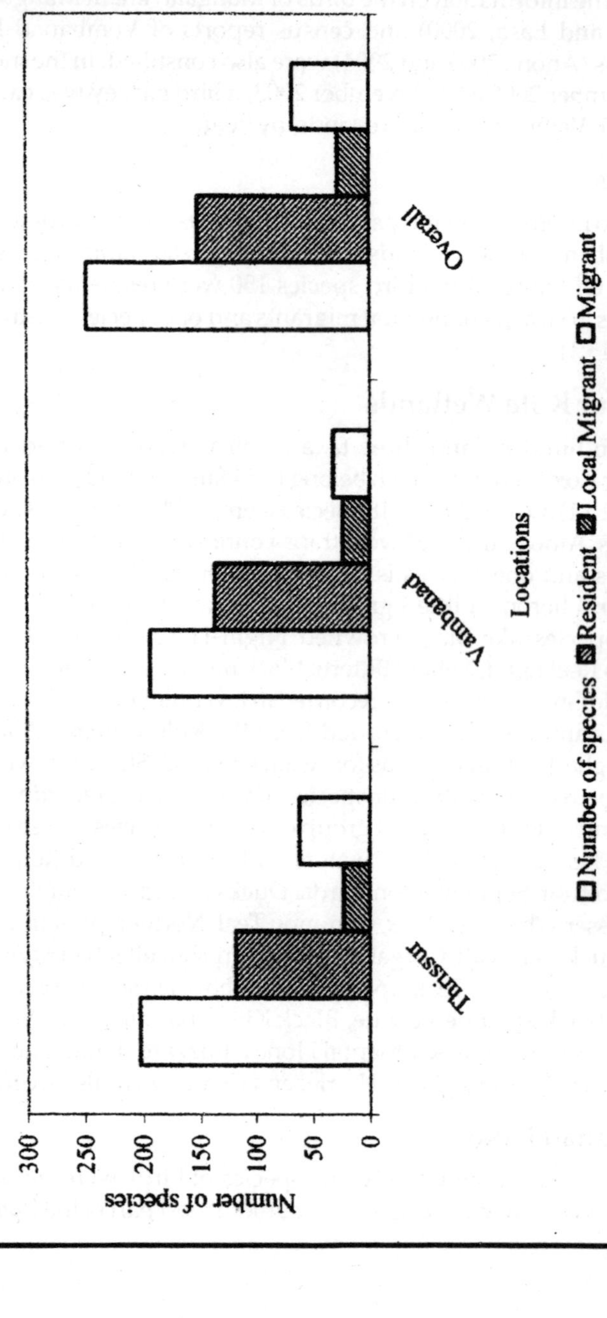

Figure 15.1: Number of Bird Species Recorded from Vembanad-Kole Wetlands

site. They belong to 56 families under 15 orders (Figure 15.1) and of these, 136 species were residents, 31 were trans-continental migrants and 24 were local migrants. Seventeen families of water birds were recorded and among them Scolapacidae (25) was the highest followed by Laridae (17), Ardeidae (13) and Anatidae (10) (Figure 15.2). The highest number of species was recorded in insectivores (29 per cent) followed by omnivores (27 per cent), aquatic feeders (18 per cent), granivores (9 per cent), carnivores (9 per cent), frugivores (5 per cent) and nectar-frugivores (2 per cent) (Table 15.1). Out of the 246 species of birds found in the Vembanad-Kole wetlands, eight species namely, Spot-billed Pelican (*Pelecanus philippensis*), Darter (*Anhinga melanogaster*), Painted Stork (*Mycteria leucocephala*), Black-headed Ibis (*Threskiornis melanocephalus*), Ferruginous Pochard (*Aythya nyroca*), Pallid Harrier (*Circus macrourus*), Wood Snipe (*Gallinago nemoricola*) and Black-bellied Tern (*Sterna acuticauda*) were listed in the threatened birds of the world (Birdlife International, 2001) (Figure 15.3).

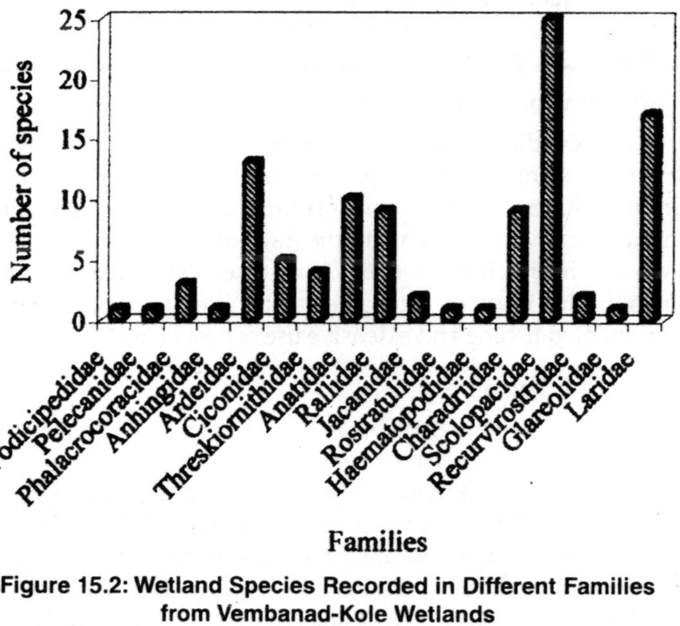

Figure 15.2: Wetland Species Recorded in Different Families from Vembanad-Kole Wetlands

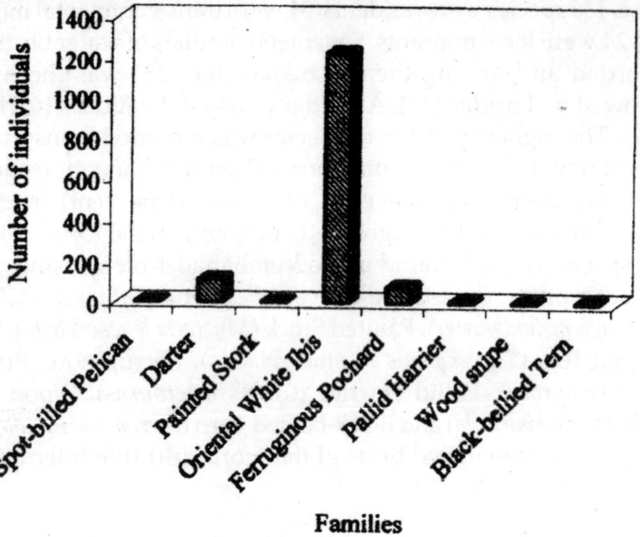

Figure 15.3: Threatened Species Recorded from the Vembanad-Kole Wetlands

Conservation Problems

Among the Ramsar sites in India, the Vembanad-Kole is the largest in extent and the area provides feeding, roosting and breeding grounds for various rare species of birds. However, with the increase in the human population and the development of industries, the avifauna in this Ramsar site is threatened. The main threats are uncontrolled tourism, extensive habitat alteration, poaching, fire, unregulated fishing and extensive use of pesticides.

Impact of Tourism

The total number of birds was low during the months of September 2000 and November 2003. This can be attributed to the heavy pressure of tourism going on this wetland and the impact of tourism on wetlands has become increasingly evident. The demands for motorized boating and land development were immediate threats on wetland birds and their habitats. Hundreds of country made houseboats are cruising in the lake. Apart from these the noise and intrusion associated with the powerboats lead to erosion, habitat destruction, and nest abandonment of birds. Migratory ducks were conspicuous due to their absence in the Vembanad Lake and this can be attributed to the heavy traffic of boats. Since such tourism is not prevalent in the Kole wetlands, thousands of migratory ducks were recorded from there.

Table 15.1: Order and Status of Birds Recorded from Vembanad-Kole Wetlands

Sl.No.	*Order*	*Status*			*Feeding Guilds*						
		R	*M*	*Total*	*A*	*I*	*G*	*N/F*	*C*	*F*	*O*
1.	Podicipediformes	1	–	1	1	–	–	–	–	–	–
2.	Pelecaniformes	5	2	7	7	–	–	–	–	–	–
3.	Ciconiiformes	20	2	22	22	–	–	–	–	–	–
4.	Anseriformes	4	6	10	10	–	–	–	–	–	–
5.	Falconiformes	10	5	15	–	–	–	–	–	–	–
6.	Galliformes	3	–	3	–	–	–	–	–	–	3
7.	Gruiformes	9	–	9	–	–	9	–	–	–	–
8.	Charadriiformes	10	48	58	–	–	–	–	–	–	58
9.	Columbiformes	4	–	4	–	–	4	–	–	–	–
10.	Psittaciformes	3	–	3	–	–	–	–	–	3	–
11.	Cuculiformes	6	–	6	–	–	–	–	–	6	–
12.	Strigiformes	7	–	7	–	–	–	–	7	–	–
13.	Apodiformes	3	–	3	–	3	–	–	–	–	–
14.	Coraciiformes	10	–	10	–	3	–	–	–	–	2
15.	Piciformes	4	–	4	–	4	–	–	–	–	–
16.	Passeriformes	80	5	85	–	63	10	6	–	3	3
	Total	**179**	**68**	**247**	**45**	**72**	**23**	**6**	**22**	**12**	**67**

R: Resident; M: Migrants; A: Aquatic feeders; I: Insectivores; C: Carnivores; G: Granivores; F: Frugivores; N/F: Nectar-frugivores; O: Omnivores.

Land Reclamation

Habitat fragmentation and loss are recognised as having three types of impacts on the fauna of remnant habitats: loss of species in fragments; changes in the composition of faunal assemblages and changes to ecological processes that involve animal species ultimately leading through a chain of process to species loss at a local and landscape level. The most important threat to the wetland birds of this Ramsar site is the last mentioned one. The mud flat of this site has been severely degraded by burning, drainage and clearing. This has affected the abundance of avifauna depended on mudflats as well as the hydrological values and functions of the mud flats.

New roads are planned through the Kole wetlands, as the paddy cultivation is not bringing in high profits. Reclamation of wetland for coconut cultivation is leading to the shrinkage of wetlands and by this practice, the wetlands are converted to dry lands, which are of no use to the wetland birds. Extensive clay mining is in vogue and paddy fields are being converted into make shift factories for manufacturing bricks. The process of reclamation of paddy field in this wetland has been accelerated by manifold human activities and is reducing the water holding capacity of the wetland. Many Islands are located in the Vembanad-Kole Ramsar site and Pathiramanal Island is facing severe threat due to the development of tourism activities. The Island used to support large number of roosting and breeding water birds namely Little Cormorant, Great Cormorant, Darter, Indian Cormorant, Indian Pond-Heron, Purple Heron and Black-Crowned Night- Heron.

Poaching

Poaching of birds for food is another problem as in other parts of the world. This is usually done with shotguns, air guns or poison and in certain cases, poaching was by professional gunman with the cooperation of farmers. In such cases, half of the birds killed were shared with the owner of the land. Groups of youngsters were also engaged in poaching the migratory and residents waterfowls as a leisure time activity. Another method used by the local people to slaughter the birds was to feed the flock of egrets with small fishes stuffed with insecticide. After consuming the poison, the dead birds were gathered from the paddy fields and consumed after removing the viscera.

Fire and Fishing

Indiscriminate fire is threatening the avifauna of the Kole wetlands. Every year, during the summer months, farmers burn the grass on the bunds by which the nests of breeding birds like Warblers (*Prinia hodgsonii, Prinia inornata* and *Acrocephalus stentoreus*),

Tailorbirds (*Orthotomus sutorius*) were destroyed. Fishing is common and usually carried out using nets of small mesh size. Catching of small fry from the wetlands is a serious problem, which reduces the food availability of birds. Using small mesh sized net, the farmers catch the fry and process it as duck feed. This practice must be stopped and the small fish should be allowed to grow, which is the source of food of wetland birds.

Pesticides and Pollution

Heavy dose of insecticides are sprayed to protect the paddy from pest attack and it was recorded that 15 commercial brands of pesticides and herbicides were used in the paddy fields. It is already known that heavy use of pesticides will cause mortality of birds. Toxic industrial wastes and agrochemicals also threaten to the Vembanad-Kole Ramsar site.

Discussion and Conclusions

The analyses showed that the Vembanad-Kole Ramsar site play a key role in hosting the migratory birds coming to Kerala. Easa and Jayson (2004) reported 19 orders, 65 families and 475 species of birds as occurring in Kerala. In the present study, 248 (52 per cent) species belonging to 60 families under 16 orders were reported from the Vembanad-Kole Ramsar site. This wetland supported 27 per cent of trans-continental migratory species. The high avian species richness recorded from the Vembanad-Kole Ramsar site is due to the presence of different microhabitats. Out of the nineteen Ramsar sites in India, the highest number of bird species was recorded from the Keoladeo National Park, Rajasthan (370 species) followed by Point Calimere Wildlife and bird sanctuary, Tamil Nadu (257 species) (Table 15.2). The Vembanad-Kole Ramsar site is having the third highest position in terms of species richness among the Ramsar sites in India. The Vembanad-Kole wetland is an ideal habitat for migratory and resident birds, especially for the winter visitors coming to the Peninsular India. This Ramsar site supported wetland birds similar to the well-known wetland habitats *viz.*, Chilka Lake, Pulicat Lake, Great Vedaranyam Swamp, Pichavaram Mangroves. It supported good habitat for one vulnerable and seven near threatened bird species and the Vembanad-Kole Ramsar site is an ideal habitat of Raptors and Passerines. Nandan and Unnithan (2004) recorded thirty-eight species of fishes and four species of prawns from this Ramsar site. Fish resources in the area are important sources of food for the wetland birds and urgent measures are needed to regulate the fishing activities. As this wetland is coming under 'Central Asian–Indian flyway', protection of the migratory species and the habitat is of highest priority in this Ramsar site.

Table 15.2: Comparison of Birds in the Ramsar Sites of India*

Sl.No.	Name of the Ramsar Site	State	Extent of Area (ha)	Number of Bird Species
1.	Ashtamudi Lake	Kerala	61,400	NA
2.	Bhoj Wetland	Madhya Pradesh	3,201	27
3.	Bhitarkanika Mangroves	Orissa	65,000	174
4.	Chilika	Orissa	1,16,500	140
5.	Deepor Beel	Assam	4,000	219
6.	East Calcutta Wetlands	West Bengal	12,500	40
7.	Harike Wetland	Punjab	2,850	167
8.	Kanjli Lake	Punjab	490	50
9.	Keoladeo National Park	Rajasthan	283	370
10.	Kolleru Lake	Andhra Pradesh	90,100	NA
11.	Loktak Lake	Manipur	31,200	116
12.	Point Calimere Wildlife and Bird Sanctuary	Tamil Nadu	38,500	257
13.	Pong Dam Lake	Himachal Pradesh	15,662	220
14.	Ropar Lake	Punjab	1,365	154
15.	Sasthamkotta Lake	Kerala	373	NA
16.	Sambar Lake	Rajasthan	24,300	45 (Water birds)
17.	Tsomoriri	Jammu and Kashmir	12,000	40 (Water birds)
18.	Vembanad-Kole wetland	Kerala	1,51,250	247**
19.	Wular Lake	Jammu and Kashmir	17,300	NA

**Source*: (WWF-India Fact Sheet); NA: Not available; **: Present study.

Acknowledgements

We are grateful to Dr. J.K. Sharma, Director, Kerala Forest Research Institute Peechi, Kerala for valuable advice and suggestions.

The authors are thankful to the Kerala Forest and Wildlife Department for sponsoring the studies at Kole wetlands of Thrissur. Dr. C. Renuka, Programme Coordinator, Division of Forest Ecology and Biodiversity Conservation is acknowledged for the encouragement.

References

Ali, S. and S. D. Ripley, 1983. *Hand Book of the Birds of India and Pakistan*. Oxford University Press, Oxford, 737 p.

Anon., 2003. *Vembanad Water Bird Count 2003: A Report*. Kerala Forest Department and Kottayam, Nature Society, 80 p.

Anon., 2004. *Vembanad Water Bird Count 2004: A Report*. Kerala Forest Department and Kottayam, Nature Society, 43 p.

Balachandran, P.V., G. Mathew and K.V. Peter, 2002. Wetland agriculture problems and prospects. In: *Wetland Conservation and Management in Kerala*, (Ed.) Jayakumar, pp. 50–68. State committee on Science Technology and Environment, Thiruvananthapuram.

Birdlife International, 2001. *Threatened Birds of Asia: The Birdlife International Red Data Book*. Birdlife International, Cambridge.

Easa, P.S. and E.A. Jayson, 2004. *Biodiversity Documentation for Kerala, Part II: Birds*. Handbook No. 17, Kerala Forest Research Institute, 45 p.

IUCN, 1971. "Ramsar Convention. The Final Act of the International Conference on the Conservation of Wetlands and Waterfowl" *IUCN Bulletin* 2 (Special Supplement): 1–4.

Jairaj, A.P. and V.K.S. Kumar, 1990. Occurrence of Spoonbill *Platalea leucorodia* Linn. in Kerala. *Journal of Bombay Natural History Society*, 87(2): 289.

Jayson, E.A., 2000. Occurrence of Black Stork *Ciconia nigra* (Linnaeus) in Kole wetlands of Thrissur, Kerala. *Newsletter for Birdwatchers*, 40(3): 39–40.

Jayson, E.A., 2002. *Ecology of Wetland Birds in the Kole Lands of Kerala*. Kerala Forest Research Institute, Peechi. Research Report No. 244. Kerala Forest Research Institute, Peechi, 100 p.

Jayson, E.A. and P.S. Easa, 2000. *Documentation of Vertebrate Fauna in Mangalavanam Mangrove Area*. KFRI Research Report No. 183, Kerala Forest Research Institute, Peechi, 42 p.

Jayson, E.A. and C. Sivaperuman, 2003. Sighting of Lesser Frigate Bird *Fregata ariel* Gray in the Kole Wetlands of Thrissur, Kerala. *Journal of Bombay Natural History Society*, 100(1): 106–108.

Manakadan, R. and A. Pittie, 2001. Standardised, common and scientific names of the birds of the Indian subcontinent. *Buceros*, 6(1): 1–37.

Ravindran, P.K., 2001. Occurrence of the white-winged Black Tern *Chilidonias leucopterus* in Kerala. *Journal of Bombay Natural History Society*, 98(1): 112–113.

Ray, J.G., 1992. The Watercock *Gallicrex cinerea* in Kuttanadu, Kerala. *Journal of Bombay Natural History Society*, 88(2): 283.

Sivaperuman, C. and E.A. Jayson, 2000. Birds of Kole wetlands, Thrissur, Kerala. *Zoos' Print Journal*, 15(10): 344–349.

Sivaperuman, C. and E.A. Jayson, 2002. Occurrence of Northern Shoveller *Anas clypeata* Linnaeus in Kole wetlands of Thrissur, Kerala. *Journal of Bombay Natural History Society*, 99(3): 517.

Chapter 16

Quantitative Behavioural Observations on Rhesus Monkey *Macaca mulatta* (Zimmermann)

S.K. Gahlawat[*#] *R.K. Gupta*[*], *R.C. Sihag*[*] *and R.C. Gupta*[**]

[*]*Department of Zoology and Aquaculture, C.C.S. Haryana Agricultural University, Hisar – 125 004*

[**]*Department of Zoology, Kurukshetra University, Kurukshetra – 136 119*

ABSTRACT

Six behavioural aspects of rhesus monkeys *Macaca mulalla viz.*, self oriented active and passive, environment oriented, social, aggressive and mother-infant of adult, sub adult males and adult, adolescent females were recorded in the present study. The aggressive behaviour was recorded maximum 1.08 to 2.22 per cent of the total activities in the adult males. The maximum activities observed were self oriented passive behaviour (27.81 to 30.78 per cent) followed by social behaviours, environment oriented and self oriented active behaviour. The adult female showed distinct tendency towards the discharge of motherly activities whereas adolescent females devoted their maximum time in social behaviours.

[#] E-mail: gahlawatsk@rediffmail.com.

Introduction

The female and male rhesus monkey take three and five years to become an adult from the birth respectively. This transformation period embodies certain definite colour as well as physical gradations. Adult males and females are members of central hierarchy and take part in mating within troops during breeding season. Infants are less than one year old; this is the period before weaving. Juvenile females are more than one year and less than 2½ years old and have a vulvar perinneal skin. Adolescent females whose age falls between 2 to 3 years, display pinkish swellings on entire hindquarter and try to form consort relation with adults. The juvenile male-I is more than one year and less than 2½ years of age and acquires from centre to subperipheral position whereas juvenile male-II is more than 2 years and less than 4 years old and acquires peripheral position. Its testes size increased without colour. Subadult males are between 4 and 5 years of age. The scrotum starts becoming pinkish and not allowed to form consort during mating period (Altmann, 1968; Neville, 1968; Lindburg, 1971 and Singh, 1981). The present study was carried out on behavioural responses of these rhesus monkeys for a period of one year in Saraswati and Veer Sonti forests of Haryana.

Materials and Methods

Six behavioural aspects of rhesus monkey were categorized by following Singh (1981). These are:

1. *Self oriented active behaviours*: Self manipulation, self grooming, autoeroticism and self reflexive.
2. *Self oriented passive behaviours*: Passive siting, dorsal lying, ventral lying and standing.
3. *Environment oriented behaviours*: Locomotion and environmental manipulations.
4. *Social behaviours*: Visual orientation, approach to others, withdrawl from others, follow, proximity and grooming.
5. *Aggression*: Threatening to or threatened by others.
6. *Mother-infant relationship*: Retrieve, reject, ventral contact, nipple contact, ventral carrying.

Five adult male, female, subadult male and juvenile female individuals were observed for 3 minutes each per visit from

5.00 A.M. to 7.30 P.M. Each individual and defined behaviours were coded and observations were recorded by using one-zero time sampling technique following Altmann, 1974. One behavioural category was recorded only once in an interval of 30 seconds, so that it permitted lower and upper limit of frequency of occurrence of a particular behaviour. The block per cent scores of each individuals (sex-wise) are summed and mean of each recorded categories were recorded and presented in Tables 16.1–16.2. All behavioural categories were converted into block percentage score using following formula:

$$\text{Block Percentage Score} = \frac{\text{Sum of Observed Frequencies}}{\text{Maximum Possible Score}} \times 100$$

Table 16.1: Showing Sum and Mean Block % Scores of Individual Behavioural Category of Different Individuals at Saraswati Forests

Sex		*Behavioural Categories*					
		I	*II*	*III*	*IV*	*V*	*VI*
AM	Sum	324.30	528.30	427.62	575.36	42.01	–
	Mean	17.07	27.81	22.51	30.28	2.21	–
AF	Sum	165.19	505.51	337.90	809.72	–	585.34
	Mean	7.18	21.98	14.69	35.21	–	25.45
SAM	Sum	101.13	203.41	140.45	150.46	–	–
	Mean	16.86	33.90	23.41	25.08	–	–
Adl.F	Sum	25.69	86.50	89.99	97.82	–	–
	Mean	8.56	28.83	29.90	32.61	–	–

I: Self oriented active behaviours; II: Self oriented passive behaviours; III: Environmental oriented behaviours; IV: Social behaviours; V: Aggressive behaviours; VI: Mother-infant behaviours; AM: Adult male; AF: Adult female with infants; SAM: Subadult male; Adl.F: Adolescent female.

Results and Discussion

The dormant or aggressive behaviour is solely evident in respect of adult males (mean aggressive behaviour value = 1.08 at Veer Sonti Torests; 2.21 at Saraswati forests), whereas, in case of all other age and sex categories (adult females, subadult males and adolescent females), the mean value is zero. The maximum chunk of the total activities is shared by self oriented passive behaviours (27.81 to

30.78 per cent) followed by social behaviours (25.08 to 40.33 per cent), environment oriented behaviours (14.69 to 29.90 per cent) and self-oriented active behaviours (6.67 to 17.07 per cent) respectively. The highest value of self-oriented passive behaviour may be attributed to the fact that the maximum quota of daily activity period is whiled away at the central core activity wherein the larger chunk of time devoted to non-productive activities perhaps in a bid to spend time only. The spatial displacement of subadult males on the peripheral position of a troop is the consequence of muscle power of adult males. Thus the transition of subadult male into adult male is more a phenomenon of gaining in size, weight and turning of reddish sexual skin rather than any predominant change in the activities. This argument is supported by similar values in (Tables 16.1 and 16.2) of all the six behavioural activities concerning adult males and subadult males.

Table 16.2: Showing Sum and Mean Block % Scores of Individual Behavioural Category of Different Individuals at Veer Sonti Forests

Sex		*Behavioural Categories*					
		I	*II*	*III*	*IV*	*V*	*VI*
AM	Sum	272.99	492.40	366.59	454.50	16.80	–
	Mean	17.06	30.78	22.91	28.72	1.08	–
AF	Sum	152.14	448.54	371.00	577.77	–	633.43
	Mean	6.92	20.39	16.86	26.26	–	28.79
SAM	Sum	50.88	91.83	73.26	83.99	–	–
	Mean	16.96	30.61	24.42	27.99	–	–
Adl.F	Sum	26.68	107.53	87.13	161.33	–	–
	Mean	6.67	26.88	21.78	40.33	–	–

I: Self oriented active behaviours; II: Self oriented passive behaviours; III: Environmental oriented behaviours; IV: Social behaviours; V: Aggressive behaviours; VI: Mother-infant behaviours; AM: Adult male; AF: Adult female with infants; SAM: Subadult male; Adl.F: Adolescent female.

Befitting a primate mother, an adult female monkey shows a distinct inclination to take sufficient portion out of self oriented active behaviours, self oriented passive behaviours and environment oriented behaviours in order to transplace this added quota of activities towards its motherly obligations. Adolescent females do not have to practise motherly obligations, and as such their behaviour seems to afford a bigger chunk of time towards social behaviour (Kumar, 1990).

References

Altmann, S.A., 1968. Sociobiology of rhesus monkeys, III. The basic communication network. *Behaviour*, 32: 17–32.

Altmann, J., 1974. Observational study of behaviour: Sampling methods. *Behaviour*, 149: 227–267.

Kumar, S., 1990. On certain behavioural and ecological aspects of free-ranging rhesus monkey, *Macaca mulatta* (Zimmermann). *M. Phil Dissertation*, Department of Zoology, K.U. Kurukshetra, pp. 1–112.

Lindburg, D.G., 1971. The rhesus monkey in north India: An ecological and behavioural study. In: *Primate Behaviour*, (Ed.) L.A. Rosenblum. Academic Press, New York, 2: 1–106.

Neville, M.K., 1968. Ecology and activity of Himalayan foothill rhesus monkeys. *Ecology*, 49: 110–123.

Singh, R., 1981. Study of behavioural patterns of age-sex classes during transational periods in free-ranging rhesus monkey (*Macaca mulatta*). *Ph.D. Thesis*, University of Rajasthan, Jaipur.

Section II

Animal Reproduction

Chapter 17

Ecology and Breeding Biology of the Earthworm *Allolobophora* (= *Aporrectodea*) *caliginosa trapezoides* (Savigny, 1826) (Oligochaeta : Lumbricidae) in Libya

*G. Achuthan Nair***# *and K. Premkumar***

**Department of Environment Sciences, Kerala University, Kariavatton, Thiruvananthapuram – 695 581, Kerala State, India*

***Department of Biology, Tuskegee University, Tuskegee, Alabama – 36088, U.S.A.*

ABSTRACT

Field and laboratory studies were conducted during 1994–2004, to determine the occurrence, density, biomass, vertical distribution cocoon production, incubation and hatchlings of *Allolobophora* (= *Aporrectodea*) *caliginosa trapezoides*, a common earthworm found in the soils of Benghazi, Libya. This peregrine earthworm formed the dominant and most widespread species of Benghazi and was present in eleven out of thirteen stations sampled. Maximum density and biomass of *A. caliginosa trapezoides* were recorded from October to December and again

[#] Corresponding Author: Sankar Compound, Sasthamangalam, Thiruvananthapuram – 695 010 Kerala State, India, E-mail: gachuthannair@yahoo.com.

during March and April, and the minimum during the summer from May to July. A juvenile dominant population observed during summer was followed by an increase in sub-adults and adults during autumn, winter and spring seasons. *A. caliginosa trapezoides* mainly confined its activities to the surface organic zone (0–10 cm depth) of the soil for most of the months studied. Cocoon production started in August and continued until April. A positive correlation between soil moisture ($r = 0.74; p < 0.05$), a negative one between and soil temperature ($r = -0.85; p < 0.05$) on cocoon production were evident. Mean incubation time was 47 days. Some cocoons contained a single worm (single hatchling) whereas the majority of them (68 per cent) contained two, the first hatchling and second hatchling. The rate of change of body mass ($F = 2.18; p > 0.05$) and the average body mass ($F = 1.25; p > 0.05$) of the single, first and second hatchlings were similar.

Keywords: *Allolobophora (= Aporrectodea) caliginosa trapezoides, Density and biomass, Vertical distribution, Cocoon, Earthworm, Incubation, Hatchlings.*

Introduction

There has been a significant increase aimed at improving the management of earthworms as a source in agricultural soils. Awareness of soil structural decline encouraged many farmers to adopt various techniques to stimulate more efficient and safer methods of transport of nutrients to plants, and the role of earthworms in this direction cannot be underestimated (Edwards and Lofty, 1977).

The farms of Benghazi (32°10′N, 20°06′E) and its neighbourhood in Libya are colonized in large numbers by the earthworm *Allolobophora* (= *Aporrectodea*) *caliginosa trapezoides* (Bennour and Nair, 1997). It is found in the root zone of moist soil and contributes much to soil fertility. It co- exists with other species such as *Allolobophora rosea* and *Microscolex dubius* in the wild (Nair *et al.*, 2005).

Reviewing the literature, one can find that the majority of information on this and other species of earthworms inhabiting North Africa comes from Libya and Egypt. The publications of Mohamed *et al.* (1995), Bennour and Nair (1997), Nair and Bennour

(1997, 1998), Alshawish *et al.* (2004) and Nair *et al.* (2005) on Libyan earthworms and the information given by Bishara and El-Kifl (1954), El-Kifl (1958), Habib and Issa (1958) and El-Duweini and Ghabbour (1964, 1965) on Egyptian earthworms are noteworthy.

The present study investigates the occurrence, density, biomass, vertical distribution, cocoon production, incubation and hatchlings of *A. caliginosa trapezoides* found in Benghazi, Libya. Such a study is important before undertaking further research on the roles played by these worms on the structure, decomposition and fertility of soils in Benghazi agroecosystems.

Material and Methods

The earthworm selected for the study was *Allolobophora* (= *Aporrectodea*) *caliginosa trapezoides.* It belongs to the class: Oligochaeta, order: Opisthopora and family: Lumbricidae.

Thirteen stations located within the municipality of Benghazi were selected for the study (Figure 17.1). These stations were:

1. Sidi Khalifa
2. El-Koefia
3. Sidi Hussain
4. Bouatni
5. Benena canal
6. Benena grassland
7. Benena farm
8. Fuwayhat
9. Garyounis
10. El-Guarsha
11. Ganfuda
12. Hawari, and
13. Nuwagia.

and they were broadly categorized into six different habitats of earthworms. These were:

1. Vegetable garden (stations 1 and 2)
2. Lemon and orange farm (station 6)

3. Green house (stations 4 and 8)
4. Plain landscape with wild grasses and plants having medium-sized trees forming canopy (stations 3, 5, 6, 9, 10, 11)
5. Rose and flower garden (station 12) and
6. Plain barren landscape without any vegetation (station 13).

Earthworms were sampled during February and March, 2002, following the procedure of Baker and Barrett (1994). A plot of 30 × 30 cm was measured within the survey site of each station which was at least 5 m away from the fence or boundary, if present. A ditch of 10 cm deep was dug in the plot and the soil was removed and spread on a white enamel tray, and hand-sorted removing earthworms as they were found and counting their total numbers. Four more ditches were dug in the same way, at least 5 m apart. The number of

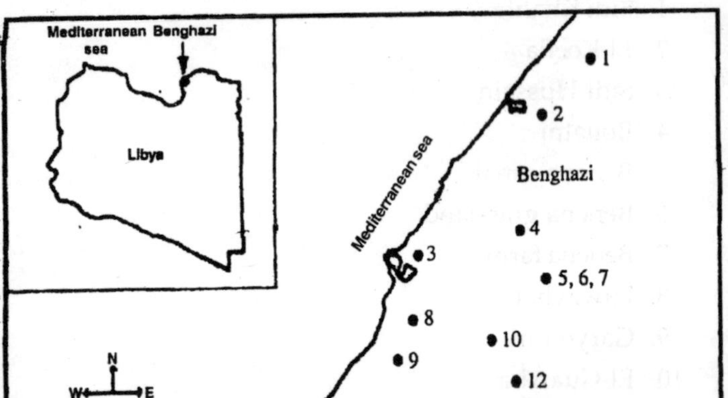

Figure 17.1: Location of Study Stations in Benghazi Municipality

Stations 1: Sidi Khalifa; 2: El-Koeia; 3: Sidi Hussain; 4: Bouatni; 5: Benina canal; 6: Benina grassland; 7: Benina farm; 8: Fuwayhat; 9: Garyounis; 10: El-Guarsha; 11: Ganfuda; 12: Hawari; 13: Nuwagia.

earthworms found in all five ditches were added together doubled and expressed as the number of earthworms m^{-2}. Special care was taken with the root mats, as earthworms tended to be concealed there. The adult worms sampled from each station were identified up to species or sub-species level and the per cent of occurrence of *A. caliginosa trapezoides* was determined.

Density, Biomass, Vertical Distribution, Cocoons and Hatchlings of *A. caliginosa trapezoids*

To study the density, biomass, vertical distribution, cocoons and hatchlings, *A. caliginosa trapezoides* were sampled from Benena farm, (Station 7; Figure 17.1) located near the airport in Benghazi, Libya. The farm has a total area of approximately 100 hectares with extensive cultivation of orange, sugarcane, alfalfa and vegetables. The farm is irrigated throughout the year from a centrally located water reservoir. The water from the reservoir is pumped to different localities through channels, and a high colonization of earthworms was observed on the banks of these channels. The site selected for sampling was the bank of one of these channels which brings water from the reservoir to the sugarcane field. The channel was 25 m long and 1.5 m wide. The soil was moist with thick grass cover throughout the year.

The earthworm populations were estimated by hand-sorting soil samples, since chemical expellants such as potassium permanganate and formalin were found impractical for heavy clay soil of the sampling site. Fortnightly sampling of *A. caliginosa trapezoides* and the cocoons laid by them began in May 1993 and continued for 12 months until April 1994. The sampling started from the bank of the channel near the reservoir and subsequent samplings were at 50 cm intervals from the previous ones towards the edge of the sugarcane field. Each sample was 0.5 m^2 in area (using quadrats) and 30 cm deep. The samples were separated into three depth fractions, 0–10 cm (within which the majority of grass roots were found), 10–20 cm and 20–30 cm. The soil was hand-sorted for earthworms and cocoons. Later the earthworms and cocoons were identified using keys (Edwards and Lofty, 1977), counted and weighed. The sampling was carried out during the second and fourth weeks of each month and the total number of worms sampled and their biomass and the total cocoons collected in the two samples were pooled. The worms were preserved in 4 per cent formation.

Freshly deposited cocoons (mean length 3.6 mm and mean width 2.9 mm) which could be easily distinguished from older ones by the light green colour of the former, were washed lightly in distilled water in order to remove debris and organic particles and then placed in small glass Petri dishes (10 ml volume) containing distilled water. The dishes were kept in the dark at 20 ± 2°C and observed every 24 hours. The sequence of emergence and the number of hatchlings were used to place the worms in three categories:

1. 'Single hatchling' (single worm emerging from a cocoon); and for two worms emerging from a cocoon,
2. 'First hatchling', and
3. 'Second hatchling', respectively.

Soon after hatchlings emerged from cocoons, water on their body surfaces was removed by rolling the worms lightly on tissue paper. They were then weighed individually on a Sartorius balance. Some were preserved in 4 per cent formalin for later measurements and some kept for growth studies. Worm length was measured with a slide caliper and the number of segments were counted under a wide stereomicroscope.

The ideal medium selected for culturing the hatchlings of *A. caliginosa trapezoides* for growth studies was the soil of the sampling site itself, a friable loam with 20 per cent to 25 per cent moisture content. A 200 ml jar half-filled with soil was chosen as a rearing container. Water was sprinkled to bring the moisture content above 20 per cent. Worms (either single, first or second hatchlings, 15 replicates of each) were introduced individually in the jar. Subsequent body mass was determined at 10-day intervals for 150 days. The jar top was covered on the outside by a net to prevent escape of the worms and was kept in the dark at 20 ± 2°C. Partly decomposed plant materials and rotten sugarcane leaves were provided as food. Every 15 days, the soil was removed and replaced by new soil brought from the field. Under these conditions, the worms thrived in the laboratory for more than 6 months to 1 year. One each of single and first hatchlings and two double hatchlings died after 130 days. Soil mixed with cow manure, as suggested by Viljoen and Reinecke

(1988), was found unsuitable for rearing the hatchlings of *A. caliginosa trapezoides* as it caused high mortality.

The temperature, pH, moisture organic matter of the soil from where the worms and the cocoons were sampled were also determined along with each sampling, using standard techniques (El-Duweini and Ghabbour, 1965; Wallwork 1970). The soil texture was determined by the hydrometer method (Foth, 1984). The rainfall data was obtained from the meteorological station located at Benena Airport, Benghazi.

Results

Allolobophora caliginosa trapezoides formed the dominant and most widespread species of Benghazi and this earthworm was present in eleven of thirteen stations. The two stations where this species was absent were El-Koefia and Benena grassland (Figure 17.1). Sidi Khalifa, Sidi Hussain, El-Guarsha, Hawari and Nuwagia stations recorded 100 per cent *A. caliginosa trapezoides,* where as in Boutani, it co-existed with *Microscolex dubius* (20 per cent) and *Allolobophora rosea* (10 per cent); in Benina canal with *M. dubius* (30 per cent); in Benena farm with *M. dubius* (20 per cent); in Fuwayhat with *A. rosea* (15 per cent) and *M. dubius* (10 per cent); in Garyounis with *A. rosea* (10 per cent) and *M. dubius* (8 per cent); and in Ganfuda with *A. rosea* (12 per cent).

Soil Analysis

The soil of Benena farm where detailed investigation on the ecology and breeding biology of *A. caliginosa trapezoides* was carried out, had more clay than silt or sand. The textural analysis of the soil of the sampling site revealed that it consisted of 36 per cent clay, 32.6 per cent silt and 31.4 per cent sand. The soil was alkaline (pH 7.3) and the organic matter averaged 4.2 per cent.

Density, Biomass, Population Structure and Vertical Distribution

Density, biomass and structure of the *A. caliginosa trapezoides* population together with environmental variables for the period from May 1993 to April 1994 are shown in Figure 17.2. Adult worms were distinguished by the presence of clitella. In subadults, the clitella were not formed but the *tubercula pubertatis* were visible. The juveniles

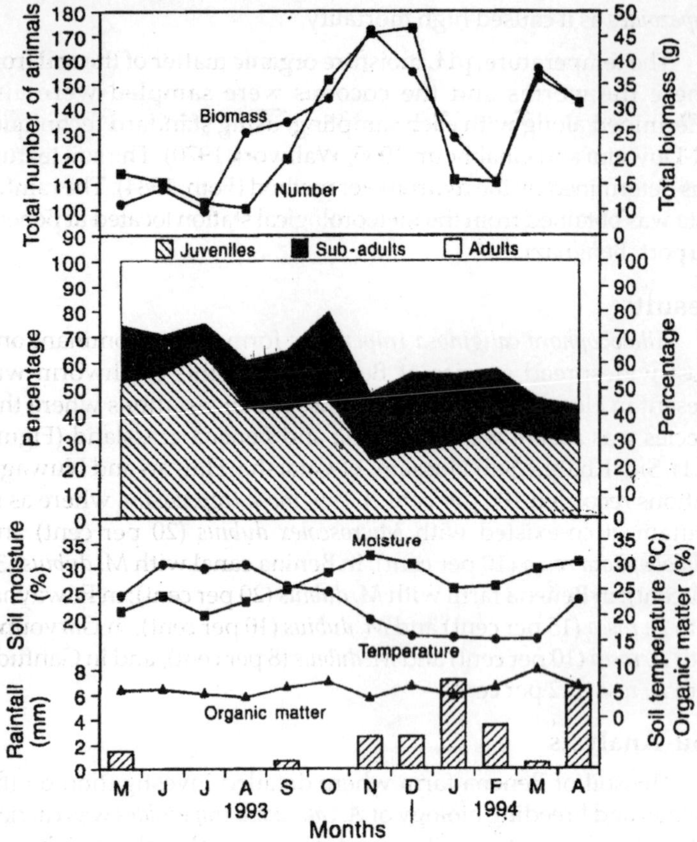

Figure 17.2: Total Number, Biomass and Percentages of Juveniles, Sub-adults and Adults of *A. caliginosa trapezoides* Together with the Mean Values of Rainfall (Benghazi), Temperature, Moisture and Organic Matter of Soil of Sampling Site

were categorized as young worms without *tubercula pubertatis* and clitella.

Number of worms sampled were low during the summer months from May (114 m^{-2}) to August (100 m^{-2}) 1993, increasing from September (120 m^{-2}) 1993 onwards, reaching a peak during December (172 m^{-2}) 1993, with a decline thereafter. During March

1994, a second peak in numbers was observed (153 m^{-2}) with a slight declining trend during April (1.14 m^{-2}) 1994.

The biomass of the worms showed more or less similar trends to population densities. However, an increase in biomass of the worms was evident from August (24.9 gm^{-2}) 1993 onwards, reaching its peak in November (46.0 gm^{-2}) 1993, with a uniform decrease until February (14.7 gm^{-2}) 1994, and subsequent increase.

More than 50 per cent of the total worms sampled during the summer months from May to July, 1993 were juveniles. This trend changed from August to October, 1993, when the percentage of subadults rose substantially (18 per cent in August and 34 per cent in October, 1993), showing that a good percentage of juveniles had reached the sub-adult stage during the late summer and early autumn seasons. A reduction in the percentage of juveniles and a subsequent increase in adults were evident from November, 1993, continuing until April, 1994. The increase in the percentage of juveniles observed from February, 1994 (35 per cent) was due to the addition of juveniles hatched from cocoons laid during the preceding winter months of December, 1993 and January, 1994. The majority of sub-adults became fully clittelate adults during January–April, 1994.

Maximum rainfall during the sampling period was between November, 1993 and April, 1994, with peaks during January and April, 1994. The highest soil temperature was recoded during July, 1993 (31°C) and the lowest during March, 1994 (15.5°C). The maximum moisture content (32.5 per cent) was noted during November, 1993, and the minimum (21 per cent) during July, 1993. The organic matter content ranged from 4.7 per cent to 9.0 per cent.

The vertical distribution of *A. caliginosa trapezoides* in three soil depths of 0–10, 10–20 and 20–30 cm, for the period from May 1993 to April 1994 are presented in Figure 17.3. It was observed that both juveniles and adults were confined to the surface organic zone (0–10 cm depth) of soil during the summer months from May to September, 1993. Later, a slow migration of the adults from the top layers of the soil to a depth of 10–20 cm was observed from October, 1993 to January, 1994. Many worms were seen copulating at 10–20 cm depth during these months and a large number of cocoons were also sampled from this depth. With the outset of spring during the latter half of February, 1994, an upward migration of adults to the top 0–10 cm was evident and most of the cocoons were also laid

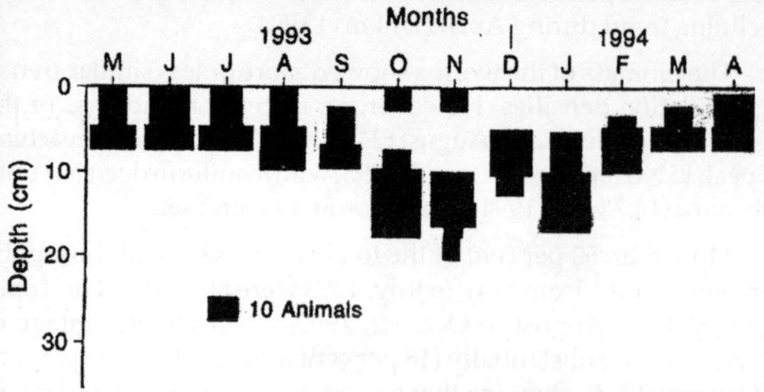

Figure 17.3: Vertical Distribution of *A. caliginosa trapezoides*

in this zone during March and April, 1994. The activities of juveniles were always confined to the top 0–10 cm soil depth, irrespective of the seasons.

Cocoons and Hatchlings

Cocoons of *A. caliginosa trapezoides* first appeared in the second half of August, 1993. With the outset of winter (end of November, 1993), the number of cocoons increased in the soil samples and this trend continued until the end of spring (March/April, 1994). No cocoons were sampled from the field during summer (May–July, 1993).

The number of cocoons per m^2 for the period May, 1993–April, 1994, was correlated with temperature, moisture and pH of soil (Figure 17.4). There was a positive correlation of cocoon production with soil moisture ($r = 0.74; p < 0.05$), and a negative one with soil temperature ($r = -0.85; p < 0.05$). There was no relationship however, between soil pH and cocoon production ($r = 0.28; p > 0.05$).

The freshly deposited cocoons were light green in colour but began to change to yellowish-green after 10 to 15 days of incubation, becoming yellowish brown after 30 days, and light brown prior to emergence of hatchlings.

The incubation period ranged from 41 to 52 days during autumn and winter, and from 38 to 48 days during spring. The mean

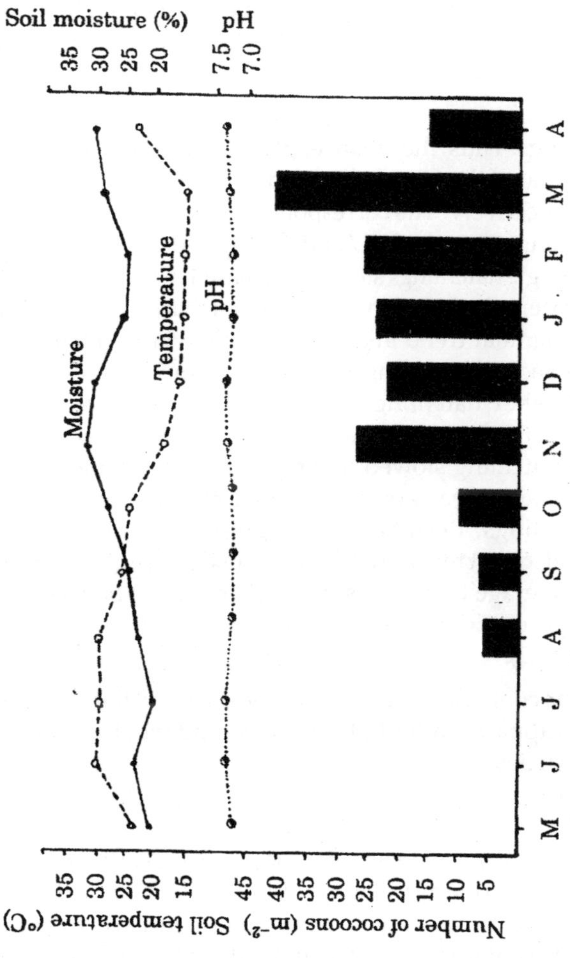

Figure 17.4: Number of Cocoons of *A. caliginosa trapezoides* per m^2 in Relation to Temperature, Moisture and pH of Soil

incubation period was 47 days. Of the 186 cocoons studied, 60 (32 per cent) contained only one worm (single hatchling) and 126 (68 per cent) had two worms (first and second hatchlings). Rapid movements of the worms inside the cocoons were observed prior to their emergence. In the two-worm cocoons, the worm having the longer body and the higher number of segments emerged first (first hatchling); the second smaller one (second hatchling) emerged 6 to 15h later.

Single hatchlings were always larger than either the first or second hatchlings. Thus, the mean length of a single hatchling was 21.2 mm, and that of the first and second hatchlings were 19.8 mm an 17.8 mm respectively. The corresponding figures for body mass and total segments were 16.1, 12.7 and 7.7 mg, and 138, 132 and 121, respectively. Single hatchlings steadily increased in body mass from 0 to 120 days after which there was a rapid increase up to 150 days (Table 17.1). A similar trend in growth was also shown by the first hatchling. The second hatchling, however, had more rapid growth rate than the other hatchlings, and by 70 days its body mass surpassed that of the first hatchling. After 130 days, however, growth of the second hatchling slowed down. In general, the increase in body mass was fairly steady over the entire 150-day period for each group of hatchlings. Results of the regression coefficient and elevation revealed that the rate of change of body mass ($F = 2.18$; $p > 0.05$) and the average body mass ($F = 1.25$; $p > 0.05$) between the single, first and second hatchlings of *A. caliginosa trapezoides* were similar (Table 17.2).

The worms continued to grow after 150 days and the tubercula pubertatis was apparent after 180 to 190 days and the clitellum was formed after 250 days.

Discussion

Allolobophora caliginosa trapezoides, the peregrine earthworm has been spread from northern and western Europe by man within the past few hundred years and it is the dominant earthworm of agricultural lands and gardens throughout temperate and some upland tropical regions of the world (Lee, 1985). Characterisation of peregrine species (Lee, 1987; Chaudhuri and Bhattacharjee, 1999) such as tolerance of environmental variability, habitat specificity, opportunism in choice of food, ability to withstand chemical stress and ecological plasticity might have helped this species to colonise in large numbers in Benghazi and surrounding regions, and possibly other parts of Libya as well.

Table 17.1: Growth Rates (mg ± SE) of Hatchlings of *A. caliginosa trapezoides* for a Period of 150 Days. Total animals = 15 from each hatchling type.

Days	*Body Mass (mg ± SE)*		
	Single Hatchling	*First Hatchling*	*Second Hatchling*
0	16.1 ± 3.0	12.7 ± 1.0	7.7 ± 0.8
10	26.7 ± 7.7	18.3 ± 1.3	16.5 ± 1.9
20	37.5 ± 4.1	24.9 ± 1.5	24.3 ± 2.1
30	42.3 ± 5.1	40.3 ± 2.5	33.1 ± 2.6
40	49.3 ± 7.2	48.6 ± 4.5	45.4 ± 4.3
50	55.3 ± 6.5	53.5 ± 4.2	47.5 ± 5.9
60	62.8 ± 7.4	55.2 ± 5.7	54.1 ± 7.1
70	67.3 ± 8.4	57.8 ± 5.7	60.4 ± 10.3
80	77.5 ± 7.3	58.9 ± 5.6	70.3 ± 13.8
90	88.4 ± 10.9	67.9 ± 6.1	82.8 ± 17.7
100	97.8 ± 13.9	88.3 ± 8.9	103.1 ± 20.6
110	111.1 ± 20.9	107.5 ± 13.8	119.9 ± 22.4
120	121.1 ± 28.1	126.3 ± 19.1	132.6 ± 16.5
130	154.3 ± 39.1	131.9 ± 20.1	154.4 ± 23.8
140	193.7 ± 42.9	154.5 ± 28.1	159.4 ± 25.5
150	224.5 ± 51.8	188.6 ± 36.23	174.8 ± 26.6

It is well known that earthworm density and biomass fluctuate during annual cycles and that there is considerable variation in size of earthworm population from one site to another; maximum densities generally occur in base-rich grassland soils and minimum densities in acid moorland and raw humus (Wallwork, 1970). Increase in population density and biomass of *A. caliginosa trapezoides* in Benghazi soils were observed from September to December and again during March and April. There appears to be a general tendency among the lumbricids for the main period of population decline to occur in the late winter when the predominant age class present is the adults (Edwards and Lofty, 1977). This aspect has been amply verified in the present study. The density and biomass of the worms during the sampling period from May, 1993 to April, 1994, ranged from 100 to 173 m^{-2} and from 9.0 to 46.0 g^{-2}, respectively. These values were higher than those recorded for other African worms from Uganda (Block and Banage, 1968) and from Nigerian grasslands (Madge, 1969) but were lower than the values recorded from Egypt (El-Duweini and Ghabbour, 1965). The timing of the peak densities in the present study was correlated with local conditions, particularly the temperature and moisture of the soil, which affected fecundity and the length of the incubation period. Nevertheless, these effects were often synchronized so that well-defined population biomass and peaks occurred in winter and spring seasons. The limiting effects of variations in temperature and moisture of the soil on the density and distribution of *A. caliginosa trapezoides* were strongly expressed since these worms were mainly surface-dwelling species. Gerard (1967), Edwards and Lofty (1977) and Reddy and Alfred (1978) also correlated peaks in population size of temperate and tropical earthworms with soil moisture and temperature.

Table 17.2: Results of Co-variance Analysis

Regression and Correlation Statistics				
Hatchling	*Regression Coefficient b*	*y-intercept a*	*Average Body Mass (mg)*	*Regression Equation*
Single	0.6988	23.0625	75.5	$y = 0.6988\ x + 23.0125$
First	0.8443	2.2007	65.5	$y = 0.8443\ x + 2.2007$
Second	1.1156	– 3.2787	80.4	$y = 1.1156\ x - 3.2787$

F-values for ANCOVA				
Source	*df.*	*SS*	*MSS*	*F-ratio*
Among *'b'*	2	3044.03	1522.02	2.18*
Sum of group deviation	42	29374.08	698.19	
Between adjusted means	2	1835.81	917.90	1.25*
Error	44	32368.11	735.64	

* Not significant

A clear pattern of juvenile dominance during summer followed by the dominance of subadults and adults during autumn and winter respectively were discernible in *A. caliginosa trapezoides* in Benghazi soils. Similar patterns of change in maturity of this species were reported by Evans and Guild (1948) from Britain and by McCredie *et al.* (1992) from Australian pasture soils. The increase in juveniles in summer was expected since their emergence from cocoons laid during the preceding winter and spring seasons started towards the end of spring and continued during the summer. Most of the juveniles which survived in the summer reached the subadult stage by autumn and the adult stage by winter. This population structure continued as an annual cycle in Benghazi soils.

A. caliginosa trapezoides which is a surface-dwelling species, mainly confined its activities to the upper organic root zone (0–10 cm depth) for most of the months. This was especially true for the juveniles. However, the migration of adults to 10–20 cm depth in the soil during the winter was observed, and this was correlated with their breeding and laying of cocoons in deep soil crevices. By spring, the upward migration of adults was again discernible. The extent to which different species of lumbricids can penetrate soil varies; consequently there are differences in vertical migration (Wallwork, 1970). Wallwork further stated that the ability of earthworms to penetrate down the soil profile is better developed in large-sized animals than in those of small sized and the character of the soil has an important effect on the vertical migration of these animals.

Cocoon production of *A. caliginosa trapezoides* in the present study in Benghazi soils lasted from August 1993 until April 1994. The optimal range of soil temperature for production of cocoons was 15°C to 22°C. Complete cessation of cocoon production was observed during the summer (from May until the first half of August, 1993) when surface temperature during the day frequently rose above 35°C. Butt (1991) reported that the life-stages of earthworms, such as cocoon production, cocoon development and growth rates of hatchlings, may be influenced by environmental temperature. He further stated that under controlled laboratory conditions, cocoon development is most rapid at 20°C, cocoon viability is not reduced significantly and growth of hatchlings to maturity is also achieved in the shorter time. On the other hand, fluctuating temperature regimes give lower annual results, but production is enhanced during periods of increased environmental temperatures. Daughberger (1988) assessed moisture and temperature preferences of several earthworms and found many species to be stenothermal. Cocoon production of many European earthworm species is highest during spring (Satchell, 1967) when temperature rises to 15°C and above. Soil moisture was never a limiting factor to the earthworms at Benena farms because of the continuous flow of water in the irrigation channels. El-Duweini and Ghabbour (1965) investigated the survival of *A. caliginosa* in relation to soil moisture content in Egypt and reported that in soils with 50-85 per cent gravel and sand, an increase in moisture content from 15 per cent to 34 per cent is associated with an increase in the number of cocoons and of cocoon production. Above 34 per cent, extra moisture has no effect.

The mean incubation period of *A. caliginosa* in Benghazi at 20 ± 2°C was 47 days which is similar to the values from Egyptian earthworms (45 to 50 days) (El-Duweini and Ghabbour, 1965). However this period is shorter when compared with the temperate worms of Europe. Thus, Butt (1993) found the incubation time of *Allolobophora longa* cocoons at 15°C to 20°C to be 61 and 56 days, respectively. Holmstrup *et al.* (1996) observed that the incubation time of lumbricid worms increases greatly with decreasing temperature. They reported a mean incubation period of 44 to 50 days at 20°C for cocoons of five earthworm species in Denmark, which increases substantially from 69 to 207 days when the temperatures are lowered to 15°C and 10°C, respectively. The ability of embryo to develop at a lower temperature should be regarded as an adaptation to the particular habitat in which the species is living, making it possible for the juveniles to emerge as soon as environmental conditions are favourable (Edwards and Bohlen, 1996).

The total number of worms emerging from cocoons of *A. caliginosa trapezoides* was either one (single hatchling) or two (first hatchling and second hatchling) (mean 1.68 ± 0.03), but never more. The number of worms emerging from cocoons differs from species to species. Thus, *A. trapezoides* found in Australian pasture soils produces two worms from a cocoon (McCredie *et al.* 1992), the African nighterawler *Eudrilus engeniae* produces up to five (Viljoen and Rejnecke, 1988), and *Eisenia foetida* and *Bismastos tumidus* produce up to six (Vail, 1974).

The large body mass, greater body length and greater number of segments of the single hatchling compared to other hatchlings are reported in many species of earthworms, *e.g. Eudrilus eugeniae, Eisenia foetida, Allolobophora caliginosa* and *Allolobophora rosea* (Vail, 1974; Reinecke and Visser, 1981; Viljoen and Reinecke, 1988). These authors postulate that the hatchlings with shorter bodies and fewer segments are still developing embryonically whereas the other worms could have utilized more energy for growth and enlargement of segments already present.

The growth rates of single, first and second hatchlings of *A. caliginosa trapezoides* showed that the second hatchling had the fastest initial growth compared with the single and first hatchlings; these differences were less marked after 120 to 130 days. However, a significant difference in the rate of change of body mass and the average body mass between the hatchlings was not discernible. Nowak (1975) studied the growth pattern of *A. caliginosa* in field conditions and observed a rapid pre-reproductive phase of growth in these worms, followed by a phase of steadily decreasing growth after sexual maturity is attained. Similar growth patterns were reported for *Allolobophora rosea* and *Eisenia foetida* by Phillipson and Bolton (1977) and Edwards and Neuhauser (1988), respectively.

Conclusion

In conclusion the soil of Benghazi is colonized by the peregrine temperate earthworm *Allolobophora caliginosa trapezoides* which forms the dominant and most widespread species in the area. The climatic and edaphic condition of Benghazi are more suitable for this species than those found in tropical and sub-tropical countries of sub-Saharan Africa. Peaks in density and biomass of *A. caliginosa trapezoides* during winter and spring together with juvenile dominance during summer and subadult and adult dominance during winter and spring show that the population structure of the species in Benghazi is more similar to that of the worms of temperate neighbouring Europe than to that of the tropical and sub-tropical worms found in other parts of Africa. The distribution of *A. caliginosa trapezoides* in the upper organic root zone during most of the months studied is expected since soil moisture and temperature are never limiting at the sampling site because of the continuous flow of water in the irrigation channel. However, this trend may not be apparent in other parts of Benghazi where the surface soil can become dry during summer. The average incubation time of 47 days for the cocoons of *A. caliginosa trapezoides* to hatch in Benghazi soils coincides with the period reported for this species in Egypt. However, it is shorter than the incubation time in Europe where the climatic conditions are more temperate than at Benghazi. The maximum of two worms per cocoon observed in the present study is also characteristic of this species. The growth rate of hatchlings are almost steady and uniform, and worms from Benghazi reach the adult stage much earlier than temperate worms but much later than tropical ones. From the data on their growth rates and from subsequent observations from the field, it can be inferred that the longevity of these animals in Benghazi soils is between 2 to 3 year.

Acknowledgements

We express our appreciation and thanks to Mrs. Shapna Nair, SAFI Institute of Advanced Study, for the help rendered in the DTP print out of the manuscript.

References

Alshawish, S.A., Mohamed, A.I. and Nair, G.A., 2004. Prolonged toxicity of sub-lethal dosages of chemical pesticides on the body mass and cocoons of *Aporrectodea caliginosa* (Savigny, 1826) (Oligochaeta : Lumbricidae) inhabiting Benghazi, Libya. *Proc. Nat. Acad. Sci.*, India, 74: 123–133.

Baker, G.H. and Barrett, V.J., 1994. *Earthworm Identifier.* CSIRO, Melbourne.

Bennour, S.A. and Nair, G.A., 1997. Density, biomass and vertical distribution of *Aporrectodea caliginosa* (Savigny, 1826) (Oligochaeta, Lumbricidae) in Benghazi, Libya. *Biol. Fertl. Soils*, 24: 102–125.

Bishara, G.M. and El-Kifl, A.H., 1954. Magnitude and distribution of earthworms in cultivated land at Giza, Egypt. *Bull. Fac. Agric. Cairo University*, 51: 21–30.

Block, W. and Benage, C.W., 1968. Population density and biomass of earthworms in some Ugandan Soils. *Rev. Ecol. Biol. Sol.*, 5: 515–521.

Butt, K.R., 1991. The effects of temperature on the intensive production of *Lumbricus terrestris* (Oligochae: Lumbricidae). *Pedobiologia*, 35: 257–264.

Butt, K.R., 1993. Reproduction and growth of three deep burrowing earthworms (Lumbricidae). in laboratory culture in order to assess production for soil restoration. *Biol. Fertl. Soils,* 16: 135–138.

Chaudhuri, P.S. and Bhattacharjee, G., 1999. Earthworm resources of Tripura. In: *Proc. Natl. A cad. Sci., India,* 69: 159–170.

Daughberger, P., 1988. Temperature and moisture preference of three earthworm species (Oligochaeta: Lumbricidae). *Pedobiologia,* 32: 57–64.

Edwards, C.A. and Bohlen, P.J., 1996. *Biology and Ecology of Earthworms.* Champman and Hall, London.

Edwards, C.A. and Lofty, J.R., 1977. *Biology of Earthworms.* Chapman and Hall, London.

Edwards, C.A. and Neuhauser, E.F., 1988. *Earthworms in Waste and Environmental Management.* SPB Academic Publications, Hague.

El-Duweini, A.K. and Ghabbour, S.I., 1964. Observations on the burrowing activities of *Allolobophora caliginosa* f. *trapezoides. Bull. Zool. Soc., Egypt,* 19: 60–90.

El-Duweini, A.K. and Ghabbour, S.I., 1965. Population density and biomass of earthworms in different types of Egyptian Soils. *J. Appl. Ecol.,* 2: 271–287.

El-Kifl, A.H., 1958. Influence of soil reaction on earthworm *Allolobophora caliginosa* (Savigny). *Bull. Fac. Agric. Cairo Univ.,* 123: 55– 72.

Evans, A.C. and Guild, W.J. Mcl., 1948. Studies on the relationships between earthworms and soil fertility IV. On the life-cycle of some British lumbricids. *Ann. Appl. Biol.,* 35: 471–484.

Foth, H.D., 1984. *Fundamentals of Soil Science.* John Wiley and Sons, New York.

Gerard, B.M., 1967. Factors influencing earthworms in pastures. *J. Anim. Ecol.,* 36: 235–252.

Habib, A. and Issa, G.I., 1958. The population density of earthworms in different cultivations. *Ann. Agr. Sci. Ain. Shams Univ.,* 3: 201–208.

Holmstrup, M., Ostergaard, I.K., Nielsen, A. and Hansen, B.T., 1996. Note on the incubation of earthworm cocoons at three constant temperatures. *Pedobiologia*, 40: 477–478.

Lee, K.E., 1985. *Earthworms: Their Ecology and Relationships with Soils and Land Use.* Academic Press, Sydney.

Lee, K.E., 1987. Peregrine species of earthworms. In: *On Earthworms*, (Eds.) A.M. Bonvicini Pagliai and D. Omodea. Modena, Italy.

Madge, D.S., 1969. Field and Laboratory Studies on the activities of two species of tropical earthworms. *Pedobiologia*, 9: 188–214.

McCredie, T.A., Parker, C.A. and Abbott, I., 1992. Population dynamics of the earthworm *Aporretodea trapezoides* (Annelida: Lumbricidae) in Western Australian Pasture Soil. *Biol. Fertl. Soils*, 12: 285–289.

Mohamed, A.I., Nair, G.A., Kassem, H. and Nuruzzaman H., 1995. Impacts of pesticides on the survival and body mass of the earthworm *Aporrectodea caliginosa* (Annelida: Oilgochaeta). *Acta Zool. Fennica*, 196: 344–347.

Nair, G.A. and Bennour, S.A., 1997. Thermal reactions of the earthworm *Aporrectodea caliginosa* (Savigny, 1826) (Oligochaeta: Lumbricidae). *Proc. Indian Natl. Sci. Academy*, 63: 53–62.

Nair, G.A. and Bennour, S.A., 1998. Cocoons and hatchlings of *Aporrectodea caliginosa* (Savigny, 1826) (Oligochaeta: Lumbricidae) in Benghazi, Libya. *J. Arid Environ.*, 40: 459–466.

Nair, G.A., Youssef, A.K., El-Mariami, M.A., Filogh, A.M. and Briones, M.J.I., 2005. Occurrence and density of earthworms in relation to soil factors in Benghazi, Libya. *Afr. J. Ecol.*, 43: 150–154.

Nowak, E., 1975. Population density of earthworms and some elements of their production in several grassland environments. *Ekologia Polska*, 23: 459–491.

Phillipson, J. and Bolton, P.J., 1977. Growth and cocoon production by *Allolobophora rosea* (Oligochaeta: Lumbricidae). *Pedobiologia*, 17: 70–82.

Reddy, M.V. and Alfred, J.R.B., 1978. Some observations on the earthworm population and the biomass in a sub-tropical pine forest soil. In: *Soil Biology and Ecology in India*, (Eds) C.A. Edwards and G.K. Veeresh. Univ. Agric. Press, Bangalore, India.

Reinecke, A.J. and Visser, P.A., 1981. Number and size of hatchlings and cocoons of the earthworm species *Eisenia rosea* and *Allolobophora trapezoides* (Oligochaeta). *Rev. de Ecologie et Biologie du Sol,* 18: 473–485.

Satchell, J.E., 1967. Lumbricidae. In: *Soil Biology,* (Eds.) A. Burges and F. Raw. Academic Press, London.

Vail, V.A., 1974. Observations on the hatchling of *Eisenia foetida* and *Biomastos timids* (Oligochaeta: Lumbricidae). *Bull. Tall Timbers Res. Stn.,* 11: 1–8.

Viljoen, S.A. and Reinecke, A.J., 1988. The number, size and growth of hatchlings of the African nightrawler *Eudrilus engeniae* (Oligochaeta). *Rev. de Ecol. et Biol. du Sol,* 25: 225–236.

Wallwork, J.A., 1970. *Ecology of Soil Animals.* Mc Grah Hill, London.

Chapter 18

Oviposition Behaviour of the Stem Borer *Apriona cinerea* Chevrolat (Coleoptera : Cerambycidae), on Poplars in Jammu Forests

Sanjay Bhatia, Baldev Sharma and Neelam Khajuria
Department of Zoology, University of Jammu, J&K, India
E-mail: siya98@sancharnet.in

ABSTRACT

Apriona cinerea Chevrolat is a serious stem borer pest of poplars in Jammu region. During studies on its biology and life-cycle, detailed observations were made on its oviposition behaviour. The most preferred site for egg laying is the main bole or the stem. Before commencing egg laying the female beetle selects the main stem usually 4.0–4.9 cm in girth, besides distinct height preference for egg laying has also been recorded. The female beetle excavates a crescent shaped niche in the main stem with its mandibles. A single egg is usually pushed inside this egg niche wherein it is glued with a colourless secretion and covered by bark fibres. Subsequently due to a host reaction this spot becomes hard and the safety of the egg is thus ensured. Significant details of the pre-oviposition behaviour, preparation of egg niche, deposition of the egg along with girth and height preference, length and width of the egg niche have been discussed in the present paper.

Keywords: *Apriona cinerea, Stem borer, Poplars, Oviposition behaviour.*

Introduction

The poplars (Genus *Populus,* Family: Salicaceae), are amongst the world's fastest growing multipurpose commercial softwoods of immense economic importance. In view of their rapid growth poplars have been planted on a large scale as agro forestry trees in the state of Jammu and Kashmir. These trees are usually raised as monoculture and are thus prone to the attack of insect pests. Of the various insect pests recorded on poplars in Jammu region (Bhatia, 1992; Sharma and Bhatia, 1995), *Apriona cinerea* Chevrolat (Figuer 18.1) has been recorded to be one of the most destructive pest. The larvae of these beetles bore and ramify galleries inside the stem and subsequently into the roots and thus hollow it completely. About 34 per cent of the poplar plantations have been recorded to be annually destroyed by the attack of *Apriona cinerea* in Jammu (Bhatia, 2004). Notes on the oviposition habits of this longicorn beetle have been made by Stebbing (1940), Beeson (1941), Singh and Prasad (1985) and Singh and Singh (1986), but no record is available on the pre-oviposition behaviour and the preferences of site for egg laying by this beetle. In the present article the selection of site for egg laying, preparation of egg niche and deposition of egg have been discussed.

Materials and Methods

Most of the studies incorporated are based on the observations of *Apriona cinerea* made in the field as well as on the stumps caged in the laboratory. Since the behaviour of the beetle under observation is nocturnal ordinary flash lights of low intensity were used to observe its oviposition behavior in the fields. The infested stumps were also brought from various forest nurseries in the region. Measurements of the egg niche were made from these stumps while girth and height preferences measurements were made in the fields. Continuous observations were made to record the oviposition habits which continued till egg laying was accomplished.

Results and Discussion

The adults of *Apriona cinerea* (Figure 18.1), start emerging in the fields during early July, the peak period being attained during the August which experience moderate to heavy rains in the area. Pairing commences a day or two after emergence and subsequently the female beetle prepares for oviposition. The various steps involved in ovipositon are:

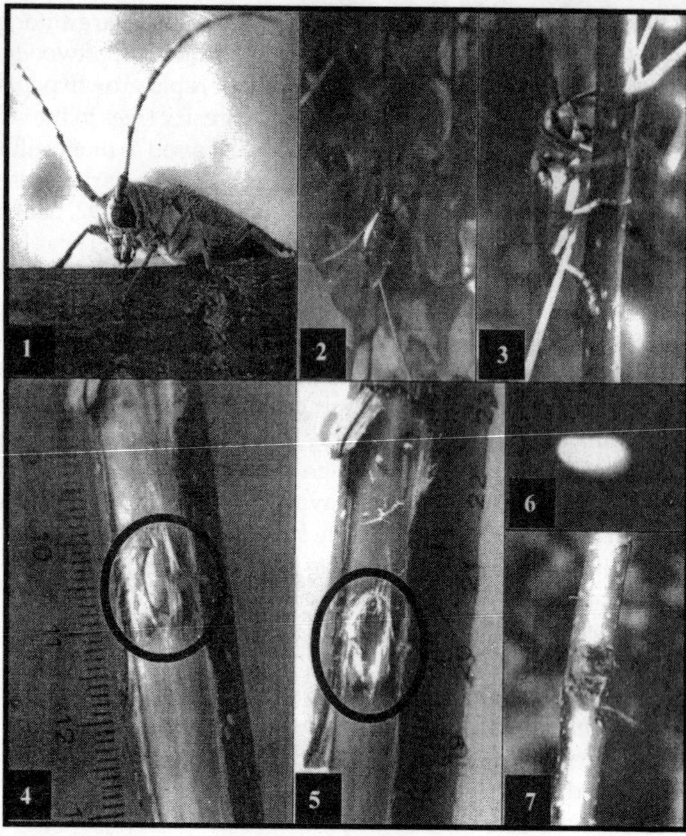

Figures 18.1–18.7

Selection of Site for Egg Laying

After mating, the inseminated female beetle feeds on the soft bark of the host plant and may even enjoy the sweet sap; which exudes out at the tips of the fresh shoots. Subsequently it starts moving over the boles and branches of the host plant, apparently for preliminary examination of the oviposition site (Figures 18.2 and 18.3). It explores the main stem or bole with the tip of its some what extended ovipositor, which evidently serve as a feeler. It appeared that the search for suitable oviposition site was not haphazard, as the female beetle takes utmost care to select the most suitable site for egg laying.

Similar preoviposition behaviour by the females in cerambycids is already on record. Banerjee and Nath (1971) for *Anoplophora versteegi*, Jolles (1932) and Saliba (1977) for *Cerambyx dux*; Wickman (1968) for *Semanotus litigiosus*, Sharifi *et al.* (1970) for *Osphranteria coerulescens*, Piper (1978) for *Dectes sayi* and Tara (1983) for *Batocera rufomaculata*.

Observations show that in *Apriona cinerea* the most suitable site for egg laying is the main bole or the stem. In order to ascertain whether there is any girth preference, observations of the girth of 50 egg laying sites were made. These have been recorded in Table 18.1 which shows that the most preferred girth for egg laying by *A. cinerea* ranges between 4.0–4.9 cm.

Table 18.1: Percentage of Egg Laying on Stems of Different Girth Range

Girth Range	*No. of Stems in Each Range*	*% age of Each Range*
1–1.9 cm	3	6
2–2.9 cm	7	14
3–3.9 cm	11	22
4–4.9 cm	29	58

Total number of stems examined–50

Out of the 50 oviposited stems examined 58 per cent (29) laying were on the stems with the girth range between 4.0–4.9 cm; while next in preference was 3.0–3.9 cm (22 per cent; 11 layings). Only 14 per cent (7) and 6 per cent (3) layings have been recorded on the stems with 2.0–2.9 cm, and 1.0–1.9 cm, girth range, respectively.

Though girth preferences for egg laying by *A. cinerea* is so far not recorded by any previous worker, but such observations for other longicorn beetles are well documented. Solomon (1977) recorded that most preferred girth for egg laying by *Goes debilis* ranges between 9–38 mm (average being 17 mm). Duffy (1946) recorded that diameter of the twigs on which *Prionus coriarius* prefers for egg laying ranges between 5–15 mm, while *Apriona germari* prefers 2½–3 cm diameter of the twig for this purpose (Stebbing, 1914). Piper (1978) observed that female beetle of *Dectes sayi* selected peripherally located plants where primary stem diameter exceeded 7.0 mm as compared to smaller narrow stemmed plants for oviposition. Tara (1983) working on *Batocera rufomaculata* observed that most preferred girth of the mulberry twig for egg laying ranged between 3.0–4.0 cm. All these observations by previous as well as present workers fit in well with the generalization of Hinton (1981) that adults of subfamily Lamiinae (Family; Cerambycidae) lay eggs on twigs and branches of 2.0–5.0 cm diameter. Forcella (1982) working on *Oncideres cingulata* in Australia observed an average of 9 mm twig diameter preferred for egg laying by this longicorn, while Raske (1973) in Albetra observed that female of *Tetropium parvulum* prefer to lay eggs on logs having eighteen inches diameter.

In sharp contrast to the egg laying behaviour of *Batocerera rufomaculata* which usually selects primary branch for egg laying (Tara, 1983), female beetle of *Apriona cinerea* selects the main stem for egg laying and very rarely the side branches for oviposition.

Height Preference for Egg Laying

Apriona cinerea, exhibits distinct height preference for egg laying. The most preferred height for egg laying ranges between 100–105 cm above the ground level (Table 18.2).

Table 18.2: Percentage of Egg Laying on Stems at Different Height from Ground Level

Height Range	*No. of Stems in Each Range*	*% age of Each Range*
85–89.9 cm	3	06
90–94.9 cm	5	10
95–99.9 cm	14	28
100–104.9 cm	20	40
105–109.9 cm	6	12
110–above	2	04
Total number of stems examined–50		

Out of 50 stems examined, 40 per cent (20) egg layings were on the stems in the height range of 100–104.9 cm while on the other lower side, only 6 per cent (3) layings were on the stems in the height range of 85–89.9 cm. Only 4 per cent (2) layings were recorded on stems beyond 110 cm. 28 per cent (14) layings were recorded on stems in the height range of 95–99.9 cm. Clubbing both ranges *i.e.* between 95–99.9 cm and 100–104.9 cm, it is concluded that preferred height for oviposition in *A. cinerea* ranges between 95–104.9 cm (40 + 28 = 68 per cent layings).

Besides girth preference for oviposition some cerambycids and some lepidopteran borers are known to exhibit height preference while others may not have for the purpose. The preferred height for oviposition by *Ziraphra rumifitgana* was recorded within 4.8–6.0 mts from the top of silver fir (Zangheri and Masutti, 1983), while *Cacoecia crataegana* female deposited its eggs between the height of 2.4–3.6 mts from ground level (Hochmat, 1959). Egrov and Solozhenikina (1963), however, observed no fixed height choice. Cheu (1941) observed significant correlation between the height of the plant and the number of the eggs layed by *Ostrinia nubilalis.* The cerambycid *Saperda carcharis* had been observed to lay eggs in the lower part of the trunks of poplars (Strot, 1962). Sanjeeva Raj (1969) observed that maximum twigs used for egg laying by *Sthenias gristator* were located between the height range of 6–10 cm above ground level.

On the contary Tara (1983) observed in *Batocera rufomaculata* that female does not show any height preference for the deposition of the eggs and that this was dependent on the availability of the preferred girth of the twigs *i.e.* if such twigs were high on the plant, they may lay egg at that height from ground level or vice versa.

Egg Laying

Many species of cerambycidae are known to oviposit in cracks and crevices of the bark on the stem or in and around the injured areas of the host plant, which requires little or no preparation of the oviposition site by the female. However, a more specialized behaviour occurs in the subfamily-lamiinae where females usually use their strong mandibles to cut out an oviposition site. Females of *Apriona cinerea,* use their mandibles to make egg niches in the living tree stem.

Having selected the preferred oviposition site (proper girth of the twig and height above the ground level) the gravid female beetle excavates a crescent-shaped niche in the stem by gnawing the bark with its mandibles. It positions its body parallel to the stem with the head directed downwards (Figure 18.3) and then proceeds to gnaw the egg niche by cutting through the stem epidermis and cortex. After completing the egg niche, it bends its abdomen downwards, extends the ovipositor slightly in order to probe the hole. Then it inserts the ovipositor and deposits a single egg in the pith parallel to the layer of the bark inside the oviposition site already prepared (Figures 18.4 and 18.5). The depth of the egg niche inside the stem has been recorded to range between 3–4 mm. In order to secure the egg inside the plant tissue, it is gummed within the slit by applying a colourless seĉretion. Subsequently the egg is covered with bark fibres aided by the secretions and ultimately the wound is closed by the tissue reaction of the host plant. Thus the oviposition site becomes hard and egg is secured. The scar which is left at the oviposition site, however, remains conspicuous (Figure 18.7).

Oviposition was observed to occur more commonly during the late afternoon and evening hours of the day. Almost identical observations with regard to oviposition are recorded for some other cerambycids such as Ritchie (1920) for *Saperds carcharias*, Piper (1971) for *Dectes sayi*, Nath and Basu (1969) for *Anoplophora versteegi*, Drouin and Wong (1975) for poplar borer *Saperda calcarata*, Nord *et al.* (1972) for *Saperda inoranta* and *Oberea schaumii* and Tara (1983) for *Batocera rufomaculata*. More references on such aspects of cerambycids are those of Ahmad *et al.* (1977), Duffy (1946, 1968), Starzyk (1977, 1978) Wickman (1968), Chowdhaury (1954), Ikeda (1979), Solomon (1968) and Raske (1972). Different observations are on record as to what is the number of eggs laid at one oviposition site. Single egg is laid at a laying site by *Apriona cinerea* according to the observations of the present author, thus being comparable with *Ospharanteria coerulescens* (Sharifi *et al.* 1970) and *B. rufomaculata* (Tara, 1983) which also lay single egg.

However, cerambycids such as *Aeolesthes sarta* lay eggs in cluster of 4–13 (Ahmad *et al.* 1977) while Duffy (1946) has observed that female of *Prionus coriarius* lay eggs in batches of 8–12 at one place, 10–110 eggs (average being 43.5) in a single batch has been observed in the case of *Phorcantha semipunctata* (Bytinski Salz and Neumark 1953). In *Sthenias gristator*, Sanjeeva Raj (1959) recorded 1–4 eggs at one place. Wickman (1968) recorded 1–5 eggs at a location laid by the female of *Semanotus litigious*, wheras Solomon (1968) recorded 1–3 eggs in one egg niche by mulberry stem borer *Dorcaschema wildi* in Mississipi. Similarly *Cyriopalus wallacei* is known to deposit eggs in cluster composing of 40–87 eggs which are laid in 2–3 days in Sabah, Malaysia (Thapa, 1974).

As has been recorded in the foregoing account *Apriona cinerea* deposits one egg in one egg niche. Measurements on 50 stems for the metery of egg niche reveals that its length (Table 18.3) varies between 1.2–1.3 cm., 44 per cent (22) laying, 2 per cent (1) laying between 0.8–0.9 cm, 4 per cent (2) laying between 0.9–1.0 cm, 8 per cent (4) layings between 1–1.1 cm, 24 per cent (12) layings between 1.1–1.2 cm, 12 per cent (6) laying between 1.3–1.4 cm and 6 per cent (3) laying between 1.4–1.5 cm.

Table 18.3: Percentage Length Range of Egg Niche

Length Range	*No. of Stems in Each Range*	*% age of Each Range*
8.0–9.0 mm	1	02
9.0–1.0 cm	2	04
1.0–1.1 cm	4	08
1.1–1.2 cm	12	24
1.2–1.3 cm	22	44
1.3–1.4 cm	6	12
1.4–1.5 cm	3	06
Total number of stems examined–50		

Breadth of the egg niches (Table 18.4) ranges between 5–6 mm, 46 per cent (23) laying; 4 per cent (2) layings between 3–4 mm; 34 per cent (17) laying between 4–5 mm, 10 per cent (5) layings between 6–7 mm and 6 per cent (6) laying between 1.3–1.4 cm and 44 per cent (3) laying between 7–8 mm.

Similar studies have also been reported by Palaniswamy *et al.* (1977), Tara (1983) for *Batocera rufomaculata* and Nord *et al.*(1972) on *Oberea schaumii.* These authors have reported uniformity in the length and breadth of the egg niche. Piper (1978) recorded that a single female of *Dectes sayi* can lay upto 46 eggs. Similarly a single *Anoplophora versteegi* female may lay upto 85 eggs (Banerjee and Nath, 1971). Sharifi *et al.* (1970) observed in *Ospharanteria coerulesces* upto 75 eggs laid by single female. Ahmad *et al.* (1977) reported that on an average a single female of *Aeolesthes sarta* may lay upto 144 eggs.

Table 18.4: Percentage Breadth Range of Egg Niche

Breadth Range	*No. of Stems in Each Range*	*% age of Each Range*
3.0–4.0 mm	2	04
4.0–5.0 mm	17	34
5.0–6.0 mm	23	46
6.0–7.0 mm	5	10
7. 0–8.0 mm	3	44
1.3–1.4 cm	6	6

Total number of stems examined–50

Egg

Eggs of *Apriona cinerea* (Figure 18.6) are pale white in colour with scattered light brown streaks on it. These are elongate ovoid, almost opaque, smooth and elliptical tapering towards either end. Micropylar end is broader. Chorion is thick with very dim light brown impressions. Length of the egg ranges between 6–7 mm and breadth between 3–3.3 mm.

Acknowledgements

The authors are thankful to the Head, Department of Zoology, University of Jammu for providing the necessary facilities during the course of this work. The help and assistance rendered by the J&K State Forest Department is also gratefully acknowledged.

References

Ahmad, M.I., I.A. Hafiz and M.L. Choudhury, 1977. Biological studies on *Aeolesthes sarta* Solsky attacking poplars in Pakistan. *Pakistan Journal of Forestry*, 77: 123–129.

Banerjee, S.N. and D.K. Nath, 1971. Life history, habits and control of the trunk borer of orange, *Anoplophora versteegi* (Ritsema) (Coleoptera : Cerambycidae). *Indian J. Agric. Sci.*, 41(9): 765–771.

Beeson, C.F.C., 1941. *The Ecology and Control of the Forest Insects of India and the Neighbouring Countries*. Govt. of India Publication, Dehradun, India, pp. 1–767.

Bhatia, S., 1992. Studies on the insect pests of poplars in Jammu forests of J&K state. *Ph.D. Thesis*, University of Jammu.

Bhatia, S., 2004. Screening of poplar varieties for relative susceptibility and resistance to the stem borer *Apriona cinerea* Chevrolat (Coleoptera : Cerambycidae) in Jammu forests, (J&K), India. *IUFRO*, 2004, New Zealand.

Bytinski-Salz. H., 1953. Two important tree borers in Israel. FAO. Pl. Prot. Bull., 1(3): 38–39.

Cheu, S.P., 1941. Correlation of corn borer damage with growth condition of corn and its significance on corn breeding work. *Kwangsi Agric.*, 2: 126–133.

Choudhury, S. 1954. Control of citrus trunkborer *Monohammus versteegi* Ritsman (Coleoptera : Cerambycidae). *Sci. Cult.*, 19: 457–458.

Drouin, J.A. and H.R. Wong, 1975. Biology, damage and chemical control of Poplar borer *Saperda calcarata* Say. (Coleoptera : Cerambycidae) in the junction of the root and stem of the Balsam poplar in western Canada. *Candian J. For. Res.*, 5(3): 433–439.

Duffy, E.A.J., 1946. A contribution towards the biology of *Prionus coriarius* Linnaeus (Coleoptera : Cerambycidae) *Trans. R. Ent. Soc., London*, 97: 419–442.

Duffy, E.A.J., 1968. A monograph of the immature stages of oriental timber beetle (Cerambycidae), pp. 1–667. British Museum (Nat. Hist.), Publ., London.

Egrov, N.N. and T.N. Solzhenikin, 1963. *Caoecia crataegana* Hubner (Coleoptera : Cerambycidae) a mass pest of oak stands in the Voronezh region. *Zool. Zh.*, 42: 1501–1512.

Forcella, F., 1982. Why twig girdling beetles girdle twigs. *Naturwissenschaften*, 69: 398.

Hinton, H.E. 1981. On the homology and nomenclature of the setae of lepidopterous larvae with some notes on the phylogeny of the lepidoptera. *Trans R. Ent. Soc., London*, 97: 1–37.

Hochmut, R., 1959. A contribution to the study of the morphology, bionomics and population dynamics of *Archis crataeganus*. *Prace. Vyzk. Ust. Lasn. CSR*, 16: 23–58.

Ikeda, K. 1979. Consumption and food utilisation by individual larva and the population of a wood borer *Phymatodes maoki* (Coleoptera : Cerambycidae). *Oecologia*, Berlin, 40: 287–298.

Jolles, 1932. A study of life-history and control of *Cerambyx dux* Fald. (Coleoptera : Cerambycidae), a pest of certain stone fruit trees in Palestine. *Bull. Ent. Res.*, 23: 251–256.

Nath, D.K. and A.G. Basu, 1969. Observations on the behavior of orange trunk borer, *Anoplophora versteegi* (pits) (Coleoptera : Cerambycidae) in Darjeeling Hills, West Bengal. *Indian J. Agric. Sci.*, 39(7): 69–78.

Nord, J.C., G.G. David and B.E. Knight. 1972. Biology of *Oberea schaumii* (Coleoptera : Cerambycidae) in trembling aspen, *Populus tremuloides*. *Ann. Ent. Soc., America*, 65(1): 114–119.

Palaniswamy, M.S., P.C. Sundrababu and T.R. Subramaniam. 1979. Studies on the biology of mango stem borer, *Batocera rufomaculata* DeGeer (Coleoptera : Cerambycidae). *South Ind. Hort.*, 27(3 and 4): 100–106.

Piper, G.L. 1978. Biology and immature stages of Dectes sayi Dillon and Dillon (Coleoptera : Cerambycidae) *Coleopt. Bull.*, 80: 1–113.

Raske, A.G., 1972a. Immature forms, genitalia and notes on the biology of *Anelaphus albofasciatus* Linnell (Coleoptera : Cerambycidae). *Pan-Pac. Ent.*, 48: 21–26.

Raske, A.G., 1972b. Immature forms, genitalia and notes on the biology of *Coenopoeus palmeri* (Coleoptera : Cerambycidae). *Candian Ent.*, 104: 121–128

Raske, A.G., 1973. Notes on the biology of *Tetropium parvulum* (Coleoptera : Cerambycidae) in Alberta. *Canadian Ent.*, 105: 757–760

Ritchie, W., 1920. The structure, bionomics and economic importance of *Saperda carcharias* Linneus, the large poplar longicorn. *Ann. Appl. Biol.*, 7: 299–343.

Saliba, L.J., 1977. Observations on the biology of *Cerambyx dux* Faldermann (Coleoptera : Cerambycidae) in the maltese island. *Bull. Ent. Res.*, 67: 107–117.

Sanjeeva Raj, P.J., 1959. The bionomics of the stem girdler, *Sthenias gristator* fabricius (Coleoptera : Cerambycidae) from Tambaram, South India. *Indian J. Ent.*, 21: 163–166.

Sharifi, S., I. Javedi and J.A. Chemsak, 1970. Biology of the rosaceae branch borer *Osphranteria coerulescens* (Coleoptera : Cerambycidae). *Ann. Ent. Soc., America*, 63(6): 1515–1520

Singh, P. and G. Prasad, 1985. Poplar stem borer *Apriona cineria* Chevrolat (Coleoptera : Cerambycidae) its biology, ecology and control. *Indian For.*, 3(7): 517–524.

Singh, P. and S. Singh, 1986. *Insect Pests and Diseases of Poplars*. pp 1–74. FRI Press, Dehradun, India.

Solomon, J.D., 1968. Cerambycid borer (Coleoptera) in mulberry. *J. Econ. Ent.*, 65: 1023–1025.

Solomon, J.D., 1972. Biology and habits of the living Beech borer in red oaks. *J. Econ. Ent.*, 65(5): 1307–1310.

Solomon, J.D., 1977. Biology and habits of the oak branch borer (*Goes debilis*). *Ann. Ent. Soc. America*, 70(1): 53–59.

Starzyk, J.R., 1977. Morphology, biology and life-history of *Carilia vivginea* L. (Coleoptera : Cerambycidae). *Z. Angew Ent.*, 83(3): 269–281.

Starzyk, J.R. and M. Lessaer, 1978. Studies on the distribution, morphology and biology of *Tetrops starki* Chevrolat (Coleoptera: Cerambycidae). *Z. Angew. Ent.*, 86 (1): 35–46.

Starzyk, J.R. and M. Lessaer, 1984. Studies on the distribution, morphology, biology and ecology of *Pronocera angusta* (Kriecb) (Coleoptera : Cerambycidae). *Z. Angew Ent.*, 97 (4): 347 – 359.

Stebbing, E.P. 1914. *Indian Forest Insects of Economic Importance: Coleoptera*. pp. 1–648. Jain Brothers Publ., Bhopal, India

Strot, M. 1962. Contribution to knowledge of the bionomics of the poplar borer *Saperda carcharia* L. Pr. Vyzk. Ust. lesn. 25: 85–114.

Tara, J.S. 1983. Investigations on the insect pests of mulberry (*Morus* sp.) in Jammu region of J&K state. *Ph.D. Thesis*, University of Jammu.

Thapa, R.S., 1974. The biology and ecology of the borer *Cyriopalus wallecei* P. (Coleoptera : Cerambycidae). *Sabah. For. Rec.*, 11: 1–33.

Wickman, E., 1968. The biology of the fir tree borer, *Semanotus litigiosus* (Coleoptera : Cerambycidae) in California. *Canadian Ent.*, 100(2): 208–220.

Zangheri, S. and L. Masuti, 1983. Appunti sulla biologia della *Ziraphra rumifitgana* altri lepidolteri tortricidi dannosi all abeto bianco in alto adigi. *Monti Bosshi.*, 14: 147–157.

Chapter 19

Mode of Action of Quinones as Potential Insect Pest Control Agents

Smita Banerjee* and Sujata Magdum**

**Department of Zoology, University of Sagar, Sagar – 470 003 (M.P.)*

***Department of Zoology, S. Nijalingappa College, Bangalore – 560 010 (Karnataka)*

ABSTRACT

The present article is a part of our work on growth and reproduction-inhibiting effects of plumbagin (5-hydroxy-2-methyl-1,4-naphthoquinone), a naturally occurring quinone in plants belonging to the genus *Plumbago*. In order to gain insight into the structure-activity relation, the studies have been rationally extended to other analogues of plumbagin and related quinonoid compounds namely, juglone (5-hydroxy-1,4-naphthoquinone) menadione (2-methyl-naphthoquinone), two benzoquinones *i.e.* 2,3-dimethyl and 2,3,6-trimethyl benzoquinone and a hydroquinone *i.e.* 2,6-dimethyl-hydroquinone. Juglone is also a naturally occurring quinone in the plant *Juglans regia* and a dimethyl derivative of plumbagin. The compounds were obtained from BARC Bombay (*Courtesy:* Dr. A. Banerji, former head, bio-organic division). The work was a first systematic attempt, as far as the authors know, to investigate effects of quinones on different aspects of reproduction. This part of the work deals with mode of chemosterilant activity of the two naturally occurring quinones in the red cotton bug *Dysdercus koenigii*. The results point to inhibition of neuroendocrine function leading to hormonal imbalance which in turn, would interfere with gonadal development. Gradual gonadal necrosis reveals effects on spermatogenesis and egg development. Moreover, metabolism is altered leading to disturbance in the feed back mechanism known to exist between insect neuroendocrine system, fat body and gonads.

Keywords: *Chemosterilant, Growth inhibitor, Insect pest management, Mode of Action, Phytochemical, Quinones.*

Introduction

The need to protect economically important crops from ravages of phytophagous insects by ecologically acceptable methods has led to the development of alternative strategies of insect control using an array of targets (Van Beek *et al.*, 1994). Research in chemical ecology show that plants have developed ingenious strategies employing secondary metabolites (Meinwald *et al.*, 1978; Bell, 1980; Agrawal, 1998). In addition to toxins they produce an armoury of chemicals which make them unsuitable for utilisation by phytophagous insects by imparting repellency, behavioural or biochemical incompatibility. Bowers (1992) has reviewed discovery of increasing number of phytochemicals in the recent past, influencing bioaccumulation, oviposition, feeding, metamorphosis, growth and reproduction in insects. Noteworthy contribution in this field of research has been by Bowers (1985, 1992), Saxena and Tikku (1988, 1990), Banerji (1988, 1990, 1994) and Rembold (1988, 1994). Gujar and Mehrotra (1988), Banerji (1991), Rao *et al.*, (1996), Saxena *et al.*, (1996) etc. have shown that plumbagin, a naphthoquinone naturally occurring in plants belonging to the genus *Plumbago* effects growth and development in insects. This quinone provided a lead structure for development of more effective growth inhibitors in order to gain insight into the structure-activity relation. These compounds were prepared in and obtained from bioorganic division laboratory, BARC, Bombay (*Courtesy:* Dr. A. Banerjee, former head) and we have shown growth regulating effects of plumbagin, another naturally occurring quinone juglone (in *Juglans regia*) and synthetic analogues of plumbagin in the red cotton bug *Dysdercus koenigii* (Banerjee *et al.*, 2001). Joshi and Sehnal (1989) reported adverse effects of plumbagin on egg laying and Magdum *et al.* (2001) have given an elaborate account of reproduction inhibition activity of plumbagin and its analogues.

Inspite of a lot of work on natural products as insect growth regulatory agents, study on mode of their action is scant. A mere knowledge of the type of a phytochemical or its derivative and its nomenclature does not serve the purpose. What is actually required is an understanding of the effect it exerts on growth, development and reproductive capacity of pest insect and pinpointing its mode of action at the cellular/physiological and biochemical level because only then, a plant product or its analogue can be brought into use for practical purpose. Elaborate study on mode of action of azadirachtin "the wonder drug" in Rembold's laboratory is worth mentioning here (Subrahmanyam and Rembold, 1989; Banerjee and Rembold, 1992; Annadurai *et al.*, 1992; Banerjee *et al.*, 1998).

Mode of action of plumbagin has been attributed to inhibition of production of ecdysoids by Mitchell and Smith (1988) and Joshi and Sehnal (1989). La Font (1991) has also shown that plumbagin lowers the ecdysteroid level of the adult insect, thus effecting the ovarian development. Inhibition of chitin synthetase and thus, chitin formation, has been describe by Kubo *et al.*, (1983), Gujar and Mehrotra (1988), and Krishnayya and Rao (1995) to be the mechanism by which plumbagin disrupts; growth and metamorphosis of young stages.

In the present study, emphasis has been laid on the mode of action of the two naturally occurring quinones plumbagin and juglone in reproduction of *Dysdercus koenigii.* The two compounds drastically reduce fecundity and fertility in the bug (Magdum *et al.*, 2001), therefore histopathological condition of the gonads of treated insects have been explored. Egg maturation, vitellogenesis and oviposition in many insects are controlled variously by hormones from neuroendocrine organs or ecdysone produced by ovaries. As in vertebrates, in insects also it appears that the neurosecretory substance consists of a carrier protein with which hormonal peptide is bound (Friedel and Loughton, 1983). Release of hormone is associated with the carrier protein and according to Prabhu (1994) the stainable material in the insect brain/associated glands may be the carrier protein with or without hormone. With this in mind, the histochemical changes in brain/corpora cardiaca of treated bugs have been investigated. Limited information on physiological effect of plumbagin or any other quinone in insects promoted the need to undertake study on their effect on proteins which are the most important physiological macromolecules. In insects, fat body is the major organ of metabolism synthesising/releasing proteins which are sequesteied by gonads through haemolymph for their development. The processes are controlled by hormones which in turn are under the influence of trophic factors from brain. This feed back mechanism is described by Banerjee (1977 a, b; 1979). Considering all the above, it was planned to explore the mode of action of the two quinones through their effect on total protein contents in the organs mentioned above. Carbohydrates also play an important role in maturation of insect gonads and their metabolism is also regulated through hormones. In the present study total glucose content of the fat body, brain and gonads has been estimated in quinone treated bugs and taken as the measure to test changes caused by these compounds in carbohydrate metabolism.

Materials and Methods

The test insect *Dysdercus koenigii* was reared in the laboratory on soaked cotton seeds and okra pods at 27 ± 2°C. The test compounds included plumbagin (5-hydroxy-2-methyl-1, 4-naphthoquinone), juglone (5-hydroxy-1, 4-naphthoquinone), menadione (2-methyl-napthoquinone), 2,3-dimethylbenzoquinone, 2,3,6-trimethyl benzoquinone and 2,6-dimethylhydroquinone. These were dissolved in acetone (1 mg/ml) and topically administered to the bugs using a Hamilton Bonaduz microlitre syringe as in Magdum *et al.*, (2001).

For histopathological changes, 4–6μ thick paraffin sections of gonads were stained using Eosin/Haematoxylin (Davenport, 1961) and those of brain-corpus-cardicum complex using Paraldehyde/Fuchsin (Ewen, 1962).

Protein and glucose were estimated using Biuret and Anthrone reagents respectively following, Plummer (1985). Samples were prepared using chilled Ringer's solution for dissection (Joshi and Muraleedharan, 1993) and proceeding according to Shibko *et al.* (1967). Five males and five females were used for each sample. Results were calculated from three replicates and expressed as average mg protein/glucose per ml of the sample. Students t-test (Ethance, 1977) was applied to determine the significance of changes.

Results and Discussion

LD_{50} of Plumbagin was found to be 28.50 µg and that of Juglone 5.86 µg per adult bug. Gradual necrosis of ovary was seen with increasing amounts of the quinones upto seven days post-treatment period, the period during which egg maturation and first laying is completed. In general, these cytotoxic effects included thickening of the follicular epithelium, pycnosis of nuclei, loosening of the epithelial cells leading to appearance of spaces between follicular epithelium and oocytes, gradual loss of the follicular epithelium cell membrane resulting in a syncytium, detachment of the follicular epithelium from oocytes, appearance of small vacuoles in ooplasm which increased into spaces with increased doses, contraction of ooplasm which in extreme cases shrunk to lie at one peripheral side, gradual loss of differentiation of oocytes in germarium and loosening in its cells giving it a spongy appearance and pycnosis of germinal vesicle at high doses like 30–40 µg plumbagin and 5–10 µg Juglone (Figures 19.1–19.8). In some cases follicular epithelial bridges between oocytes were broken (Figure 19.8). In control females each ovariole contained 10–11 oocytes whereas that of quinone treated females had only 4–6 of them. Karnavar *et al.* (1994) have shown adverse effects on ovarian development during first oocyte cycle in azadirachtin treated *Teleogryllus mitratus.* Similar results were noted in azadirachtin treated ovaries of *Labidura riparia* in a dose dependent manner by Sayah *et al.* (1996). According to Saxena *et al.* (1996) follicular epithelial cells of plumbagin treated *Musca domestica* changed their shape from cuboid to columnar in a usual way and yolk was also deposited, though a marked vacoulisation in cytoplasm was seen in the cells. Thus, according to these authors, yolk deposition, though depleted continues to some extent. The results of the present study are similar to those obtained for chemosterilised *Poecilocerus pictus* by Sahai and Banerjee (1986) and Banerjee and Sahai (1986, 1987).

The follicular epithelium in insect ovary is a very important organ whose function includes transport of nutrients and other substances to the oocyte thereby assuring its proper development. In the red cotton bug, it normally forms a smooth cellular layer of columnar cells. Degeneration of this layer as described above would prevent it from functioning properly and thus uptake of nutrients from haemolymph would be affected. Degeneration of follicular epithelial bridges between oocytes most probably makes them

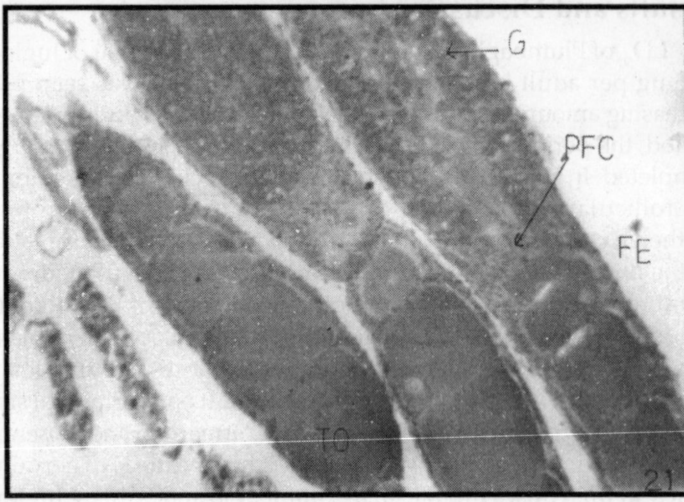

Figure 19.1: L.S. of Ovary of Control *Dysdercus koenigii* Showing Three of the Ovarioles. × 200

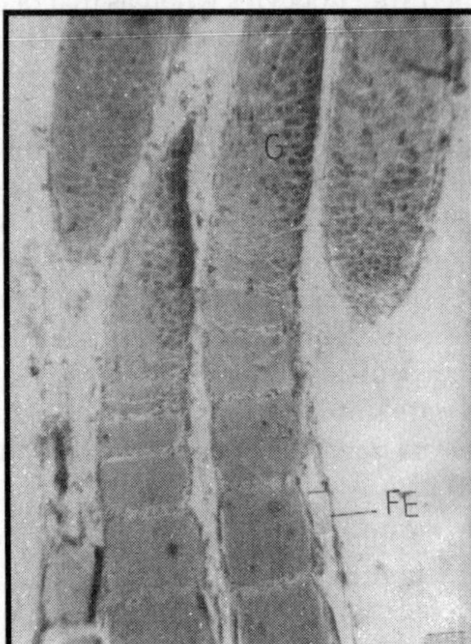

Figure 19.2: L.S. of Ovary After One Day of Treatment with 10 µg Plumbagin Showing Shrinkage of Germarium and its Pycnotic Cells. × 200

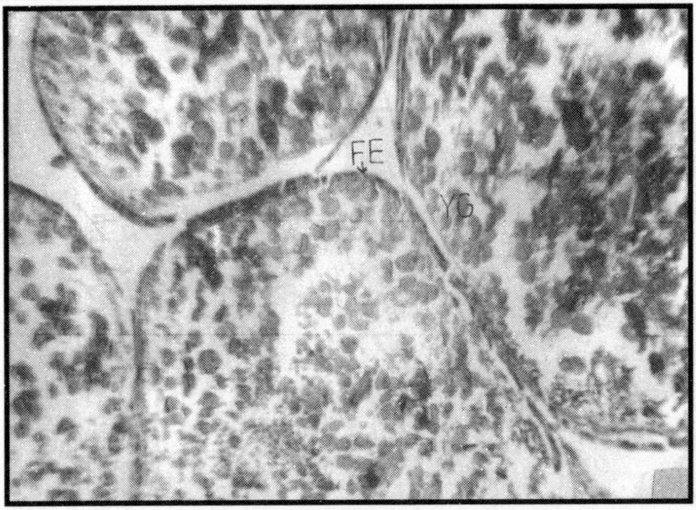

Figure 19.3: T.S. of Ovary of a Control Female Showing Gravid Oocytes with Yolk Granules. × 450

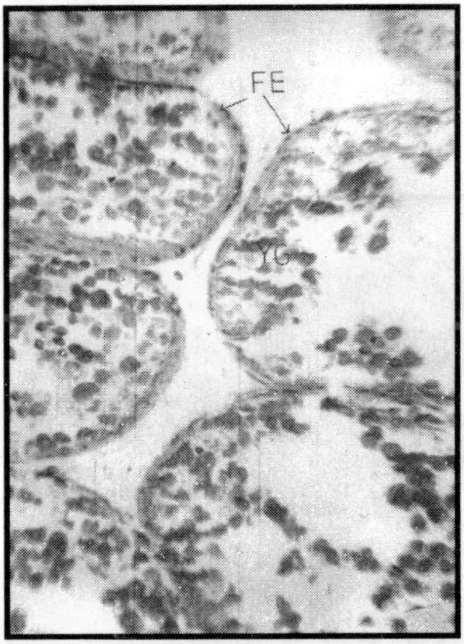

Figure 19.4: The Same of a 20 µg Plumbagin Treated Female after Seven Days Showing Smaller Size of Oocytes, Syncytial Follicular Epithelium and Displaced Yolk Granules Resulting in Spaces in the Ooplasm. × 450

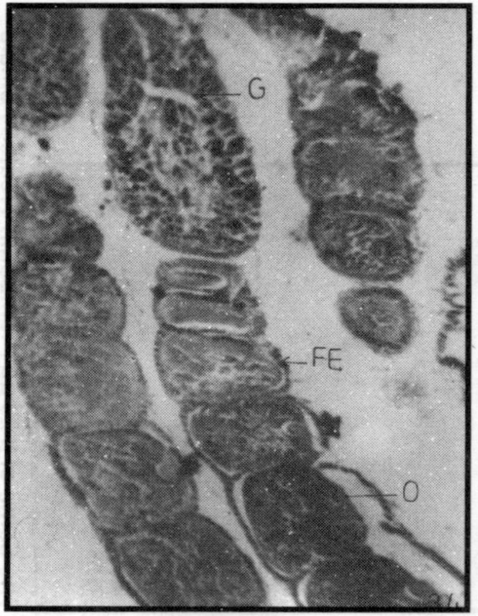

Figure 19.5: L.S. of Ovary after One Day of Treatment with 40 µg Plumbagin Showing Pycnotic Cells and Spaces in the Germarium, Loosening of Follicular Epithelial Cells and its Detachment, Degenerating Ooplasm. × 200

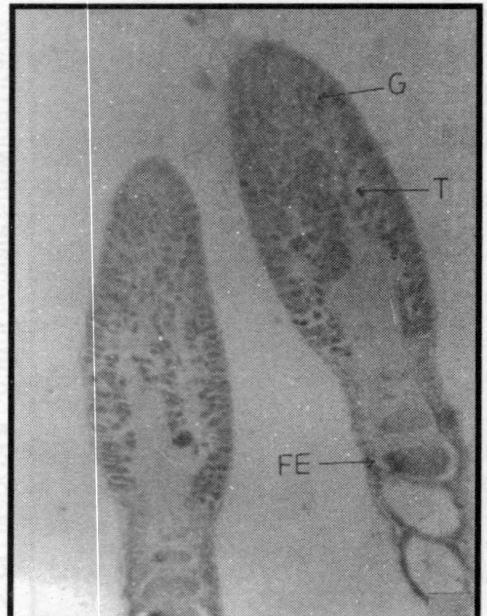

Figure 19.6: L.S. after Five Days of Treatment with 2 µg Juglone Showing Two Ovarioles with Degenerated Germarium, Smaller Developing Oocytes with Shrunken Ooplasm and Degenerating Follicular Epithelium. In some Oocytes Ooplasm is Completely Dissolved. × 200.

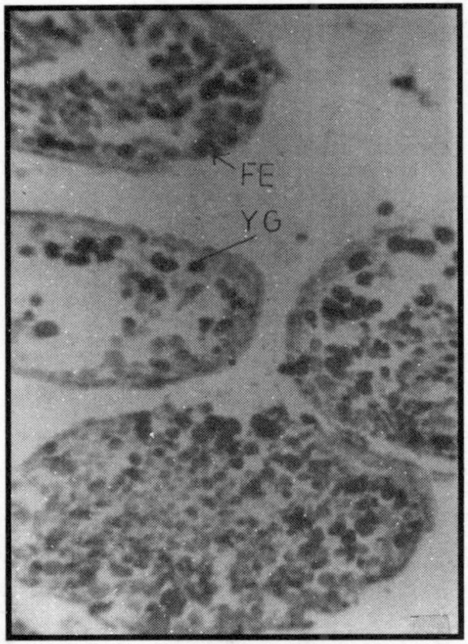

Figure 19.7: T.S. of the Same as Figure 19.6, After Seven Days of Treatment. × 450.

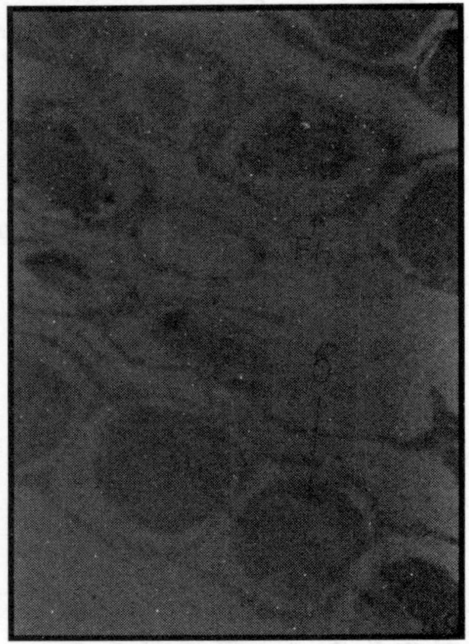

Figure 19.8: L.S. of Ovarioles Five Days After Treatment with 10 µg Juglone. Contracted ooplasm or no ooplasm at all in some oocytes, detached follicular epithelium and broken FE bridge. × 200.

dependent on each other for nutrition and protection. Appearance of vacuoles and progressive formation of large spaces and absence of yolk granules in plumbagin and juglone treated ovaries indicates gradual absorption of oocytes as in chemosterilised *P. pictus* in contrast to observations of Saxena *et al.* (1996) for plumbagin treated *Musca domestica.* Pathological changes described above, during the first reproductive cycle, indicate that the quinones damage the ovaries in an irreversible manner leading to sterility in female bugs.

Pathological changes leading to interference with spermatogenesis were not necessarily dependent on amounts of the quinones. These changes revealed damage to different extent in different stages of germ cell development alongwith damage to germinal epithelium. Testes of treated bugs were even morphologically reduced in size as reported in chemosterilised insects by LaBrecque and Fye (1978) and azadirachtin treated locusts by Linton *et al.* (1997). In general, the cytotoxic effects seen due to quinones in the present study were thickening of outer connective tissue, weak construction of the testicular mass, shrinkage of the whole gonad, appearance of vacuoles in testicular lumen, presence of pycnotic spermatogonial nuclei at high doses, improper development of germ cells and presence of few abnormal hypertrophied sperms and spermatozoa (Figures 19.9–19.13). Saxena *et al.* (1996) have also shown that sperm production in plumbagin treated *Musca domestica* is hamperd due to distortion of cells. The central core of testicular compartment of the treated flies is occupied with early stages of spermatogenesis *i.e.* spermatogonia and spermatocytes, pointing towards delay in cell division. In *M. domestica* Saxena *et al.* (1996) have concluded that at lower concentrations, plumbagin behaves like a spindle poison colchicine by effecting mitosis and at higher quantities it exhibits nucleotoxic and cytotoxic effects. According to Taber and Borkovec (1969) and Champion (1972) an insect chemosterilant exerts its effects directly on gametes or indirectly through interference with the tissue supporting/nourishing gametes. In the present case, the two quinones studied seem to effect the gonads both by direct and indirect ways.

Both the quinones have been found to cause accumulation of paraldehyde-fuchsin positive material in median as well as lateral neurosecretoy cells of pars intercerebralis as compared to uniform granular distribution of this substance in brain of control insects

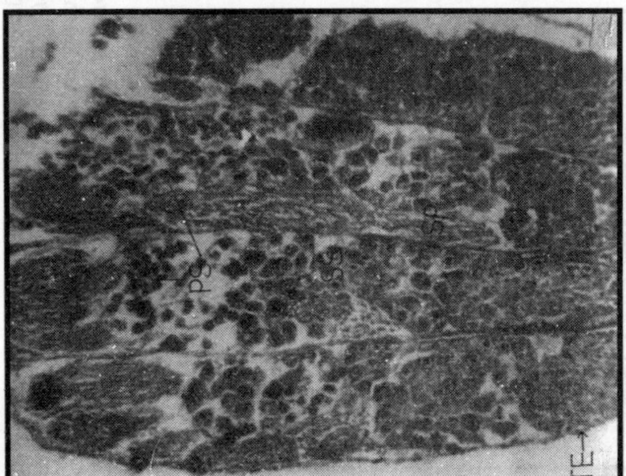

Figure 19.9: Section of Testis of a Control *D. koenigii* Showing Spermatogonia, Primary and Secondary Spermatocytes, Spermatids. × 200

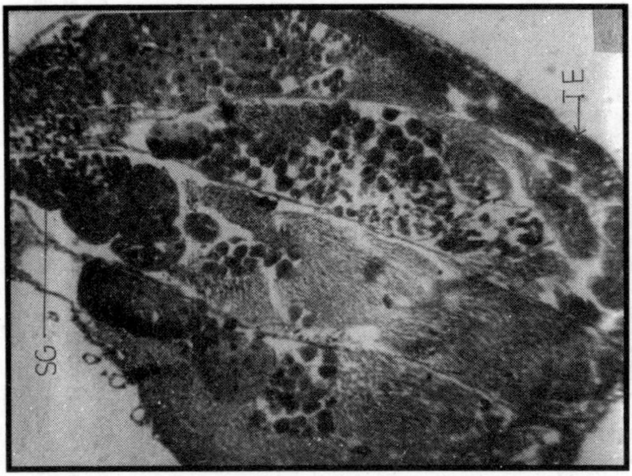

Figure 19.10: Section of Testis Three Days After Treatment with 10 µg Plumbagin Appearance of Spaces, Pycnotic Spermatogonia. × 200

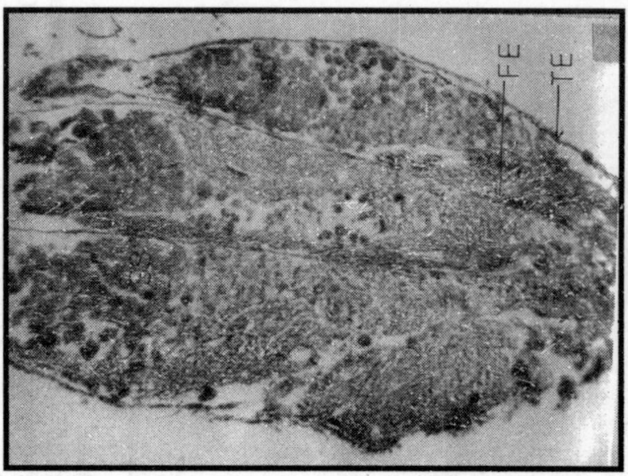

Figure 19.11: The Same After Three Days of Treatment with 20 μg Plumbagin. Loosely arranged germ cells, less number of spermatocytes which are pycnotic and detachment of testicular epithelium. × 200

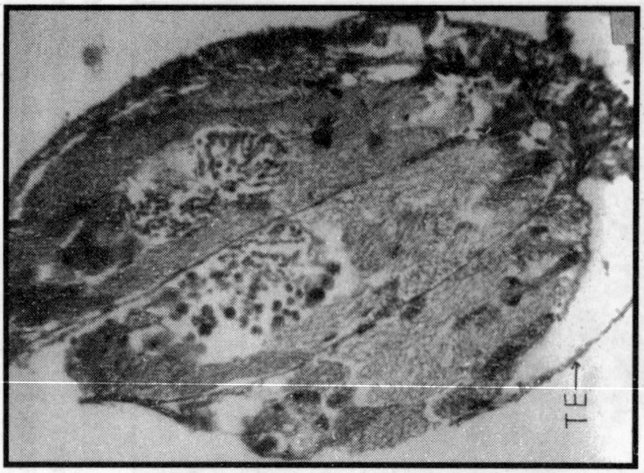

Figure 19.12: Section of Testis Showing Core of Follicles Filled with Loosely Arranged Pycnotic Early Spermatocytes/Spermatogonia, Broken Testicular Epithelium. × 200

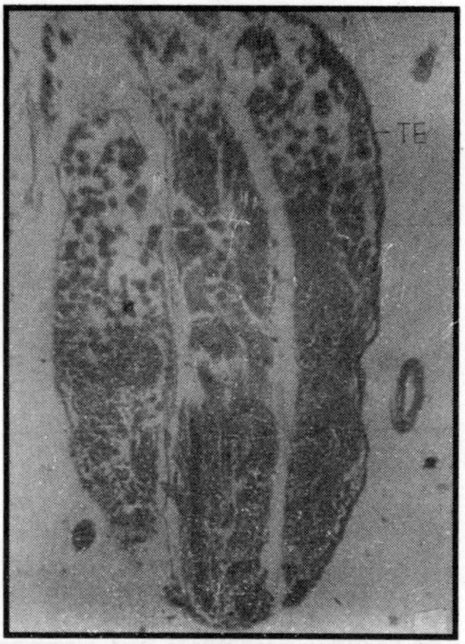

Figure 19.13: Section After Five Days of Treatment with 40 µg Plumbagin. Thick testicular epithelium, loosely disposed degenerating germ cells causing large spaces in the lumen, hypertrophied spermatozoa. × 200.

(Figure 19.14–19.18). This accumulation was not dose dependent but the whole organ became distorted with increased dosage of both the compounds. Loosening of the brain tissue, appearance of space and detachment of neural sheath were the prominent effects noted with both the compounds. Treatment with juglone caused more accumulation of neuorsecretory material in all types of cells of the pars intercerebralis (Figure 19.19) whereas plumbagin caused less accumulation in 'A' type cells only. Juglone also caused accumulation of neurosecretory material at the lowest dose of 1 µg/insect whereas the lowest dose of plumbagin *i.e.* 10 µg/insect did not show any effect.

It is speculated that accumulation of paraldehyde fuchsin positive neuorsecretory material, which according to Friedel and Loughton (1983) and Prabhu (1994), is the carrier protein for brain hormones as described earlier (Introduction), is due to inhibition of synthesis and/or release of neurosecretory material. Similar

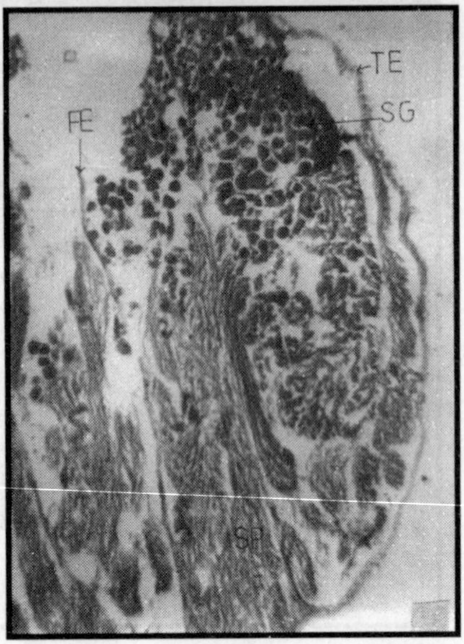

Figure 19.14: The Same After Seven Days of 10 µg Juglone Treatment. Broken follicular epithelium in addition to the anomalies described in Figures 19.10×13. × 200.

Figure 19.15: Section of Brain of Control *D. koenigii*. × 450

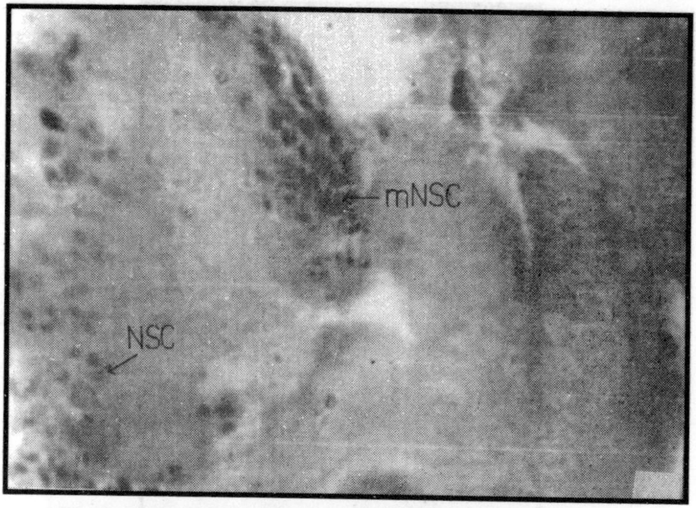

Figure 19.16: The Same of a 30 µg Plumbagin Treated Bug After Three Days Showing Distorted Tissue and Accumulation of Neurosecretory Material in Median and Lateral Neurosecretory Cells. × 450

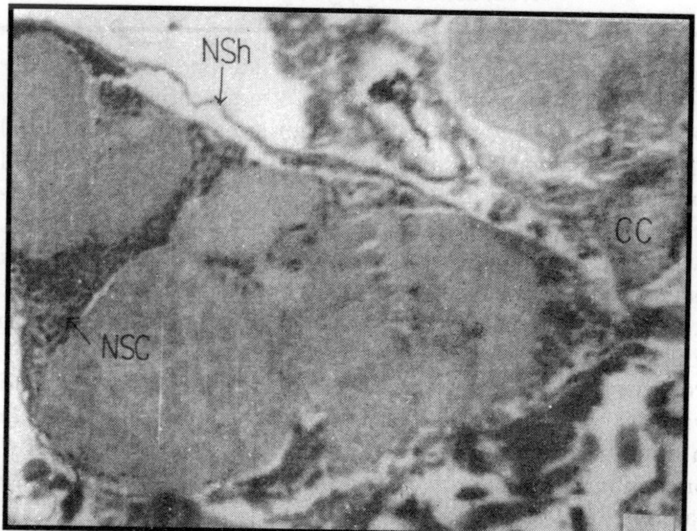

Figure 19.17: The Same After Three Days of Treatment with 1 µg Juglone. Wavy and fragmented neural sheath and deposition of neurosecretory material. × 200.

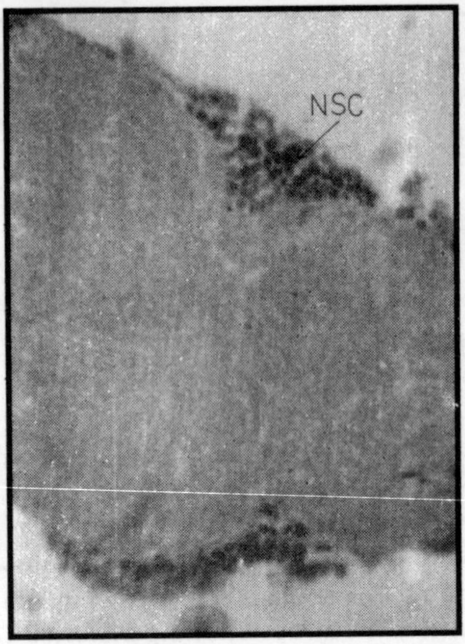

Figure 19.18: Section of Brain After Five Days of Treatment with 2 μg Juglone. × 450

observations were noted in case of ecdysterone treated *Mylabris pustulata* (Kamalakannan *et al.*, 1984), azadirachtin treated *Earias fabia* (Banerjee, 1989) and locusts (Subrahmanyam and Rembold 1989). Krishnayya and Rao (1995) have mentioned delayed build up of neurosecretory material in the brain of plumbagin treated *Helicoverpa armigera* larvae leading to interference at the hormonal level. Thus, anomalies seen in the brain tissue of treated insects in general and in the neurosecretory cells in particular, points towards a restricted release of the trophic factors from brain which in turn, would interfere with release of hormones necessary for physiological processes including reproduction. The blockade of synthesis and/ or release of trophic factors has been explained on the basis of either an inhibition of the feedback control of hormone titres or by a concomitant blockade of release or synthesis of neurosecretory material (Rembold, 1989).

The results of total protein estimation (Tables 19.1–19.4) revealed that in normal (control) bugs there was a gradual decrease in fat body protein contents upto six days of age followed by a rise during

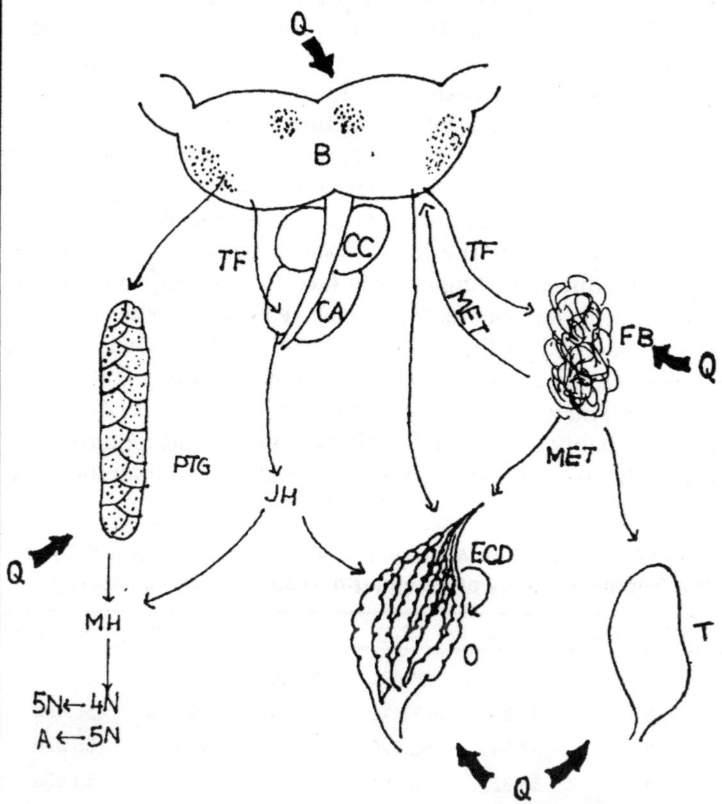

Figure 19.19: Vulnerability of the Vital Organs: Neuroendocrine system, fat body and gonads of a *Dysdercus koenigii* to quinones as derived from present investigation and a possible mode of action.

A: Adult; B: Brain; CA: Corpus allatum; CC: Corpus cardiacum; ECD: Ecdysone; FB: Fat body; JH: Juvenile hormone; MET: Metabolite; MH: Moulting hormone; O: Ovary; PTG: Prothoracic gland; Q: Quinone; T: Testis; TF: Trophic factor; 4N: 4th instar nymph; 5N: 5th instar nymph.

the following days. The decrease in first few days of the reproductive phase has been attributed to utilisation of the metabolite by the gonads. The rise during the consequent days could be due to further synthesis of protein required for the second reproductive phase.

Statistically significant decrease in protein content of treated bugs occurred in dose dependent manner with all the doses of plumbagin and juglone. Krishnayya and Rao (1995) have also shown decline in protein content of fat body as well as haemolymph of plumbagin treated *Helicoverpa armigera.* According to them, disturbance in haemolymph and fat body protein of plumbagin treated insects was also due to lesser feeding since fat body is a multifarious tissue, being a storage organ for nutrients as well as a biosynthetic organ. This might also hold true for *D. koenigii* because we have also witnessed reduced feeding in the quinone treated bugs. Of the two phytochemicals, juglone was more effective. Also, the female fat body was more susceptible, the reason for which is not known. However, presence of more amount of protein in female fat body (Table 19.1, 19.3 control) than in the male counterpart (Table 19.2, 19.4 control) is most probably due to the fact that ovaries would require more proteins during vitellogenesis and egg maturation than the male gonads.

Table 19.1: Changes in Average Protein Content (mg/ml ± S.D) of the Female Fat Body of Plumbagin Treated *Dysdercus koenigii*

Dose PTP (Days)	*Control*	*10 µg*	*20 µg*	*30 µg*	*40 µg*
1	59.25	58.49c	58.26b	57.89b	57.56b
2	57.99	57.71c	57.34c	57.11b	56.89b
3	57.24	56.82c	56.22b	55.65b	55.27b
4	55.92	55.15b	54.70a	54.10a	53.53a
5	49.90	29.63a	29.20a	28.81a	28.56a
6	15.20	14.93c	14.26b	13.66a	12.96a
7	17.61	17.02c	16.67b	16.27b	15.77a

S.D. (Standard deviation): Less than 1 in all cases; PTP: Post treatment period. Values marked 'a' significant at $p < 0.0001$; Values marked 'b' significant at $p < 0.001$; Values marked 'c' significant at $p < 0.01$.

Protein content in ovaries is at much higher level than in testes (Tables 19.5 and 19.7) throughout the life cycle. A gradual increase in total protein was seen upto five days of emergence in control insects because the ovaries attain maturity during this period, proteins are synthesised and incorporated into the oocytes. Although fat body is the primary site of vitellogenic synthesis in insects (Wyatt *et al.*, 1987), Chang (1993) has shown that ovary itself may be the site of vitellogenin synthesis. A drop in protein content was observed in control *D. Koenigii* ovary at 6–7 days of age which is obvious since egg laying in this bug takes place exactly during this age. In both plumbagin and juglone treated insects there was a gradual fall in total protein content of ovaries with increased dosage and post treatment period indicating inhibition of protein synthesis in the organs alongwith interference in absorption from haemolymph.

Table 19.2: Changes in Average Protein Content (mg/ml ± S.D) of the Male Fat Body of Plumbagin Treated *Dysdercus koenigii*

Dose PTP (Days)	*Control*	*10 μg*	*20 μg*	*30 μg*	*40 μg*
1	34.70	34.23c	33.68c	33.13a	32.81a
2	33.28	32.74c	32.36b	32.11b	31.74b
3	32:61	32.01a	31.24a	30.82a	30.32a
4	28.86	28.11c	27.66a	27.56a	24.75a
5	24.50	24.28d	23.89e	23.66c	23.33c
6	08.91	07.79c	07.41a	06.82a	06.22a
7	15.82	15.17c	14.68b	13.83c	13.08b

S.D. (Standard deviation): Less than 1 in all cases; PTP: Post treatment period.

Values marked 'a' significant at $p < 0.0001$; Values marked 'b' significant at $p < 0.001$; Values marked 'c' significant at $p < 0.01$; Value marked 'd' significant at $p < 0.05$; Value marked 'e' significant at $p < 0.1$.

Table 19.3: Changes in Average Protein Content (mg/ml ± S.D) of the Female Fat Body of Juglone Treated *Dysdercus koenigii*

Dose PTP (Days)	*Control*	*10 μg*	*20 μg*	*30 μg*	*40 μg*
1	69.20	62.36a	59.48a	52.71a	49.98a
2	62.24	36.05a	35.00a	29.55a	27.56a
3	54.80	27.84a	27.41a	24.65a	23.43a
4	53.06	25.27a	25.02a	24.55a	23.21a
5	51.17	26.04a	25.57a	25.05a	24.78a
6	21.74	19.55a	18.71a	12.64a	10.00a
7	59.40	17.87e	17.34a	11.72a	07.59a

S.D. (Standard deviation): Less than 1 in all cases; PTP: Post treatment period.

Values marked 'a' significant at $p < 0.0001$; Value marked 'e' significant at $p < 0.1$.

Table 19.4: Changes in Average Protein Content (mg/ml ± S.D) of the Male Fat Body of Juglone Treated *Dysdercus koenigii*

Dose PTP (Days)	*Control*	*1 μg*	*3 μg*	*5 μg*	*10 μg*
1	35.72	34.25a	33.68a	32.61a	32.26a
2	32.06	31.09a	30.75a	29.40a	26.74a
3	25.03	24.60b	24.40b	23.73a	22.69a
4	24.53	23.13a	22.84a	22.51a	22.24a
5	15.72	15.62e	15.30c	15.17a	14.30a
6	08.53	08.01a	07.94c	07.34b	06.74d
7	13.46	10.85a	10.60a	10.10a	09.98a

S.D. (Standard deviation): Less than 1 in all cases; PTP: Post treatment period.

Values marked 'a' significant at $p < 0.0001$; Values marked 'b' significant at $p < 0.001$; Values marked 'c' significant at $p < 0.01$; Value marked 'd' significant at $p < 0.02$; Value marked 'e' significant at $p < 0.5$.

Table 19.5: Changes in Average Protein Content (mg/ml ± S.D) of the Ovary of Plumbagin Treated *Dysdercus koenigii*

Dose PTP (Days)	*Control*	*10 μg*	*20 μg*	*30 μg*	*40 μg*
1	14.35	14.05d	13.71c	13.33c	12.74b
2	18.31	16.84c	15.97b	15.55b	15.07b
3	24.43	19.80a	19.28a	18.76a	18.23a
4	36.37	24.53a	23.93a	23.46a	22.91a
5	45.57	34.01a	33.38a	33.04a	32.69a
6	21.52	21.29d	20.72c	20.37b	20.03c
7	09.13	07.39b	06.77b	06.37b	05.80b

S.D. (Standard deviation): Less than 1 in all cases; PTP: Post treatment period.

Values marked 'a' significant at $p < 0.0001$; Values marked 'b' significant at $p < 0.001$; Values marked 'c' significant at $p < 0.01$; Value marked 'd' significant at $p < 0.05$.

In testes of control bugs also, there was a gradual increase in protein content upto four days of emergence. (Tables 19.6 and 19.8) followed by a drop on 5–7 days. This may be due to the reason that proteins are required in large amounts during maturation. Total protein content was always less in quinone treated insects indicating inhibition of uptake by the male gonads. While studying effects on gonads, it was difficult to determine which of the two phytochemicals was more effective as the per cent change in comparison to controls was almost of the same level.

Table 19.6: Changes in Average Protein Content (mg/ml ± S.D) of the Testis of Plumbagin Treated *Dysdercus koenigii*

Dose PTP (Days)	*Control*	*10 μg*	*20 μg*	*30 μg*	*40 μg*
1	09.78	07.14b	06.37b	05.30b	05.15b
2	10.10.	09.60c	08.73b	08.56c	07.76b
3	11.17	10.03a	09.88a	09.28a	07.96a
4	27.51	24.53a	23.91a	23.43a	23.13a
5	15.97	14.43b	13.88b	13.09b	12.69b
6	09.63	09.11d	08.56c	07.66b	07.39b
7	09.36	08.76c	08.16b	07.46b	07.14b

S.D. (Standard deviation): Less than 1 in all cases; PTP: Post treatment period.

Values marked 'a' significant at $p < 0.0001$; Values marked 'b' significant at $p < 0.001$; Values marked 'c' significant at $p < 0.01$; Value marked 'd' significant at $p < 0.02$.

Table 19.7: Changes in Average Protein Content (± S.D) of the Ovary of Juglone Treated *Dysdercus koenigii*

Dose PTP (Days)	*Control*	*1 μg*	*3 μg*	*5 μg*	*10 μg*
1	10.54	10.39c	09.98a	09.90a	09.25a
2	12.94	11.79a	10.97a	10.22a	09.52a
3	22.47	20.69a	20.20a	15.45a	13.78a
4	25.57	25.07b	24.68a	21.54a	15.00a
5	31.02.	27.79a	27.76a	24.78a	19.38a
6	20.45	20.36c	20.45b	18.46a	14.88a
7	06.49	06.29c	06.04a	05.52b	04.64a

S.D. (Standard deviation): Less than 1 in all cases; PTP: Post treatment period.

Values marked 'a' significant at $p < 0.0001$; Values marked 'b' significant at $p < 0.001$; Values marked 'c' significant at $p < 0.01$.

Table 19.8: Changes in Average Protein Content (mg/ml ± S.D) of the Testis of Juglone Treated *Dysdercus koenigii*

Dose PTP (Days)	*Control*	*1 μg*	*3 μg*	*5 μg*	*10 μg*
1	00.57	00.52d	00.47c	00.27b	00.09a
2	01.57	01.42e	01.32c	00.93b	00.45a
3	02.26	02.04c	01.77a	01.44a	00.67a
4	07.39	06.99b	05.57b	04.88a	02.44a
5	03.91	03.65c	03.45b	02.79b	02.36b
6	02.36	01.99e	01.82b	01.32a	00.35a
7	01.87	01.77c	01.34c	00.70a	00.27a

S.D. (Standard deviation): Less than 1 in all cases; PTP: Post treatment period.

Values marked 'a' significant at $p < 0.0001$; Values marked 'b' significant at $p < 0.001$; Values marked 'c' significant at $p < 0.01$; Value marked 'd' significant at $p < 0.02$; Value marked 'e' significant at $p < 0.5$.

Table 19.9: Changes in Average Protein Content (mg/ml ± S.D) of the Brain of Plumbagin Treated *Dysdercus koenigii*

Dose PTP (Days)	*Control*	*10 μg*	*20 μg*	*30 μg*	*40 μg*
1	28.11	24.63a	24.28a	23.68a	23.29a
2	27.49	24.23a	23.63a	23.01a	22.69a
3	25.05	23.93b	23.26b	22.76a	22.39a
4	24.48	19.50a	18.76a	18.56a	17.66a
5	21.89	18.13b	17.51a	17.14a	16.77a
6	14.38	11.02b	12.19a	09.88a	08.66a
7	18.83	13.53a	13.21a	12.79a	12.41a

S.D. (Standard deviation): Less than 1 in all cases; PTP: Post treatment period.

Values marked 'a' significant at $p < 0.0001$; Values marked 'b' significant at $p < 0.001$.

The brain of insects is known to regulate gonadal development in insects through the endocrine sequence (Banerjee, 1977 a, b; '79; Sahai and Banerjee, 1984). In recent years information on the endocrine system of several insects has established more than one fact which has a direct bearing on oogenesis and folliculogenesis. This may be extended to the fat body metabolism under insect hormonal regulation as already mentioned. Adverse effects of plumbagin on ecdysoids in developmental stages of insects is now well established (Banerjee *et al.*, 2001) and a similar action of juglone is envisaged. Thus, quinones may also cause disturbance in levels of ecdysone in adult insects. In recent years several functions for ecdysteroid function in insect ovarian development has been demonstrated (La Font, 1991). Fallon *et al.* (1974) and Hagedom *et al.* (1975) have shown that ecdysone is synthesised and released from ovaries in mosquitoes. When secreted, this ecdysone stimulates the fat body to synthesise yolk protein (Fallon *et al.*, 1974). The follicular cells are identified as source of hormones in locusts (Goltzene *et al.*, 1978). Less is known about involvement of ecdysone in male reproduction. According to Happ (1992) ecdysteroids cause several diverse effects in males like increased mitotic levels in spermatogonia of *Rhodnius prolixus*, initiation of meiosis in diapausing lepidopteran larvae and mitosis of male accessory glands in beetles. Loeb's laboratory (Chang, 1993) has demonstrated that the testes of some insects can secrete ecdysteroids. Secretion of ecdysone is under the control of brain hormone. Similarly juvenile hormone synthesised by the corpora allata is involved in vitellogenesis and is under the control of neuropeptide neurohormones secreted by brain.

Table 19.10: Changes in Average Protein Content (mg/ml ± S.D) of the Brain of Juglone Treated *Dysdercus koenigii*

Dose PTP (Days)	*Control*	*1 μg*	*3 μg*	*5 μg*	*10 μg*
1	26.05	24.25a	23.26a	16.97a	11.72a
2	25.82	22.19a	17.79a	15.00a	11.69a
3	21.69	18.28a	15.72a	14.18a	09.80a
4	16.22	14.85a	13.58a	11.74a	07.41a
5	16.02	14.65a	11.89a	09.86a	05.03a
6	11.52	09.73a	09.38a	07.49a	02.91a
7	12.81	11.92a	10.40a	10.12a	02.94a

S.D. (Standard deviation): Less than 1 in all cases; PTP: Post treatment period.

Values marked 'a' significant at $p < 0.0001$.

An age wise decrease in total protein content was noted in brain of control bugs from one to seven days (Tables 19.9 and 19.10). This might be due to the fact that the insect undergoes progressive developmental changes during the first reproductive phase and total proteins of brain which include a multiple array of polypeptides like enzymes, neurohormones, carrier proteins etc. are constantly released during this period. Both the quinones induced decrease in protein content in comparison to controls, indicating their interference with synthesis of brain proteins. The rate of decline was more in juglone treated bugs confirming once again that it was more effective than plumbagin and exerted its effects at very low doses. According to Krishnayya and Rao (1995) decrease in neurosecretory material in normal *Helicoverpa armigera* larvae is due to cessation of feeding and slow build up of neurosecretory material induced by plumbagin treatment was again, due to less feeding. Also, according to them, a delayed build up of neurosecretory substance in larvae and lesser corpora cardiaca volume are due to interference of plumbagin with hormonal level of the insects. Similar observations were reported earlier by Subrahmanyam and Rembold (1989) and Subrahmanyam *et al.*, (1989) for azadirachtin treated *Schistocerca gregaria* and by Gupta and Rao (1990) for *Spodoptera litura*.

Spectrophotometric observations showed a dose dependent decrease in total glucose content in different organs mentioned earlier, after treatment with both the quinones. In the control group a gradual increase in glucose content was noted in the fat body of both female and male *D. koenigii* during the first reproductive phase (Tables 19.11–19.14). This was due to accumulation of carbohydrates in insects which is directly proportional to feeding habit of an individual as well as due to gonadal development during this period. It was also found that female fat body possess lesser amount of glucose as compared to male fat body. May be sperm maturation requires more carbohydrates as described by Sugai (1967); Banerjee (1977 b). With increased amounts of quinones administered, glucose content of the fat body decreased at a given post treatment period in comparison to that of the controls, juglone being more effective than plumbagin in this case also (Table 19.11–19.14). This decrease could be due to low feeding rate after treatment with quinones.

Ovarian glucose content in the control group of insects showed progressive increase upto four days of emergence and a sharp rise was noted after five days (Tables 19.15-19.18). This may be because the glucose level is associated with morphological ooplasmic events during egg maturation. Similar results have been reported by Cummings *et al.* (1971) by ultrastructural and cytochemical studies of ovaries of *Drosophila melanogaster.* Both plumbagin and juglone caused depletion of glucose content in the female as well as male gonads, in comparison to those of control bugs. However, testes were more susceptible than the ovaries because almost at each dose and post-treatment period, per cent change in relation to controls was more in testes than in ovaries (Tables 19.15–19.18).

Table 19.11: Changes in Glucose Content (mg/ml ± S.D) of the Female Fat Body of Plumbagin Treated *Dysdercus koenigii*

Dose PTP (Days)	*Control*	*10 μg*	*20 μg*	*30 μg*	*40 μg*
1	10.65	09.96b	09.45b	09.00a	08.55a
2	12.68	10.73c	10.58b	10.43c	10.13e
3	13.32	12.78b	12.54b	12.18b	11.94b
4	16.20	15.42c	15.00b	14.70b	14.57b
5	16.92	19.08b	16.08b	15.48b	15.06b
6	25.28	21.76b	21.60b	21.88b	21.28b
7	27.29	25.93e	23.89a	23.29a	22.87a

S.D. (Standard deviation): Less than 1 in all cases; PTP: Post treatment period.

Values marked 'a' significant at $p < 0.0001$; Values marked 'b' significant at $p < 0.001$; Values marked 'c' significant at $p < 0.01$; Value marked 'e' significant at $p < 0.1$.

Table 19.12: Changes in Glucose Content (mg/ml ± S.D) of the Male Fat Body of Plumbagin Treated *Dysdercus koenigii*

Dose PTP (Days)	*Control*	*10 μg*	*20 μg*	*30 μg*	*40 μg*
1	11.58	11.22b	10.86b	10.56b	10.26a
2	13.62	13.20b	12.90b	12.60b	12.12a
3	15.18	14.88c	14.70b	14.46b	14.04b
4	16.80	16.38d	16.02a	15.54a	15.30a
5	19.14	18.96d	16.92b	16.68b	16.38b
6	22.92	20.82b	20.58b	20.46b	20.22b
7	27.94	27.23b	26.59a	26.32b	24.90a

S.D. (Standard deviation): Less than 1 in all cases; PTP: Post treatment period.

Values marked 'a' significant at $p < 0.0001$; Values marked 'b' significant at $p < 0.001$; Values marked 'c' significant at $p < 0.01$; Value marked 'd' significant at $p < 0.02$.

Table 19.13: Changes in Glucose Content (mg/ml ± S.D) of the Female Fat Body of Juglone Treated *Dysdercus koenigii*

Dose PTP (Days)	*Control*	*1 μg*	*3 μg*	*5 μg*	*10 μg*
1	14.00	13.60c	13.20a	13.00a	12.40a
2	14.20	13.70a	13.60c	13.20f	12.90a
3	14.50	13.90f	13.70c	13.40c	12.90b
4	15.20	14.40c	13.90b	13.60c	13.20b
5	16.10	14.90b	14.60b	14.40b	14.10b
6	16.90	15.40a	15.00b	14.60a	14.20a
7	17.00	16.40b	16.00b	15.60a	15.00a

S.D. (Standard deviation): Less than 1 in all cases; PTP: Post treatment period. Values marked 'a' significant at $p < 0.0001$; Values marked 'b' significant at $p < 0.001$; Values marked 'c' significant at $p < 0.01$; Value marked 'f' significant at $p < 0.5$.

Table 19.14: Changes in Glucose Content (mg/ml ± S.D) of the Male Fat Body of Juglone Treated *Dysdercus koenigii*

Dose PTP (Days)	*Control*	*1 μg*	*3 μg*	*5 μg*	*10 μg*
1	16.10	15.80d	15.50c	14.90b	14.60b
2	16.40	16.00b	15.50b	15.10b	14.80a
3	16.90	16.40d	16.20b	15.60b	14.90b
4	17.40	16.80b	16.40b	15.70b	15.20a
5	17.70	17.10c	16.70b	16.20b	15.60b
6	18.20	17.40b	16.90b	16.50b	16.10b
7	18.40	17.80c	17.50b	17.00b	16.70b

S.D. (Standard deviation): Less than 1 in all cases; PTP: Post treatment period. Values marked 'a' significant at $p < 0.0001$; Values marked 'b' significant at $p < 0.001$; Values marked 'c' significant at $p < 0.01$; Value marked 'd' significant at $p < 0.02$.

Carbohydrate (glucose) content of the brain in control insects increased with age upto five days, but after six days a fall was noted followed by another rise. The reason for this is not known. The quinones caused depletion of glucose content with increased doses and post-treatment period, just as in fat body and gonads and juglone being more effective than plumbagin (Tables 19.19–19.20). The results indicate disturbed metabolism due to these compounds.

Table 19.15: Changes in Glucose Content (mg/ml ± S.D) of the Ovary of Plumbagin Treated *Dysdercus koenigii*

Dose PTP (Days)	*Control*	*10 μg*	*20 μg*	*30 μg*	*40 μg*
1	04.28	04.13c	03.93b	03.85b	03.73b
2	04.38	04.23d	04.03b	03.88a	03.75a
3	05.05	04.38a	04.28a	04.08a	03.90a
4	05.40	05.23a	05.23a	05.25a	05.00a
5	07.90	07.05a	06.93a	06.75a	06.48a
6	11.68	07.95a	07.38b	07.03a	06.75a
7	12.40	08.02a	07.90a	07.13a	06.90a

S.D. (Standard deviation): Less than 1 in all cases; PTP: Post treatment period.

Values marked 'a' significant at $p < 0.0001$; Values marked 'b' significant at $p < 0.001$; Values marked 'c' significant at $p < 0.01$; Value marked 'd' significant at $p < 0.02$.

Table 19.16: Changes in Glucose Content (mg/ml ± S.D) of the Testis of Plumbagin Treated *Dysdercus koenigii*

Dose PTP (Days)	*Control*	*10 μg*	*20 μg*	*30 μg*	*40 μg*
1	04.50	04.40f	04.27b	04.58b	04.80c
2	05.20	04.45a	04.58a	04.38a	04.28a
3	08.48	05.13a	05.00a	04.88a	04.83a
4	09.85	07.16a	06.98a	06.83a	06.67a
5	11.15	07.20a	07.00a	07.83a	06.75a
6	13.93	08.15a	07.13a	06.98a	06.80a
7	14.23	08.33a	07.43b	07.03a	06.85a

S.D. (Standard deviation): Less than 1 in all cases; PTP: Post treatment period.

Values marked 'a' significant at $p < 0.0001$; Values marked 'b' significant at $p < 0.001$; Values marked 'c' significant at $p < 0.01$; Value marked 'f' significant at $p < 0.5$.

Table 19.17: Changes in Glucose Content (mg/ml ± S.D) of the Ovary of Juglone Treated *Dysdercus koenigii*

Dose PTP (Days)	*Control*	*1 µg*	*3 µg*	*5 µg*	*10 µg*
1	07.55	07.45e	07.30e	07.10c	06.65c
2	07.65	07.54c	07.40c	07.15c	06.90b
3	08.11	07.65a	07.55a	07.35a	07.25a
4	08.30	08.10a	07.75c	07.60c	07.35a
5	08.50	08.30a	08.00c	07.70a	07.45a
6	08.70	08.40c	08.20c	08.00c	07.55a
7	08.75	08.40a	08.25a	08.00a	07.90a

S.D. (Standard deviation): Less than 1 in all cases; PTP: Post treatment period.

Values marked 'a' significant at p < 0.0001; Values marked 'b' significant at p < 0.001; Values marked 'c' significant at p < 0.01; Value marked 'e' significant at p < 0.10.

Table 19.18: Changes in Glucose Content (mg/ml ± S.D) of the Testis of Juglone Treated *Dysdercus koenigii*

Dose PTP (Days)	*Control*	*1 µg*	*3 µg*	*5 µg*	*10 µg*
1	07.55	07.45f	07.00c	06.65c	06.10b
2	08.15	07.70d	07.20c	07.05b	06.90b
3	08.45.	07.85b	07.30a	07.05a	07.00a
4	08.70	08.00b	07.70b	07.55a	07.15a
5	11.90	08.15a	07.90a	07.80a	07.25a
6	12.65	08.55a	08.30a	08.10a	07.50a
7	14.55	09.65a	09.15a	08.80a	08.55a

S.D. (Standard deviation): Less than 1 in all cases; PTP: Post treatment period.

Values marked 'a' significant at p < 0.0001; Values marked 'b' significant at p < 0.001; Values marked 'c' significant at p < 0.01; Values marked 'd' significant at p < 0.02; Value marked 'f' significant at p < 0.5.

The present investigation was planned to study the mode of naturally occurring quinones as reproduction inhibiting agents. Thus, it is a novel approach to insect pest control in which interference with reproduction, the process which in fact gives the pest status to an insect, could play a key role in its control. At the end of the investigation, the objective has been satisfactorily fulfilled, though a lot more of the possibilities are still there to set up various experiments, specially to study the mode of action and follow up studies. This could provide ecofriendly insect control agents in insect pest management (IPM) programme.

Table 19.19: Changes in Glucose Content (mg/ml ± S.D) of the Brain Plumbagin Treated *Dysdercus koenigii*

Dose PTP (Days)	*Control*	*10 μg*	*20 μg*	*30 μg*	*40 μg*
1	15.40	14.80b	14.50b	14.10b	13.90b
2	15.80	14.70b	14.40b	13.90b	13.60a
3	16.00	14.40a	14.10a	13.50a	13.30a
4	16.30	14.20a	13.90a	13.30a	13.10a
5	16.50	14.20a	13.50a	13.00a	12.10a
6	16.10	12.90a	10.30a	09.10a	08.70a
7	16.20	09.30a	08.80a	08.50a	07.90a

S.D. (Standard deviation): Less than 1 in all cases; PTP: Post treatment period.

Values marked 'a' significant at $p < 0.0001$; Values marked 'b' significant at $p < 0.001$.

Table 19.20: Changes in Glucose Content (mg/ml ± S.D) of the Brain of Juglone Treated *Dysdercus koenigii*

Dose PTP (Days)	*Control*	*1 μg*	*3 μg*	*5 μg*	*10 μg*
1	15.60	15.30e	15.10e	14.80c	14.50b
2	16.00	15.20c	14.90c	14.80c	14.40c
3	16.40	15.00a	14.70a	14.30a	13.90d
4	16:60	14.90a	14.40c	14.10c	13.90d
5	16.70	14.50a	14.20c	13.80a	13.50d
6	15.10	14.40c	14.10c	13.60c	13.20d
7	17.50	13.30a	12.60a	12.10a	11.60d

S.D. (Standard deviation): Less than 1 in all cases; PTP: Post treatment period.

Values marked 'a' significant at $p < 0.0001$; Values marked 'b' significant at $p < 0.001$; Values marked 'c' significant at $p < 0.01$; Value marked 'd' significant at $p < 0.05$; Value marked 'e' significant at $p < 0.10$.

Acknowledgement

Sanction of a BRNS (DAE) research project to Smita Banerjee and a Junior Research fellowship to Sujata Magdum is gratefully acknowledged. The authors are very much thankful to Dr. A. Banerji (former head, bioorganic division, BARC, Bombay) for the quinone compounds and for his valuable discussions.

Key to Lettering

CC: Corpus cardiacum; FE: Follicular epithelium; G: Germarium; mNSC: Neurosecretory material; NSC: Neurosecretory cell; NSh: Neural sheath; O: Ooplasm; PFC: Prefollicular cells; PS: Primary spermatocyte; SG: Spermatogonia; SP: Spermatocyte; SS: Secondary spermatocyte; T: Trophocytes; TE: Testicular epithelium; TO: Terminal oocyte; YG: Yolk granules.

References

Agrawal, A.A., 1998. Induced responses to herbivory and increased plant performance, *Nature*, 270: 1201–1202.

Annadurai, R. S., Oberthuer, W. and Rembold, H., 1992. High resolution two-dimensional mapping of tissue specific polypeptides in the desert locust *Schistocerca gregaria*. *Electrophoresis*, 13: 456.

Banerjee, S., 1977a. Observations on the neuroendocrine complex and neurosecretory cells of castrated *Poecilocerus pictus* Fab. *Zool. Jb. Anat. Bd.*, 975: 45–53.

Banerjee, S., 1977b. Effect of the gonads on lipid and glycogen deposition in the fat body of *Poecilocerus Pictus*. *Zool. Anz.*, 167.

Banerjee, S., 1979. A histochemical study on the control of lipid metabolism in the fat body of *P. pictus*. *Curr. Sci.*, 48(24): 1090–1092.

Banerjee, S., 1989. A possible mode of action of azadirachtin as a growth regulator in *Earias fabia* Stahl. In: *Environmental Risk Assessment and Management*, (Eds.) Dalela, R.C., Deshmukh, P.B., Mathai, T. and Pillai, K. S. Academy of Environmental Biology, Muzaffarnagar, p. 65–75.

Banerjee, S. and Sahai, Y.N., 1983. Neuroendocrine control of oocyte development in *P. pictus* (Acrididae : Orthoptera). *Curr. Sci.*, 52: 432–434.

Banerjee, S. and Sahai, Y.N., 1986. Histopathological effects of Tepa and Metepa on the ovaries of *P. pictus*. *Folia Morphol.*, 34: 206–212.

Banerjee, S. and Sahai, Y.N., 1987. Histological effects of apholate on the ovary of *P. pictus.* In: *Environment and Ecotoxicology,* (Eds.) Dalela, R.C., Sahai, Y.N. and Gupta, S. Academy of Environmental Biology, Muzaffarnagar, p. 239–242.

Banerjee, S. and Rembold, H., 1992. Azadirachtin 'A' interferes with serotonin pools in the neuroendocrine system of locusts. *Naturwissen.,* 79: 81–84.

Banerjee, S., Oberthuer, W. and Rembold, H., 1998. Effect of Azadirachtin on GABA pools in neuroendocrine organs of locusts. In: *Current Topics in Pollen Spore and Biopollution Research,* (Eds.) Dutta, N.M.; Gupta-Bhattacharya, S.Y., Mandal, S. and Bhattacharya, K.N. Research Periodicals and Book Publishing House, USA/UK/Taiwan, p. 173–177.

Banerjee, S., Magdum, S., Kalena, G.P. and Banerji, A. 2001. Insect growth regulatory activity of naturally occurring quinones and their derivatives in *Dysdercus koenigii* Fabr. (Hem., Pyrrhocoridae). *J. Appl. Entomol.,* 125: 25–30.

Banerji, A., 1988. Alternate strategies for insect control: Natural products as potential biocides. Technical report, BARC–1416, BARC, Bombay, p. 1–62.

Banerji, A., 1991. Insect control by biochemical procedures. Lessons from the plant kingdom. In: *Insect Chemical Ecology,* (Ed.) Hardy I. Academia, Praha, p. 235–244.

Banerji, A., 1994. Insect-plant interactions: Natural insect control agents. In: *Recent Advances In Insect Physiology and Toxicology,* (Ed.) Gujar, G.T. Agricole Publishing Company, New Delhi, p. 41–47.

Bell, E.A., 1980. The possible significance of secondary compounds. In: *Encyclopaedia of Plant Physiology, New Series, Vol. 7: Secondary Plant Products,* (Eds.) Bell, E.A. and Chartwood, B.V. Springer-Verlag, New York, p. 11–19.

Bowers, W.S., 1985. Phytochemical disruption of insect development and behaviour. ACS Symposium series No. 276, *"Bioregulation for Pest Control,"* (Ed.) Hedin P.A.

Bowers, W.S., 1992. Insecticidal compounds from plants. In: *Phytochemical Resources for Medicine and Agriculture,* (Eds.) Nigg, H.N. and Seigler, D. Plenum Press, New York, p. 227–235.

Champion, D.G., 1972. Insect Chemosterilants: A review. *Bull Entomol. Res.*, 61: 577–590.

Chang, E.S., 1993. Comparative endocrinology of moulting and reproduction: Insects and crustaceans. *Ann. Rev. Entomol.*, 38: 161–180.

Cummings, M.R., Robins, M.S. and Ganetzky, B., 1971. Biochemical aspects of oogenesis in *Drosophila melanogaster*. *J. Insect Physiol.*, 17: 2105–2118.

Davenport, N.A., 1961. *Histological and Histochemical Techniques*. W.B. Saunders Co., Philadelphia.

Elhance, D.N., 1977. *Fundamentals of Biostatistics*. Kitab Mahal, Allahabad, p. 796–809.

Ewen, A.B., 1962. An improved aldehyde-fuchsin staining for neurosecretory products in insects. *Amer. Micro. Soc.*, 81(1): 94–96.

Fallon, A., Hagedorn, H., Wyatt, G.R. and Laufer, H., 1974. Activation of vitellogenin synthesis in the mosquito *Aedes aegypti* by ecdysone. *J. Insect Physiol.*, 20: 1815–1823.

Goltzene, F., Lagueuz, M., Charlet, M. and Haffman, J., 1978. The follicle cell epithelium of maturing ovaries of *Locusta migratoria:* A new biosynthetic tissue for ecdysone. *J. Physiol. Chem.*, 359: 1427–1434.

Gujar, G.T. and Mehrotra, K.N., 1988. Toxicity and morphogenetic effects of plumbagin on *Dysdercus koenigii* F. *J. Appl. Entomol.*, 105: 460–470.

Gupta, S. and Rao, P.J., 1990. Effect of azadirachtin on neuroendocrine system of *Spodoptera litura*. *Indian J. Ent.*, 52: 589–594.

Hagedorn, H.H., O'Conner, Fuchs, J.d., Sage, M.S. and Schlaeger, D.A. 1975. The ovary as a source of α-ecdysone in an adult mosquito. In: *Proc. Nat. Acad. Sci., USA*, 72: 3240–3259.

Happ, G.M., 1992. Maturation of the male reproductive system and its endocrine regulation. *Ann. Rev. Entomol.*, 37: 303–320.

Joshi, A. and Muraleedharan, D., 1993. Hormonal modulation of glycogen reserves in the fat body of castor semilooper *Achaea janata* Linn. (Lepid., Noctuidoe). *J. Biosci.*, 18(2): 261–270.

Joshi, N.K. and Sehnal, F., 1989. Inhibition of ecdysteroid production by plumbagin in *Dysdercus* koenigii. *J. Insect Physiol.*, 35(10): 737–741.

Kamalakannan, D.R. and Jeyachandran, K.P.S., 1984. Inhibition of neurosecretory activity by ecdysone in the blister beetle, *Mylabris pustulata* Thumb. *Curr. Sci.*, 53(16): 881–883.

Karnavar, G.K., Strambi, C., Strambi, S. and Kumar, M.A., 1994. Haemolymph ecdysteroid level and oocyte development in *Teleogryllus mitratus* (Orth., Gryllidae). *Entomon.*, 19(3–4): 149–152.

Krishnayya, P.V. and Rao, P.J., 1995. Plumbagin effects on *Helicoverpa armigera* final instar haemolymph protein, amino acids and fat body protein content. *Indian J. Exp. Biol.*, 33(3): 217–221.

Kubo, I., Uchida, M. and Klochek, J.A., 1983. An insect ecdysis inhibitor from the African medicinal plant *Plumbago capensis:* A naturally occurring chitin synthetase inhibitor. *Agric. Biol. Chem.*, 47: 911–913.

La Brecque, G.C. and Fye, R.L., 1978. Cytogenetic and other effects of chemosterilants tepa, metepa, apholate and hempa in insects (a review). *Mut. Res.*, 47: 99–113.

Lafont, R., 1991. Reverse endocrinology of hormones secretion functions. *Insect Biochem.*, 21: 697–721.

Linton, Y.M., Nisbet, A.J. and Mordue, A.J., 1997. The effects of azadirachtin on testes of the desert locust *Schistocerca gregaria* F., *J. Insect Physiol.*, 43(2): 1077–1084.

Magdum, S., Banerjee, S., Kalena, G.P. and Banerji, A. 2001. Chemosterilant activity of naturally occurring quinones and their analogues in the red cotton bug *Dysdercus koenigii*, (Het., Pyrrhocoridae). *J. Appl. Entomol.*, 125: 589–596.

Meinwald, J., Prestwich, G.D., Nakanishi, K. and Kubo, L., 1978. Chemical ecology: Studies from East Africa. *Science*, 199: 1167–1173.

Mitchell, M.J. and Smith, S.L., 1988. Effect of chitin synthetase inhibitor plumbagin and its 2-demethyl derivative juglone on insect ecdysone, 20-monoxygenase activity. *Experientia*, 40: 990–991.

Plummer, D.T., 1985. *An Introduction to Practical Biochemistry*, Tata McGraw Hill Book Co. (UK) Ltd., TMH edition, p. 144–145; 183–184.

Prabhu, V.K.K., 1994. Insect neuroendocrine system. In: *Recent Advances in Insect Physiology and Toxicology*, (Ed.) Gujar, G.T. Agricole publishing Academy, New Delhi, p. 48–58.

Rao, J.V., Sreenivasan, C. and Makkapati, A.K., 1996. Plumbagin effect on growth and metamorphosis of housefly *Musca domestica* L. (Dipt., Muscidae). *Intn. Pest Control*, p. 24–27.

Rembold, H., 1989. *Insecticides of Plant Origin*, (Eds.) Arnason, J.T., Philogene, J. and Morand, P. ACS Symposium Series, 387. Washington D.C.

Rembold, H., 1994. Secondary plant products in insect control with special reference to azadirachtin. In: *Advances in Insect Reproduction*, (Ed. Engels, W.E. Elsevier, Amsterdam, p. 481–491.

Sahai, Y.N. and Banerjee, S., 1986. Histopathological and cytochemical effects of hempa on ovaries of *Poekilocerus pictus* Fabr. (Orth., Acrididae). *Folia Morphol.*, 34: 201–205.

Sayah, F., Fayet, C., Idaomar, M. and Karlinsky, A., 1996. Effect of azadiarachtin on vitellogenesis of *Labidura riparia* (Derm., Insecta). *Tissue and Cell*, 28(6): 741–749.

Saxena, B.P. and Tikku, K., 1988. Exploitation of lacunae by some allelochemicals in insect-plant interactions. In: *Dynamics of Insect Plant Interactions: Recent Advances and Future Trends*, (Eds.) Ananthakrishnan, T.N. and Raman, A. Oxford and IBH publications Co. Ltd., New Delhi, p. 105–122.

Saxena, B.P. and Tikku, K., 1990. Impact of natural products on the physiology of phytophagous insects. In: *Proc. Indian Acad. Sci. (Animal Sci.)*, 99: 185–198.

Saxena, B.P., Tikku, K. and Thappar, R.K., 1996. Plumbagin, a sterilant for *Musca domestica* males. A light and electron microscopic study of its effects on its testes. In: *Proc. Nat. Sci. Acadi.*, 62(3): 199–206.

Saxena, B.P., Thappa, R.K., Sharma, A. and Suri, O.P., 1996. Effect of plumbagin on gonadotropic cycle of the housefly *Musca domestica* L. *Indian J. Exp. Biol.*, 34: 739–749.

Shibko, S., Koivistoines, P., Tratnyek, C.A., Newhall, A.R. and Friedman, L., 1967. A method for sequential quantitative separation and determination of protein, RNA, DNA, lipid and glycogen from a single rat liver homogenate or from a subcellular fraction. *Anal. Biochem.*, 9: 514–528.

Subrahmanyam, B. and Rembold, H., 1989. Effect of azadirachtin 'A' on neuroendocrine activity in *Locusta migratoria*. *Cell Tissue Res.*, 256: 513–517.

Subrahmanyam, B., Mueller, T. and Rembold, H., 1989. Inhibition of turnover of neurosecretion by azadirachtin in *Locusta migratoria*. *J. Insect. Physiol.*, 35: 493–500.

Sugai, E., 1967. Cytological studies of male sterility of silkworm *Bombyx mori*. *Agri. Tech.*, 10: 36–39.

Taber, S. and Borkovec, A.B., 1969. Chemical sterilisation of honeybee spermatozoa *in vitro*. *Nature*, 224: 1217–1218.

Van Beek, Blaakameer, A., Griepink, F.C., Van Loon, J.A., Visser, J.H. and DeGroot, A.E., 1994. Chemical ecology as a lead for the development of environmentally safe insect control agents. In: *Advanced Chemistry of Insect Control*.

Wyatt, G.R., Cook, K.E., Firko, H. and Dhadialla, T.S., 1987. Juvenille hormone action on locust fat body. *Insect Biochem.*, 17: 107–173.

Chapter 20

Ecobiological Studies on *Zygogramma bicolorata* Pallister (Coleoptera : Chrysomelidae) a Biocontrol Agent on *Parthenium hysterophorus* L. in Jammu

Sanjay Bhatia, Rajani Choudhary and Manjot Singh

Department of Zoology, University of Jammu, J&K – 180 006, India

E-mail: siya98@sancharnet.in

Introduction

Parthenium hysterophorus L. (Figure 20.1) is a weed of national significance. It is commonly called in India as parthenium weed, congress grass, carrot weed, white top, starweed, chatak chandani etc. It is a herbaceous, erect and annual plant belonging to the family Asteracae (compositae), tribe Heliantheae and subtribe Ambrosinae.

This weed is very aggressive in growth and usually reaches upto 1–2 mts in height. Several aspects of the ecology of parthenium weed contributes to its aggressiveness. These include the size and persistence of the soil seed bank, longevity of seed when buried; fast germination rate, and the innate dormancy mechanism of seed, which makes it well adapted to varied types of environmental conditions. In addition, this weed releases allelopathic chemicals that inhibit germination and growth of pastures grasses, legumes, cereals, vegetables, other plants and even trees and can cause total habitat change in native vegetation. It is also a serious human health hazard. This weed has gained notoriety mainly due to the health hazards, such as nasobranchial allergy and eczematous contact dermatitis, it causes to humans (Towers *et al.*, 1977; Mohd *et al.*, 1978; Towers and Subba Rao, 1992; MacFadyen, 1995).

Parthenium is an exotic deadly weed, native to the area around the Gulf of Mexico including the West Indies and to the Central South America. Within the last 100 years it has found its way to Africa, Australia and Asia. It is widely adventitious and has become a major problem in both India and Australia, where it has achieved the status of "worst weed".

It is a serious problem in India with the weed spreading across majority of states including the State of Jammu and Kashmir overrunning waste lands, pastures and agricultural fields. The overall spread is extensive and thus it has acquired a status of one of the most prominent invasive plants and has been listed as one of the seven worst weeds in the world. In Jammu region of J&K state this weed has spread to all the six districts of the region and it is now moving fast and far in other high altitude areas of the region, including the Kashmir valley. However, its intensity has been estimated to be less in high elevation, as compared to plains; the weed so far has not been reported from Ladakh region of the State of Jammu and Kashmir.

Ever since this plant became a menace in India and other countries, efforts are on to control and manage this weed by different methods. However, biological control of the weed is the only cost effective, environmentally safe and ecologically viable method available.

During the last few years much emphasis has been given to control *Parthenium hysterophorus* through various biological agents like insects, pathogen and competitive displacement of it by other plants. Biological control trials against this weed were initiated in India in 1983 by the Indian Institute of Horticultural Research, Bangalore, by introducing the leaf-feeding beetle *Zygogramma bicolorata* Pallister (Coleoptera : Chrysomelidae) from Mexico, its native home and later introduced on a trial basis in few pockets in and around Jammu city in the early nineties by the J&K State Council for Science and Technology. However, no subsequent studies on its establishment, dispersal, biology and life-cycle and the efficacy of the beetle in controlling the weed in Jammu region was carried out.

This present work embodies a detailed observations on the ecobiology and life-cycle of *Zygogramma bicolorata* and the nature of damage it causes to *Parthenium hysterophorus* in Jammu region of J&K state.

Materials and Methods

Observations regarding various aspects of the life-cycle and biology of *Z. bicolorata* were carried out both in the laboratory, as well in the fields. For the study of the life-cycle and its associated biological observations the beetle was reared in clear plastic jars, with wire meshed windows on their lids for aeration. Studies on development were carried out under laboratory conditions between April–June (Pre-monsoon period); July–September (monsoon period) and October–December (Post-monsoon period). All along this period both temperature and humidity records were maintained. The various aspects of the life-cycle studied were:

Copulatory Behaviour

The experiments were initiated with the collection of newly emerged beetles from the fields. In the laboratory, the male and the female beetle were segregated on the basis of their abdominal sternites as suggested by McClay (1980). Single pair in 5 replicates were introduced into plastic jars containing fresh shoots of parthenium plants, which were rolled in cotton strips and then dipped in glass bottles containing water. Observations with regard to precopulatory behaviour, pairing and copulation and duration of mating was recorded.

Oviposition

After copulation, the inseminated female beetles were introduced into cages containing bouquets of parthenium shoots. Pre-oviposition and oviposition behaviour was recorded.

Incubation Period

For determining the incubation period, batches of freshly laid eggs were collected from some of the field cages and from laboratory cages. The egg batches were then placed separately in petridishes for determining the incubation period. Effect of temperature and humidity on the incubation period was also recorded.

Determination of Larval Instars and Larval Duration

For determining the number of larval instars and larval duration, newly hatched larvae were reared individually in glass vials. Each Larva was examined daily in the morning to see the moulting which was confirmed by the presence of exuviae in the vials. The moulted larvae were transferred with a fine camel hair brush to a separate glass vials. This process was continued till the larvae reached fully matured stage. The total larval duration was also recorded.

Pupation

Fully grown larvae were released in the cages containing 2–3 cm layer of soil for the purpose of pupation as the larvae of these beetles pupate in the soil. Water was sprinkled uniformly on the soil daily to keep the soil moist. Observations regarding the prepupal and pupal duration were made.

Sexual Dimorphism

For studies on sexual dimorphism, pupae were collected from the soil and placed individually in numbered vials after recording their body length and width. When adults emerged, the sex of the individual was recorded by observing the last abdominal sternite, as described by McClay (1980). The sex ratio of the beetle was continuously monitored, using at least 75 adults collected from field every month for one year.

Adult Longevity

10 pairs of freshly emerged adults were used for studying fecundity and adult longevity. The pairs were maintained in separate cages and eggs counted on alternate days until the death of the female beetle. The mortalities of the males was also recorded. Observations were made on the hatchability of the eggs.

Observations and Discussion

Adults of *Zygogramma bicolorata* (Figure 20.2) are small in size and attractively coloured. The elytra are marked with undulating dark brown lines that run longitudinally over an off white

Figures 20.1–20.8

background. The rest of the body and appendages are dark brown, except that the pronotum has off white patches at the anterior edges. Female beetles are generally larger than males and measures about 7 to 7.5 mm in length and 5 to 5.2 mm in width at mid point while male measures about 6 to 6.5 mm in length and 4 to 4.2 mm in width. Head with pair of prominent black eyes on the lateral side. Antennae are situated on the lateral side of the eyes and are comprised of 11 well marked segments. First 5 segments of antennae are light brown in colour while rest 6 segments are dark brown in colour. Mouth parts are prominent and are of biting and chewing type. Thorax is dome shaped and the texture is smooth and shiny. Elytra is closely applied to the abdomen. Scutellum is triangular in shape. The body on the whole is studded with granular punctures.

Pairing and Copulation

Freshly emerged adults of *Zygogramma bicolorata* were observed to mate only from the fifth to sixth day after emergence. No special precopulatory behaviour was shown by the beetle. Just approaching the female and initial body contact appears to be stimuli for the male beetle for mounting. Male mounted the female from behind by moving directly forward on to its dorsum (Figure 20.3). Male lie at 45° angle to the female. Both the prothoracic and mesothoracic legs rest upon the female's elytra and the metathoracic legs lying over the posterior ventral portion of the female. During the process of copulation both the sexes either remain motionless or sometimes female starts moving in paired condition carrying male partner on its back.

Mating was found to last as long as 1 hr 45 mins to 3 hrs and copulatory pairs were seen from morning till evening in the field. Mating pairs remain in mounting position for long time and sometimes for the whole day and intromission and extrusion of the aedaegus often occurs during the process of mating. Male remain in mounting position till insemination was accomplished. Repeated mating between the same male and female occurred frequently in the laboratory both before and after the commencement of oviposition.

Male was found capable of mating with different females and a female with number of males. Observations show that slight disturbance does not hinder the process of mating. Mating pairs were observed from the month of March till October in the field conditions.

Oviposition

After mating the inseminated female feeds on parthenium leaves for a week or so prior to egg laying. Jayanth and Geetha Bali (1993 a) reported that pre-oviposition period in *Zygogramma* lasted for 10–70 days at Bangalore. Oviposition occurred throughout the day and prior to oviposition, the female beetle has been seen moving over the stem, branches and leaves of the plant, apparently for selection of the oviposition site. Eggs are generally laid on the ventral surface of the both young and older leaves singly or in clusters of 6–10 eggs in most of the cases (Figure 20.4). However occasionally they may lay eggs on the upper surface of the leaves, stems and even on the flowers.

Eggs are pale yellow in colour and placed in juxtaposition to each other but not in overlapping manner (Figure 20.4). They are small and oblong tapering towards either end and measures about 1.21 mm long and 0.56 mm wide. In field, first generation of eggs were seen in the month of April soon after the emergence of the beetle. Jayanth and Geetha Bali (1993a) reported that oviposition continued for 89–138 days while post-oviposition period was short and adults died within 1–21 days after the termination of egg laying.

Hatching

The first instar larva emerges from the egg case by gnawing away an irregular hole with the help of its mandibles and wriggles out by enlarging the slit through peristaltic movements. Small setae present on the body of the larvae may facilitate the emergence. Emergence results into irregular split at least halfway down the egg capsule length. Occasionally the egg capsule is completely shredded into fragments. The newly hatched larvae are bright yellow in colour, turning to creamy white as they grow.

Post Hatching and Feeding Behaviour of Larvae

Immediately after hatching, the first instar larvae start feeding on the area adjoining to the oviposition site and after sometime start migrating towards the auxiliary buds as newly hatched larvae congregate and feed on the terminal and axilliary buds. Initial instar larvae were observed to be feeding in groups of 6–7 larvae on the buds. As they grow they move towards the peripheral leaves. Observations shows that later instar larvae do not feed in groups. The mature larvae feed on both the young and older leaves and result into extensive defoliation of the plant (Figure 20.5). Newly hatched larvae are yellowish in colour turning creamish white as they grow. Fully grown larva measures about 6 mm in size. Body slightly curved with protrusible proleg like structure at the posterior end, which is used for locomotion. Eight abdominal segments are visible both from dorsal and ventral side. A prominent grey coloured streak passes mid-dorsally from the thorax to the last abdominal segment.

Number of Larval Instars

The larvae moulted thrice and as such four larval instars could be differentiated. Jayanth and Geetha Bali (1993a) also reported 4 larval instars during their studies carried out at Bangalore. Under laboratory conditions, maximum length attained by first to fourth instar larvae were: 2.6 ± 0. 04 mm, 3.2 ± 0.3 mm, 4.5 ± 0.3 and 6.5 ± 0.04 respectively. The first two larval stages looked alike, except for size. Like wise third and fourth instar larvae had similar appearance. The last two instars could be differentiated from the earlier stages by their spiracles which were seen as clear dark spots after the second moult.

The total larval duration was recorded during pre-monsoon, monsoon and post monsoon seasons of the year and the findings have been listed in the Table 20.1.

Table 20.1: Showing Total Larval Duration of *Zygogramma bicolorata*

Sl.No.	*Season*	*Larval Duration (in days)*	*Temperature*		*Humidity (%)*
			Maximum (°C)	*Minimum (°C)*	
1.	Pre/monsoon (April–June)	14	39.2	23.0	45
2.	Monsoon (July–September)	12	37.4	26.0	71
3.	Post-monsoon (October–November)	9	31.4	19.20	72.50

Fecundity

Jayanth and Geetha Bali (1993 a) reported that total number of eggs laid by a female beetle ranged between 1695–3368 (mean 2520.90) at Bangalore. The females were capable of laying upto 45 eggs per day and 299.60 eggs per week. Mcfadyen and Mcclay (1981), recorded an average of 836 eggs per female with a maximum of 1786. During the present studies, a single female beetle in the laboratory laid up to 100 eggs in a day. However, generally they lay upto 60–70 eggs per day. On an average a female beetle lays upto more than 800 eggs.

Incubation Period

The chorion of the egg which is pale yellow when freshly laid changes to deep yellow during the course of incubation. Prior to hatching, colour of the egg changes to orange. Colour changes on the chorion of the egg during the course of incubation was also earlier recorded by Mcclay (1980), but no such colour changes were reported by Jayanth and Geetha Bali (1995).

Incubation period of the eggs of *Z. bicolorata* was recorded during Pre-monsoon (April–May), monsoon (July–September) and Post-monsoon (October–November) seasons of the year and the findings have been listed in the Table 20.2.

Table 20.2: Showing Incubation Period of Eggs of *Zygogramma bicolorata*

Sl.No.	*Season*	*Duration of Hatching (in days)*	*Temperature*		*Humidity*
			Maximum (°C)	*Minimum (°C)*	*(%)*
1.	Pre/monsoon (April–June)	05	35.5	23.0	35
2.	Monsoon (July–September)	03	38.6	25.3	45
3.	Post-monsoon (October–November)	3.5	32.30	20.70	63

The incubation period was shortest (3 days) during monsoons, 5 days during Pre-monsoon and 3.5 days during Post-monsoon periods. Eggs in the same cluster usually hatched concurrently. Jayanth and Geetha Bali (1993 b) while studying the temperature tolerance of *Zygogramma bicolorata* reported that embryonic development was noticed only at temperature ranging between 15°C and 35°C.

Within this range, the duration of development decreased with the increase in temperature. They recorded 52 per cent hatching between 20–25°C, while 48 per cent of them hatched at 30°C. The percentage of hatching declined at 35°C (16 per cent) and none of the eggs hatched at 40°C.

On the contrary, highest percentage of hatching was recorded when the temperature ranged between 32–38°C in Jammu region. Besides, 90 to 95 per cent eggs hatched in the laboratory conditions, whereas Jayanth and Geethabali (1993 b) reported only 30 to 53 per cent eggs hatching under laboratory conditions at Bangalore. Further they also reported that hatching percentage was higher earlier in the breeding season (May–July), as compared to later months (August–October). However no such difference in the egg hatching percentage was observed in Jammu conditions.

Table 20.1, shows that the total larval duration was recorded to be of 14 days during pre-monsoon season, 12 days during monsoon and only 9 days during post-monsoon period. Jayanth and Geetha Bali (1993a), reported an average of 11–13 days as larval duration of this beetle at Bangalore. In another study Jayanth and Geetha Bali (1993b) recorded that extreme temperatures affected the development of larvae. At 5°C and 10°C, they were found incapable of feeding.

They further recorded that when newly hatched larvae were placed at 40°C, most of them died during the first instar, while 10 per cent developed upto the third instar and later died. The duration of larval development reduced with the increase in temperature from 15°C to 35°C. Thus at 15°C, 46.23 days were required and at 35°C the development was completed in 8.32 days.

Prepupal Behaviour of the Larvae

Between the fourth and fifth day of development of last instar, the larvae stop feeding, shrinks in size, become sluggish and sometime changes to creamish yellow or dull cream in colour and drops down on the soil and burrowes itself into the soil for pupation (Figure 20.6).

Having reached suitable depth, usually few centimeters below the soil, the larvae start making a rounded pupal chamber (Figure 20.7). Full grown larvae were found incapable of burrowing into the dry soil, and as such soil moisture is very important as it helps in making the pupal chamber and in burrowing into the soil. Larvae remain as prepupa for an average of about 4 days during pre-monsoon, 5 days during monsoon and 4 days during post-monsoon season before metamorphosing into the pupal stage.

Jayanth and Geetha Bali (1995) while studying the effect of soil moisture on pupation also reported that full grown larvae of *Zygogramme bicolorata* were incapable of burrowing into dry soil for pupation, nor could they pupate on the surface of the soil, although they survived upto 10 days when water equivalent to 0.5 mm of rainfall was added. Soil moisture equivalent of 1.5 mm of rainfall was required for both burrowing in of full grown larvae and emergence of adults from pupal chambers.

Pupation

Pupation occurred within an earthern pupal chamber. Pupa is creamish in colour and about 6.5 ± 0.03 length mm in length. Pupa is off white in colour (Figure 20.8). It is an exarate type and measures about 6.5–6.8 mm in length. The duration of pupal period last between an average of about 4 to 7 days during different seasons of the year. That pupal stage lasted for 2 weeks was observed by McFadyen and McClay (1980) and Jayanth (1987, 1993).

During pupal period the colour of the pupa changes and turns to dark cream and eyes also become pigmented. In the laboratory conditions, the pupation was also observed on the soil surface. Complete metamorphosis from pupal stage to adult beetle take place in pupal chamber.

Jayanth and Geetha Bali (1993b), recorded that the beetle was capable of pupating only at temperature ranging from 15 to 35°C. The duration of development decreased with increase in temperature. They reported that 50 days are required at 15°C, while only 7.40 days were needed at 35°C.

In an another study, Jayanth and Geetha Bali (1995) while studying effect of soil moisture on adult emergence, showed that when the pupal cells are formed very close to the soil surface, adult could come out even if the soil was dry. But when pupation took place at a depth of 2–3 cm from the soil surface as is usually the case, they were unable to break through and emerge. Pupa is almost quiescent but sometimes move its abdominal tip when disturbed.

Emergence of the Beetle

As soon as the teneral adult beetles becomes sufficiently sclerotized within the earthen pupal chambers in the soil, they make their way out from the soil to the exterior. Newly emerged adults are white in colour and normal coloration and complete sclerotization were attained within 10 to 12 hours. Emergence of the over wintering adult beetles of *Zygogramma bicolorata* from the soil begins during middle of March in Jammu region and they remain active in the field up to December.

Two peaks of beetle population have been recorded one during April–May and second during July–November. Highest density of the beetle is recorded between July and November. This is apparently synchronized with the onset of monsoons in the area of this study. In Jammu conditions this beetle has been recorded to remain active in the field from March–December.

This observation is contrary to Jayanth and Geetha Bali (1993c) and Jayanth (1996) who reported that the beetle remained active under field conditions in Bangalore between May and October. Besides, these authors also recorded that *Z. bicolorata* undergo diapause within the soil as adults during the remaining months and emerged with the onset of rains. A comparative emergence period of this is given in the Table 20.3.

Table 20.3: Showing Comparative Emergence and Activity Period of *Zygogramma bicolorata* in Bangalore and in Jammu

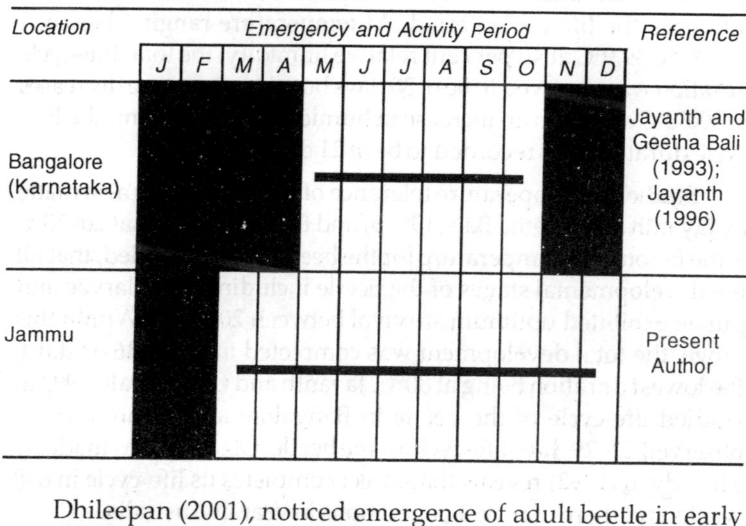

Dhileepan (2001), noticed emergence of adult beetle in early summer, reaching peak population in middle of summer but declines subsequently with no larval activity from early autumn in Australia.

Life Cycle Duration

Total life-cycle duration of the beetle was studied in three seasons (Table 20.4). During pre-monsoon season the life-cycle duration lasted for an average of 30 days, while during monsoon period it was recorded to be of 24 days and an average period of 21 days during post-monsoon season was recorded in Jammu.

Table 20.4: Showing Total Life-cycle Duration of *Zygogramma bicolorata* in Different Seasons

Sl.No.	*Season*	*Duration of Hatching (in days)*	*Temperature*		*Humidity*
			Maximum (°C)	*Minimum (°C)*	*(%)*
1.	Pre/monsoon (April–June)	30	39.9	24.3	26
2.	Monsoon (July–September)	24	37.6	25.8	53.5
3.	Post-monsoon (October–November)	21	30.80	19.25	61.5

The observations shows that temperature and humidity directly influence the life-cycle period. At temperature ranging between 24.3°C to 39.9°C at 40 per cent relative humidity, the total life-cycle duration was observed to be of 30 days but as temperature decreases to 30.80°C and with an increase in humidity (61.5 per cent) the life-cycle duration was recorded to be of 21 days.

Studies on temperature tolerance of the beetle were also made by Jayanth and Geetha Bali (1993b) and they recorded that 20–30°C is the favourable temperature for the beetle. They recorded, that all the developmental stages of the beetle including eggs, larvae and pupae exhibited optimum survival between 20–30°C. Within this range, the total development was completed in 20.75–46.63 days, the lowest duration being at 30°C. Jayanth and Geetha Bali (1993 a) studied life-cycle of the beetle in Bangalore and on an average observed 27–29 days life-cycle of the beetle. Observations made by McFadyen (1992) reveals that insect completes its life-cycle in 6–8 weeks and there may be 4 generations in year in Australia.

Number of Generations

Beetle was continuously reared in the laboratory from March onwards *i.e.*, from emergence up to the time it undergo diapause and 6–7 generations in an year have been recorded in Jammu conditions. During extreme summer periods, the beetles do not breed and bury themselves in the soil and become active again with the onset of rain. Thus it was observed that during the month of May when temperature exceeded 45°C the number of the beetles decline considerably in the fields. During winter months (December to February), these beetles undergo diapause and remain inactive in the soil.

Observations made in Bangalore showed that *Z. bicolorata* completes 5–6 generations under field conditions (Jayanth and Bali, 1992).

Sex-Ratio and Adult Longevity

The sex-ratio was found to be in favour of females. Samples collected from the field at monthly intervals showed that the mean sex-ratio was 1 female : 0.40 males.

The males of *Z. bicolorata* lived for an average of 220 days in laboratory conditions while the female has a shorter life span of an average of 135 days. As the sex ratio favoured females, with the males constituting only about 40 per cent of the population, longer survival of males would be advantageous in ensuring fertilisation of all the females, since the adults were capable of mating repeatedly throughout their life-span.

Nature of Damage

Both the adult beetles and their larvae feed voraciously on the parthenium leaves and also on the flowers thereby affecting the vigour, growth and reproduction of the plant (Figure 20.9). Adults generally feed on the leaves and cause complete defoliation of the plant. The newly hatched larvae feed initially on the area adjoining the oviposition site and later move to the terminal and axillary buds, where they congregate and feed, leading to stunted growth, altered branching pattern and also prevents emergence of flowers.

During peak infestation 7–8 early instar larvae can be seen feeding on the terminal and axillary buds, where they congregate and feed (Figure 20.10). As the larvae grow, they migrate towards the peripheral leaves and in severe infestation many mature larvae can be seen feeding voraciously on the plants. Feeding was observed on both fresh as well as the older leaves and the last two larval instars *i.e.*, cause maximum damage to the plant.

Bhumannavar and Balasubramanian (1998) reported that among the larval and adult stages, the maximum amount of food

Figures 20.9–20.10

ingested was seen in the third instar, besides relative consumption and growth rate were high during first and third instar larvae and in the adults during the oviposition stage.

Defoliation of parthenium both by adults and larvae reduce the number of leaves, plant height, plant density and reduction in flowers. In the completely defoliated plants, the beetle even oviposits on the flower heads and emerging larvae feeds on the flowers.

In Jammu conditions, the beetle remaining active in the fields from late March up to December, with peak activity period observed during July to late November. During this period, it causes maximum damage to the plant and completely strips it off the leaves and large stands of leafless plants can be seen all around (Figure 20.11), which in due course of time turn brown and dry, resulting in significant reduction in plant density, growth and flower production, thereby resulting in reduction in seed production and subsequent affecting the soil seed bank which in turn can result in reduced density of parthenium.

Acknowledgements

The authors are grateful to the Head, Department of Zoology, University of Jammu for providing the necessary facilities during the course of this work.

References

Bhumannavar, B.S. and Balasubramanian, C., 1998. Food consumption and utilization by the Mexican beetle, *Zygogramma bicolorata* Pallister (Coleoptera : Chrysomelidae) on *Parthenium hysterophorus* L. *Journal of Biological Control*, 12(1): 19–23.

Dhileepan, K., 2001. Effectiveness of introduced biocontrol insects on the weed *Parthenium hysterophorus* (Asteraceae) in Australia. *Bulletin of Entomological Research*, 91: 167–176.

Jayanth, K.P., 1987. Introduction and establishment of *Zygogramma bicolorata* on *Parthenium hysterophorus* at Bangalore, India. *Curr. Sci.*, 56(17): 310–311.

Jayanth, K.P., 1993. Report of the fact finding committee meeting on parthenium. *Curr. Sci.*, 65(12): 902–905.

Jayanth, K.P., 1996. Status of biological control trials against *Parthenium hysterophorus* by *Zygogramma bicolorata* in India. *Madras Agricultural Journal*, 83(11): 672–678.

Jayanth, K.P. and Geetha Bali, 1992. Estimation of number of generations of the Mexican beetle, *Zygogramma bicolorata* Pallister (Coleoptera : Chrysomelidae) by measurements of thermal units. *Journal of Entomological Research*, 16(4): 273–276.

Jayanth, K.P. and Geetha Bali, 1993a. Biological Studies on *Zygogramma bicolorata* Pallister (Coleoptera : Chrysomelidae): A potential biocontrol agent of *Parthenium hysterophorus* L. (Asteraceae) in Bangalore, India. *Journal of Biological Control*, 7(2): 93–98.

Jayanth, K.P. and Geetha Bali, 1993b. Diapause behaviour of *Zygogramma bicolorata* (Coleoptera : Chrysomelidae): A biological control agent for *Parthellium hysterophorus* (Asteraceae) in Banglore, India. *Bulletin of Entomological Research*, 83: 383–388.

Jayanth, K.P. and Geetha Bali, 1993c. Temperature tolerance of *Zygogramma bicolorata* (Coleoptera : Chrysomelidae) introduced for biological control of *Parthenium hysterophorus* (Asteraceae) in India. *Journal of Entomological Research*, 17(1): 27–34.

Jayanth, K.P. and Geetha Bali, 1994. Biological control of *Parthenium hysterophorus* by the beetle *Zygogramma bicolorata* in India. *FAO Bulletin of Plant Prot.*, 42(4): 207–213.

Jayanth, K.P. and Geetha Bali, 1995. Effect of soil, moisture on population and adult emergence of *Zygogramma biocolorata* Pallister (Coleoptera : Chrysomelidae). *Journal of Entomology Research*, 19(2): 183–185.

McClay, A.S., 1980. Studies on some potential biocontrol agents for *Parthenium hysterophorus* in Mexico. In: *Proc. 5th Int. Symp. Biological Control of Weeds*, Brisbane, Australia, pp. 471–482.

McFadyen, R.E., 1992. Biological control against parthenium weed in Australia. *Crop Protection*, 11: 400–407.

McFadyen, R.E., 1995. Parthenium weed and human health in Queensland, Australia. *Family Physician*, 24: 1455–1459.

McFadyen, R.E.C. and McClay, A.R., 1981. Two new insects for the biological control of parthenium weed in Queensland. In: *Sixth Australian Weeds Conference*, (Eds.) B.J. Wilson and J.D. Swarbrick. Weed Science Society of Queensland, Australia, pp. 145–149.

Mohd, A.S., Singh, R. and Sharma, R.C., 1978. Contact dermatitis due to *Parthenium hysterophorus*. *Indian Journal of Medical Research*, 68: 481–484.

Towers, G.H.N., Mitchell, J.C., Rodriguez, E, Bennett, F.D. and Subbarao, P., 1977. Biology and chemistry of *Parthenium hysterophorus*. L. problem weed in India. *Journal of Scientific and Industrial Research*, 36: 672–684.

Towers, G.H.N and Subba Rao, P.V., 1992. Impact of the pan tropical weed, *Parthenium hysterophorus* L. on human affairs. In: *Proc. First Int. Weed Control Congr.*, Melbourne, 1992, (Eds.) H.H. Combellock, K.J. Levick, J. Parsons and I.G. Chardson, Vol. I, pp. 134–138.

Chapter 21

Reproductive Performance: A Deciding Factor for the Hunter Reduviids Utilization in Biological Control

K. Sahayaraj

Crop Protection Research Center, Department of Zoology, St. Xavier's College, Playamkottai – 627 002, Tamil Nadu, India

E-mail: ttn_ksraj@sancharnet.in

ABSTRACT

Hunter reduviids constitute an important predator found almost in all the ecosystems of the biosphere. Though most of them are general predators, some of them are specific in their predation. Naturally occurring reduviids have the capability to control the pest population under the economic threshold level. Hence now-a-days reduviids have been augmented under laboratory conditions, released in the fields and fruitful results have been achieved by many agriculture entomologists. Evaluating the reproduction and multiplication capacity of the reduviids both in the field and laboratory conditions are very crucial aspect in the field of biological control. Here the reproduction capacity of various reduviids in relation to their density, prey density, prey rearing medium, rearing arena, composition of male and female, female parental age, and substratum has been discussed. The data revealed that reproduction is a deciding factor while integrating hunter reduviids in the biological control programme.

Introduction

Heteropteran behaviour, governed by natural instincts, includes patterns that enable individual species to live within specific environments. The patterns are so similar for all members of a family that each family can be characterized as essentially phytophagous or predatory, and terrestrial, semi-aquatic, or aquatic. Reduviids are an important group of insects belonging to the order heteropertera. Heteropteran predators are important biological control agents of many agricultural insect pests. Of this group, reduviids are one of the least researched and most poorly understood families, although evidence in support of their potential in biological control is accumulating. More than 6250 species have been described throughout the world. They have been distributed in all types of habitats ranging from semi-arid zones to evergreen forest. There are two groups of reduviids identified: one is haematophagous and another one is hunter reduviids. The later group has an important role in the field of insect pest management. Reduviids constitute the largest group of predaceous land bugs. Economic loss to agricultural crops resulting from insect damage can be reduced by development of ecologically sound, environmentally friendly pest management techniques. New methods of pest insect control are urgently needed to reduce and replace the use of chemical pesticides in farming. Biological control techniques using beneficial insects can be used for insect pest control. Mass production of reduviids has not been available in the literature. Mass production purely depends upon the reproductive ability of an organisms.

The production of gametes, the maturation of various tissues and the expression of behaviour related to mating and birth or oviposition characterize the reproductive development of animals. Morphological and functional changes of the neuro-muscular system are considered a prerequisite for animals to show adequate sexual behaviour, which in turn is necessary for successful reproduction. Although evolutionarily remote, similar processes occur in the adult insect, albeit regulated by a different hormone named juvenile hormone (JH). JH, first discovered by Wigglesworth (1934), regulates several aspects of insect development and reproduction. In cooperation with 20-hydroxyecdysone (20-HE), JH governs metamorphosis by inhibiting the development of adult characters. In the adult insect, by contrast, JH function has clearly been associated with reproductive development.

The metamorphosis of insects is truly one of nature's most amazing phenomena. This beautifully photographed program documents the life cycles of various insects from reproduction to adult, and shows how the insects adapt at each stage. It provides vividly detailed support for Life Science and Biology units on insect reproduction and both complete and incomplete metamorphosis. Included are the large diving beetle, swallowtail butterfly, fruitfly, locust and dragonfly. Also explored are the changing food requirements and common habitats of each of the insects. The quickly evolving physical characteristics of each insect at each stage of development are expertly depicted and thoroughly discussed. Prey type and density; predator density; polymorphism of the predator; rearing arena (nature and type) and temperature having an influence on the reproductive performance of any predator have been discussed.

Materials and Methods

The material and methods have been explained elsewhere in details (Sahayaraj, 1991, 1994, 2004). In addition to our own publications, informations were also collected from the literature and presented in this review.

Results and Discussion

Ecotypic Heterogenecity on Reproduction

Ambrose (1980) studied the ecotypic heterogenecity on the biology of four reduviids *viz., Rhynocoris marginatus, S. pubinotum, Acanthaspis pedestris* and *A. siva*. Sahayaraj (1991) studied the ecotypic heterogenecity on the biology of *N. therasii* and *Eciomocoris tibialis;* and postulated that reduviids living in scrub jungle has higher reproductive capacity than semi-arid zones. *Acanthaspis quinquespinosa* Fabricius is an alate, warning colored, crepuscular, entomosuccivorous, polyphagous and multivoltine-assassin bug found in the tropical evergreen forests, scrub jungles, semiarid zones and agroecosystems of peninsular India (Sahayaraj, 1991). Intensive collection in Tamil Nadu, India shows that *A. quinquespinosa* distributed in Tropical rain forests (Keeriparai, Patchaimalai, Surlitheertham), scrub jungles (Marunthuvazhmalai, Sivanthipatti), semiarid zones (Kanyakumari, Pollachi, Tambaram, Usarathukudieruppu) and agroecosystems (cotton and groundnut respectively). *A. quinquespinosa* inhabiting Patchaimalai had shorter developmental, pre-ovipositional and incubation periods and longer adult longevity. This category also had higher fecundity and hatchability and larger size (Sahayaraj unpublished data). Each reduviid has its own ability in reproduction (Table 21.1) and it purely depends on the nature of ecotype where it is, living. Furthermore, particular ecotype is favourable for only the particular reduviid. For instance ecotype Courtallam is favourable for *Eugores plagiatus* but not for *Sphedanolestes pubinotum*. Similarly Maruthuvazhmalai is supporting the reproduction of *Ectomocoris tibialis* and not *Acanthaspis siva*. These kinds of studies are very useful to select the viable reduviid for the mass production and incorporate them in biological control programme.

Table 21.1: Ecotypic Heterogenecity on Reproduction of Reduviids

Reduviids	*Name of the Ecotype*	
	Maximum Egg Laid	*Minimum Egg Laid*
Neohaematorrhopus therasii	Melapattam	Sunkankadai
Sphedanolestes pubinotum	Kodayar	Courtallam
Sphedanolestes minusculus	Sunkankadai	Aralaymozhi
Eugores plagiatus	Courtallam	Perunchani
Ectomocoris tiobilais	Maruthuvazhmalai	Chandrapuram
Acanthaspis pedestris	Mettupalayam	Chandrapuram
Acanthaspis siva	Salem	Maruthuvazhmalai
Acanthaspis quinquespinosa	Patchaimalai	Usarathukudieruppu

Reproductive Performance of Reduviid: Cluster Analysis

Very little information is available about the cluster analyses of reduviids reproductive behaviour including pre-oviposition period, number of egg batches, number of eggs and incubation period with adult longevity, pre-oviposition period, number of egg batches, total number of eggs laid, average number of eggs and oviposition index. Cluster analysis of reproductive parameters such as pre-ovipostion period, number of egg batches laid, number of total eggs laid by a female, average number of eggs per batch and oviposition index (per cent of egg-laying days in female adult longevity) were consided. The analyses revealed that both the *A. quinquespinosa* and *A.*

biannulipes and *A. quadrisignatusm* and *E. annulata* (Figure 21.1) were closely related species. Similar analyses were also performed for 26 harpactorine reduviids. The analyses revealed that *E. inornatus* had a unique reproductive behaviour when we compare it to other harpactorine reduviids (Figure 21.2). The reproductive performance of the former is very close when we compare to the latter. Among the nine reduviinae subjected for this cluster analysis, *E. slateri* is a unique species than other eight reduviinae. Another cluster analysis was performed with reproductive indices such as pre-oviposition period, number of egg batches, number of eggs and incubation period with adult longevity, pre-oviposition period, number of egg batches, total number of eggs laid, average number of eggs and oviposition index revealed that *E. tibilais* has better reproductive performance than other Piratinae reduviids like *E. vishnu* and *E. xavierei* (Figure 21.3).

Reproduction: Life Table Studies

Arilus cristatus (Linn.) laid in an average 64 eggs/female and eggs hatched within 60 to 61 days. Studies on the life history of a emesinae reduviid *Emesaya brevipennis* (Say) under field and laboratory conditions have revealed that this bug laid eggs during February and March. Similarly *Pseudometapterus umbrosus* (Blatchley) from southern Illinois laid eggs from early May to late October. *Coranus nodulosus* laid 43 eggs in 13.5 batch (Sahayaraj, 2001).

Prey Density

Reduviids individually required more and consumed greater number of preys than other predatory insects. Furthermore they killed more number of preys than they need to satiate it. One of the potential factors that need to be taken into account to understand field situation is the effect of prey scarcity or rarity. Prey scarcity caused the predator to compromise reproductive output for metabolic maintenance until sufficient prey is available to provide the necessary energy for reproduction. A life table was constructed in relation to the predator density. The net reproductive rate was higher in the 50 predator group and it decreased with an increase in the predator density. *R. marginatus* population doubling time (14.9 days) was found to be higher in the 100 predator category and increased as the predator density decreased (Sahayaraj, 2001a). Sahayaraj (2001b) also studied the influence of *Corcyra cephalonica* larval density on the development and reproduction of *N. therasii.* The results revealed that prey density has great impact on the preoviposition period

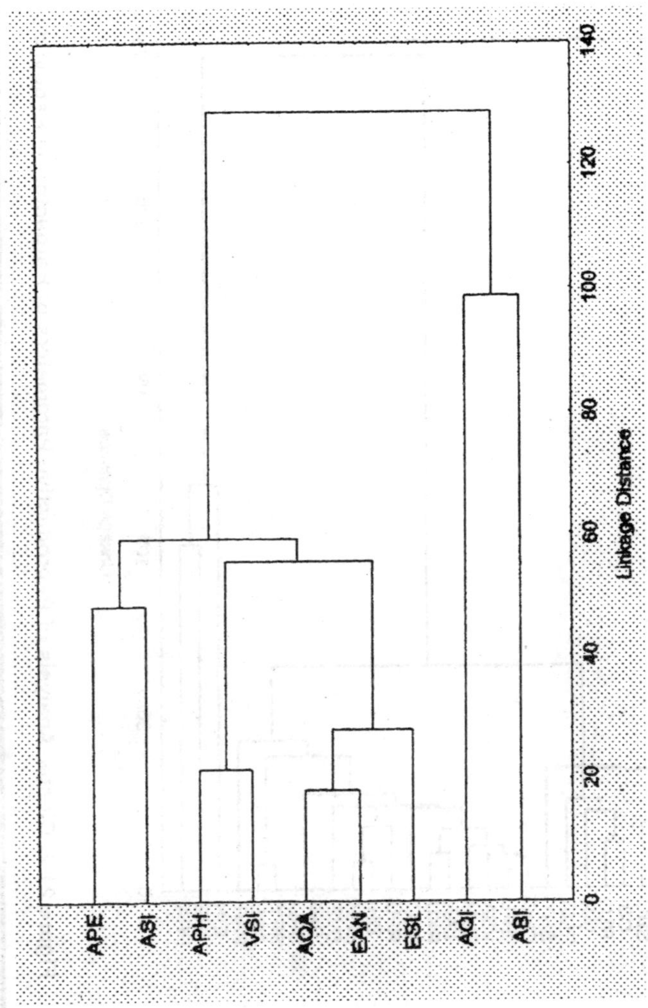

Figure 21.1: Cluster Analyses of Reproductive Parameters of Acanthaspidinae Reduviids

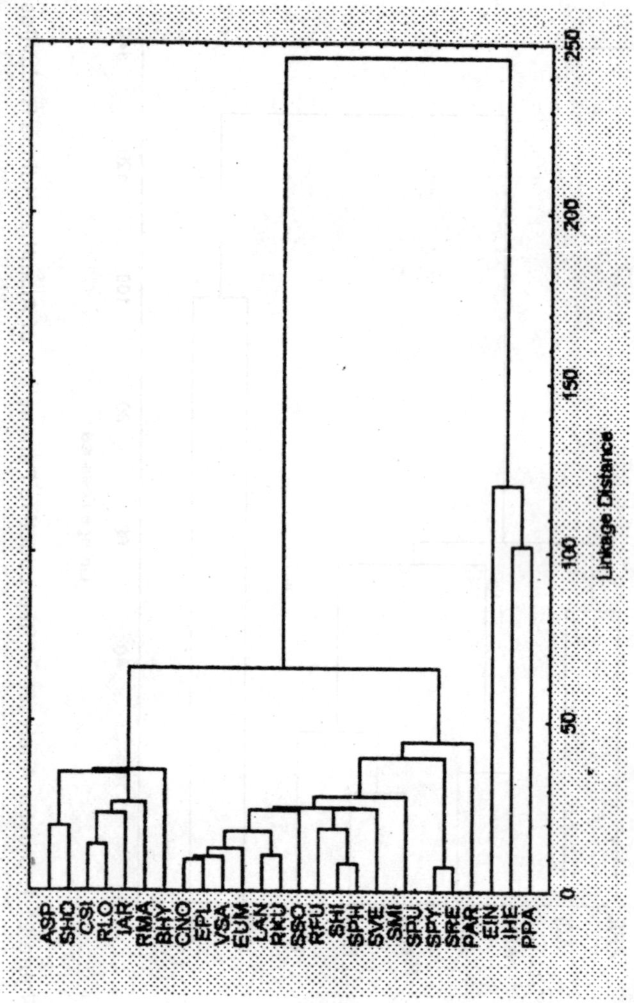

Figure 21.2: Cluster Analysis of Reproductive Parameters of Harpactorine Reduviids

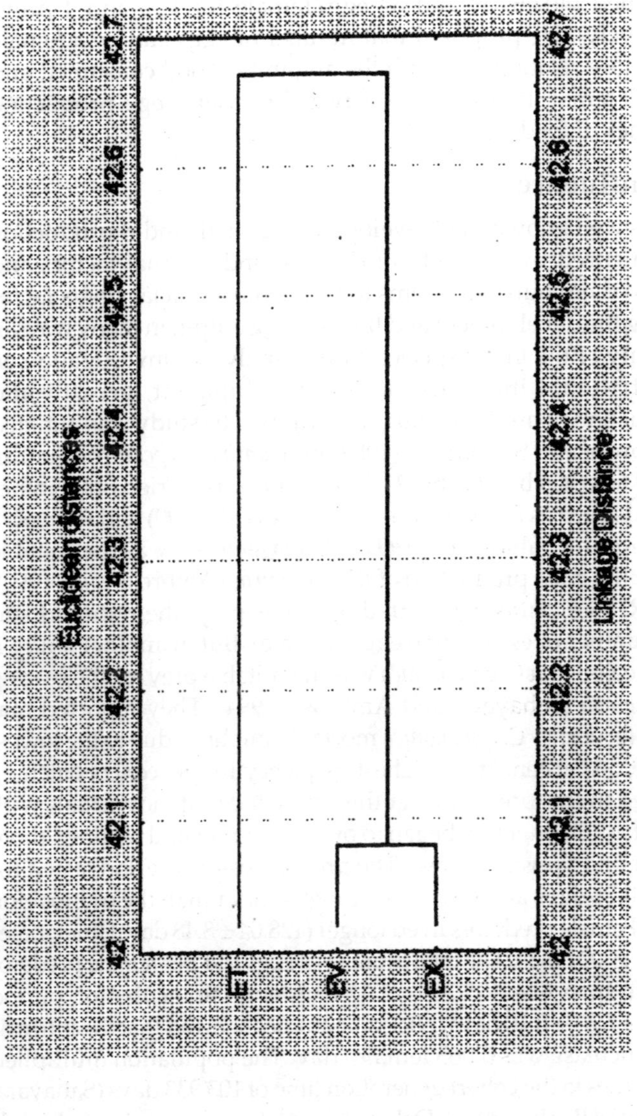

Figure 21.3: Cluster Analysis of Reproductive Parameters of Piratinae Reduviids

(15.66, 12.33, 9.5 and 8.16 days for 1, 2, 4 and 8 prey density respectively). Egg laying capacity was also higher in 8 prey density (68 egg/female) and diminished when the density of the prey diminished (40.66, 59 and 62 egg/female for 1, 2 and 4 preys respectively). He proposed that number of eggs laid by a female shows close relationship with the amount of food consumed and the adult female may cause a delay in the onset of egg-production due to lack of food.

Prey Influence

It is well known that development, growth and reproduction are influenced by a variety of external and internal factors and amongst them nutrition seems to be the most crucial single factor affecting the development and the total egg output in many ways in a majority of the insect species. Surprisingly, the investigation on the nutritional influence on the development, growth and reproduction of predatory bugs are scanty. The study was initiated by Ambrose and Subburasu (1988) in *Acanthaspis pedestris* Stal. an oriental assassin bug (Table 21.2). Latter it was carried out in other reduviids such as *R. marginatus* (Ambrose *et al.*, 1990) and *R. kumarii* (Ambrose and Sahayarani, 1991). *Chrotogonus* sp. was the suitable prey for the mass production of *R. margiantus* (Ambrose *et al.*, 1990). When *R. margiantus* was reared with this prey, the reduviid laid more number of eggs (210.0 eggs/female). But in another study it was revealed that *C. cephalonica* was the suitable prey for the rearing of *N. therasii* (Sahayaraj and Ambrose, 1994). They reported that highest fecundity *C. cephalonica* mortality might be due to the lowest frequency 0 per cent of the highest frequency 100 per cent of hatching observed in this prey. Further they reported that individuals with shortest adult longevity began to oviposit earlier and retained more number of oviposition days. The pre-oviposition period lasted for 18.64 ± 0.76 days and it was 6.31 times shorter than the oviposition period. Female predators lived longer (128.04 ± 8.48 days) than males (82.84 ± 11.09 days). The total number of eggs laid per female predator in its life span was 405.28 ± 22.15 eggs and the incubation period lasted for 6.81 ± 0.10 days. Life-table studies showed that the intrinsic rate of increase was 0.063 female/day. The population multiplied 292.28 times in the cohort generation time of 103.933 days (Sahayaraj and Gabriel Paulraj, 2001). Delayed ovipostion observed in reduviids might be due to the predatory stress developed and the consequent shortage of food.

Table 21.2: Prey Influence on the Development, Fecundity and Life Table of Reduviids

Prey	*TDP*	*SUR*	*FAL*	*FC*	*HA*	*NRR*	*WM*	*DT*
			Acanthaspis siva					
S. litura	116.70	–	66.47	114.12	–	33.78	1.23	23.45
E. vittella	126.43	–	59.12	098.11	–	29.48	1.21	25.11
C. cephalonica	132.00	–	56.92	083.87	–	20.62	1.28	29.45
			A. pedestris					
C. compresus	122	55	35	07	45	–	–	–
M. domestica	98	78	68	23	44	–	–	–
O. obesus	97	63	25	09	38	–	–	–
C. cephalonica	70.47	86.00	56.0	173.5	84.3	–	–	–
			Rhynocoris marginatus					
S. litura	46.71	–	128.04	148.6	92.45	292.29	–	11.15
S. litura	70.38	–	79.87	191.89	–	65.95	1.399	14.49
E. vittella	76.93	–	68.47	151.44	–	41.35	1.314	17.67
C. cephalonica	83.47	–	65.29	121.75	–	27.90	1.278	20.15
			Rhynocoris fuscipes					
C. cephalonica	–	63.0	43.71	–	–	69.86	6.66	1.109
S. litura	42.00	–	47.17	114.34	–	32.88	1.48	12.54
E. vittella	48.19	–	42.33	98.49	–	25.05	1.42	13.68
C. cephalonica	50.02	–	38.11	87.14	–	18.15	1.33	16.72
			R. kumarii					
S. litura	63.45	–	111.46	221.46	–	83.80	1.419	13.87
E. vittella	70.43	–	94.16	207.18	–	66.70	1.342	16.45
C. cephalonica	74.12	–	83.33	186.32	–	56.90	1.314	18.13
Chrotogonus sp.	74.4	70.0	101.5	57.0	87.7	–	–	–
O. obersus	71.5	20.0	86.5	74.5	93.6	–	–	–
Burchus theoromae	97.6	23.3	110.0	91.5	94.4	–	–	–
			Cydnocoris gilvus					
S. litura	62.33	–	43.00	89.00	–	31.90	1.522	11.58
Oxya nitidula	65.00	–	39.00	72.00	–	20.94	1.437	13.32
Odontotesmes obesus	68.00	–	27.00	45.30	–	12.88	1.361	15.72

Contd...

Table 21.2–Contd...

Prey	TDP	SUR	FAL	FC	HA	NRR	WM	DT
				Alleocranum biannulipes				
Anagasta kuehniella	75.56	8.7	16.92	140.27	96.0	–	–	–
Corcyra cephalonica	51.62	–	20.34	247.64	81.97	–	–	–
Tribolium confusum	39.97	–	21.45	444.75	96.67	–	–	–

Refer Sahayaraj (2004).

TDP: Total developmental period; SUR: Survival rate; FAL: Female adult longevity; FC: Fecundity; HA: Hatchability; NRR, WM, DT: Indicates not observed.

The total number of eggs laid by a female was influenced by nymphal diet (F = 3.56; df = 1.63; p = 0.05). Maximum and minimum number of eggs in a batch was in favour of *S. litura* fed group. However, total number of eggs per female and number of egg batches per female were higher in *H. armigera* fed group. Within each nymphal diet regime, *H. armigera* larvae as nymphal and adult diet appeared more suitable for egg production than *S. litura*. When *R. marginatus* was fed on *H. armigera*, egg production diminished to virtually zero after 54 days, with the rest of the oviposition period contributing to a minor extent to the total egg production. Egg hatching was not significant from *H. armigera* to *S. litura*. The mean incubation period of eggs from females that fed on *H. armigera* was shorter. The difference in incubation period was statistically not significant (ANOVA, f = 2.536; df = 1, p = 0.1164). Daily rate of oviposition per female fluctuated during lifetime for the two diet regimes. The intrinsic rate of increase, the net reproductive rate, the finite rate of increase, the weekly multiplication followed the trend of total fecundity, with females on *H. armigera* demonstrating a higher intrinsic rate of increase and a net reproductive rate. The generation times were 72.283 and 76.881 days on *S. litura* and *H. armigera*, respectively. Nutritional quality of the preys revealed that *S. litura* has more amounts of proteins (0.29 ± 0.02 g/g), carbohydrates (0.26 ± 0.08 g/g) and lipids (0.15 ± 0.02 g/g) than *H. armigera* (0.23 ± 0.04, 0.21 ± 0.01 and 0.14 ± 0.01 g/g for carbohydrates, proteins and lipids respectively). *R. marginatus* consumed more amount of carbohydrates (84 per cent), proteins (95.7 per cent) and lipids (57.1 per cent) from *H. armigera* than *S. litura* (73.1 per cent of carbohydrates, 79.3 per cent of proteins and 46.7 per cent of lipids) (Sahayaraj *et al.*, 2004). The adult female body size is often influenced by the larval nutrition, which in turn affects the fecundity of the insect. Stewart *et al.* (1991) pointed out the diet relationship between female body size and number of ovarioles. This generalization is supported by the results of the present study, wherein the fecundity was higher in *R. marginatus* reared on *H. armigera,* which were heavier as compared to the smaller *R. marginatus* females reared on *S. litura.* This is in accordance with the results of Phoofolo and Ovbrycki (1997), which confirms the fact that female body size may be a good indicator of potential fecundity. The nature of the prey species consumed by the predator has a major impact on the fecundity of the predators as stated by Enkegaard *et al.* (1997) and Hansen *et al.* (1999). Thus the higher fecundity of *R. marginatus* females reared on *H. armigera,* may be due to the higher primary nutrients in the latter.

The life table statistics of *R. marginatus* varied with *S. litura* compared to *H. armigera.* The net reproductive rate was higher for *H. armigera* reared *R. marginatus* as compared to *S. litura* fed *R. marginatus.* The net reproductive rate was lower than the gross reproductive rate in both the cases. This is in accordance with the results of George *et al.* (1998a), who have attributed such finding to the sharp decline in the survivorship value of parent females. However, the net reproductive rate, innate capacity for increase, intrinsic rate of increase and weekly multiplication declined in the case of *A. siva* on *S. litura* (George *et al.*, 1998a) (Table 21.2). Similar findings were recorded by Venkatesan *et al.* (1997) in *C. gilvus,* Claver (1998) in *R. kumarii,* Singh *et al.* (2002) in *Cotesia glomeratus* (Linn.).

Kind of Prey

Ambrose (1999) reported that reduviids feed only live prey. One of the basic requisite for a successful biological control agent is the availability of a sound, low cost rearing and mass multiplication technique, which are not available for reduviids. Other natural enemies like pentatomids lacewings and ladybird beetles have been mass produced not only with the live preys, but also with frozen or heat killed preys or artificial diets. Such efforts could be undertaken in reduviids, definitely as most of the pest control practitioners utilize this efficient predators in most of the crops IPM programme. Studies on the effectiveness of alive and frozen prey on the reproductive performance of the reduviid were scanty in the literature. This lacuna in reduviid was first fulfilled by Sahayaraj and Jeyalakshmi (2002). They tried to mass-produce *R. marginatus* with alive and frozen larvae of *Corcyra cephalonica*. According to them the net reproductive rate was higher in alive prey fed group (205.48) than frozen larvae fed group (156.7). However, they don't find any difference in weekly multiplication (7.49 and 7.48 for alive and frozen larvae fed group respectively) and doubling time corrected (10.13 and 10.41 for alive and frozen larvae fed group respectively). This study postulated an idea that the fertility of a predator could vary based on the type of host larvae provided.

Prey Rearing Media

The pre-oviposition period of *R. marginatus* varied with prey rearing media. Among the four media, the shortest pre-oviposition period was observed in Jowar fed Corcyra (JFC) (62.7 ± 15 days) followed by Rice Fed Corcyra (RFC) (70.8 ± 1.3 days), Wheat Fed Corcyra (WFC) (73.3 ± 1.6 days) and SFC (68.0 ± 1.2 days). In an average the pre-oviposition period was 68.7 days which is longer than 41.4 and 38.1 days for *R. marginatus* fed on *S. litura* and *C. cephalonica* respectively further studies are necessary to confirm these results. As observed for the pre-oviposition, the oviposition period was also minimum and maximum in JFC (44.7 days) and SFC (71.6 days) respectively. The variations in the reproductive period among the four categories were due to the variations in the nymphal duration. As observed in pre-oviposition, the fecundity was also higher in JFC group (360.9 ± 13.5) followed by WFC (223.6 ± 5.5) and the minimum was observed in SFC (208.3 ± 3.9) and RFC (208.4 ± 4.4). A diet consists solely by Corcyra reared on wheat thus seems to meet the optimal requirements of *R. marginatus* egg-laying female. The low egg production found by George (2000) for females fed with *C. cephalonica* (100.97 egg/female) larvae corresponds well with the 148.74 eggs for females fed with *S. litura*. More number of batches of eggs were recorded in JFC (6.49 ± 1.84) followed by wheat (5.00 ± 1.06). The maximum fecundity of JFC group might be due to the maximum weight of female predator. Evans (1976) reported that the mean fecundity of a predator decreased depending upon the predator density. The reason for the differences in pre-oviposition and total number of eggs laid per female may be due to the different types of diet provided to the *C. cephalonica*. It may also be due to differences in the experimental setup, in that the female *R. marginatus* was present along with a male throughout its life-time (16) and in the present examination the females were together with many males. The maximum number of nymphs emerged from the *R. marginatus* fed with *C. cephalonica* reared on jower (82.88 per cent) and wheat (82.51 per cent). The minimum numbers of nymphs than those emerged from the sorghum (74.33 per cent) fed category (Sahayaraj and Jeyalakshmi, 2002).

Predator Polymorphism

Polymorphism is an adaptive feature of the insects. An individual when exists in more than one form is called polymorphism. It exists either in colour variation or in sexual variations. These are present in reduviid predators. Both variations can be used for the survival. Polymorphism has been reported in four reduviids such as *N. therasii, Ectomocoris tibialis, R. marginatus* and *A. siva*. Polymorphism of the farmer two reduviids were studied by Sahayaraj (1991). *R. marginatus* exhibited three colour polymorphic farms such as niger (N), sanguineous (S) and nigerosanguineous (NS). Among the three farms, S-laid more number of eggs (175) followed by N (154.67) and NS (101.25) (Ambrose, 1980). *A. siva* exists in four morphs with uniform development of wings but with different morphological features in prothorax and legs. They have black prothorax and legs, black prothorax and white-banded legs, white spotted prothorax and white-banded legs, and white spotted prothorax and black legs. Ambrose (1980) also reported that black prothorax and legs had higher fecundity, moderate index of oviposition days and the highest hatchability. *Neohaematorrophus therasii* exhibited different colour morphs *viz.*, entirely violaceous black males and female (B), entirely red (except abdomen and a central prothoraci spot) male and female (R) and violaceous black male with red corium (BR) (Sahayaraj, 1991). It is a polyphagous reduviid present in cotton field. This polymorphic diversity of this reduviid does not have any influence on the egg laying capacity of this reduviid. What ever it may be the morph nature, *N. therasii* laid 45 to 47 eggs during its reproductive period. Similarly the total number of batches of eggs laid by the reduviid was not also changed (11 to 12 batched/female). However, *R. marginatus* laid the more days (22 days) followed by BR (20) and B (12 to 13 days). A late male (AM), brachypterous male (BM) and female (BF), micropterous male (MM) and female (MF) are the sexually dimorphic forms observed in

E. tibialis. Mating partners of MM and MF laid more number of eggs (49.83 eggs/female) than AM and MF (37.83 eggs/female). However, the farmer category has higher index of oviposition (in days) (36.98) than the latter (32.97) (Sahayaraj, 1991).

Predator Sex Ratio

Sex ratio is the key factor for determining the fecundity, hatchability and other parameters involved in population biology. A male biased sex ratio beyond certain limits badly affects the egg production in insects. Over proportion of male's results in higher frequency of male encounters, unnecessary copulation and copulation attempts (chasing) adding up to an increased disturbance of the female. *Rhynocoris fuscipes* is an important reduviid predator predating upon more than 42 insect pests. In this study we evaluated the suitability of sex ratios (4 : 1; 2 : 1, 3 : 2, 1 : 1, 2 : 3 and 1 : 2 female : male) for the mass production of *R. fuscipes* by life table analysis. The gross reproductive rate, orbitrary value of innate capacity for increase, innate rate of increase and weekly multiplication were higher in 3 : 2 sex ratio. This category also had minimum precise generation (75.61 days) and doubling time (13.10 days) respectively. The present findings suggested that 3 female and 2 male combinations of adults could be maintained at the time of mass rearing of this predator. Sahayaraj *et al.* (2003) studied the suitability of *R. marginatus* sex ratio for mass rearing. They reported that the fecundity was higher in sex ratio of 0.6 (3 female and 2 male) than other categories such as 0.4 (2 female 3 male), 0.5 (1 female and 1 male), 0.7 (2 female and 1 male) and 0.8 (4 female and 1 male). The better group also had maximum gross weekly multiplication (1.449) and hypothetical female in F2 generation (3286.1).

Rearing Container

First observation in reduviid indicated that crowding of the predator reduced fecundity, when the size of the container was increased the productive capacity of *Ectomocoris tibialis* has not changed. For instance both 80 and 160 capacity reared *Ectomocoris tibialis* produced 25 and 24 eggs respectively. However, it reduces the pre-oviposition period from 37 days to 20 and 24 days for 80 and 160 ml capacity containers respectively.

Acknowledgements

We thank Rev. Fr. Dr. A. Antonysamy, S.J., Principal, and Prof. M. Thomas Punthan, Head, Department of Zoology, St. Xavier's college, Palayamkottai for institutional facilities.

References

Ambrose, D.P., 1999. *Assassin Bugs*. Science Publications Inc. Enfield, New Hamphne, USA, pp. 337.

Ambrose, D.P., Saju T. and Sahayaraj, K., 1990. Prey influence on the development, reproduction and size of assassin bug *Rhynocoris marginatus*. *Environ. and Ecol.*, 8(1): 280–287.

Ambrose, D.P. and Subbarasu, P.A. 1988. Prey influence on the development, reproduction and size of the assassin bug *Acanthaspis pedestris*. *Environ. and Ecol.*, 6(4): 948–955.

Ambrose, D.P. 1980. Bioecology, ecophysiology and ethology of reduviids (Heteroptera) of the scrub jungles of Tamil Nadu, India. *Ph.D. Thesis*, University of Madras, Madras, India.

Birch, S., 1948. The intrinsic rate of natural increase in an insect population. *Journal of Animal Ecology*, 17: 15–26.

Bragdon, T.H., 1951. Colorometric determination of blood lipids. *Journal of Biochemistry*, p. 190–513.

Claver, M.A., 1998. Biocontrol potential evaluation of a reduviid predator *Rhynocoris kumarii* Ambrose and Livingstone (Insecta : Heteroptdra : Reduvuiidae) against three cotton insect pests. *Ph.D. Thesis*, Manaonmaniam Sundaranar University, Tirunelveli.

Enkegaard, A., Saradar, M.A. and Brodsgaard, H.F., 1997. The predatory mite Hypoaspis mites: Biological and demographic characteristics on two prey species, the mushroom sciarid fly, *Lycoriella solani*, and the mould mite, *Tyrophagus putrescentiae*. *Entomologia Experimentalis et Applicata*, 82: 135–146.

Evans, H.F., 1976. The effect of prey density and host plant characteristics on oviposition and fertility in *Anthocoris confusus* (Reuter). *Ecol. Ent.*, 1: 157–161.

George, P.J.E., 1999. Biology and life table of *Rhynocoris marginatus* (Fab.) (Heteroptera : Reduviidae) on three lepidopteran insect pests. *Journal of Biological Control*, 13: 33–38.

George, P.J.E., 2000. Life tables and intrinsic rate of natural increase of the three morphs of *Rhynocoris* marginatus (Fab.) (Heteroptera : Reduviidae) on *Corcyra cephalonica* Stainton. *Journal of Experimental Zoology*, 3(1): 59–63.

George, P.J.E., Seenivasan, R. and Kannan, S., 1998a. Influence of prey species on the development and reproduction of *Acanthaspis siva* Distant (Heteroptera : Reduviidae). *Entomon.*, 23(4): 69–75.

George, P.J.E., Seenivasaga, R. and Kannan, S., 1998b. Biology and life table studies of *Acanthaspis pedestris* stal. (Heteroptera : Reduviidae) population on three lepidopteron insect pest. *Journal of Biological Control*, 12(1): 1–6.

George, P.J.E., Kannagi, J. and Ambrose, D.P., 2002. Nutritional influence of prey on the biology and biochemistry in *Rhynocoris marginatus* (Fab.). *Journal of Biological Control*, 16(1): 1–4.

Hansen, D.L., Brodsgaard, H.F. and Enkegaard, A., 1999. Life table characteristics of *Macrolophus caliginosus* preying upon *Trtranychus urticae*. *Entomologia Experimentalis et Applicata*, 93: 269–275.

House, H.L., 1966. The role of nutritional principles in biological control. *The Canadian Entomologist*, 98: 1121–1134.

House, H.L., 1977. Nutrition of natural enemies. In: *Biological Control of Insects by Augmentation of Natural Enemies*, (Eds.) R.L. Ridgeway and S.B. Vinson. Plenum, New York, p. 151–182.

Isikber, A.A. and Copland, M.J.W., 2002. Effects of various aphid foods on *Cycloneda sanguinea*. *Entomologia Experimentalis et Applicata*, 102: 93–97.

Jayaraman, J., 1985. *Laboratory Manual of Biochemistry*. Wiley Eastern Limited, New Delhi.

Lowrey, O.H, Rosebrough, J., Farr A.L. and Randell, R.J., 1951. Protein measurement with folin phenol reagent. *Journal of Biochemistry*, 193: 265–275.

New, T.R., 1991. *Insects as Predators*. New South Wales Univ. Press, Kensington.

Phoofolo, M.W. and Obrycki, J.J., 1997. Comparative prey suitability of *Ostrinia nubilalis* eggs and *Acyrthosiphon pisum* for *Coleomegilla maculata*. *Biological Control*, 9: 167–172.

Sahayaraj, K., 1991. Bioecology, Ecophysiology and Ethology of chosen predatory hemipterans and their potential in biological control (Insecta : Heteroptera : Reduviidae). *Ph.D. Thesis*, Madurai Kamaraj University, Madurai, India.

Sahayaraj, K., 1994. Biocontrol potential evaluation of the reduviid predator *Rhynocoris marginatus* (Fabricius) to the serious groundnut pest *Spodoptera litura* (Fabricius) by functional response study. *Fresenius Envir. Bull.*, 3: 546–550.

Sahayaraj, K., 1995. Bio-efficacy and development of reduviid predator, *Rhynocoris marginatus*, on *Spodoptera litura*. *International Arachis Newsletter*, 15: 56–57.

Sahayaraj, K., 1999a. Field evaluation of the predator, *Rhynocoris marginatus* (Fab.), on two groundnut defoliators. *International Arachis Newsletter*, 19: 41–42.

Sahayaraj, K., 1999b. Effect of prey and their ages on the feeding preference of *Rhynocoris marginatus* (Fab.). *International Arachis Newsletter*, 19: 39–40.

Sahayaraj, K. and Paulraj, M.G., 2001. Rearing and life table of reduviid predator *Rhynocoris marginatus* Fab. (Het., Reduviidae) on *Spodoptera litura* Fab. (Lep., Noctuidae) larvae. *J. Appl. Ent.*, 125: 321–325.

Sahayaraj, K. and Jeyalakshmi, T., 2002. Mass rearing of *Rhynocoris marginatus* Fab. on live andf rozen larvae of *Corcyra cephalonica* Stainton. *Entomol. Croat.*, 6(1–2): 35–49.

Sahayaraj, K. and Martin, P., 2003. Assessment of *Rhynocoris marginatus* (Fab.) (Hemiptera : Reduviidae) as augmented control in groundnut pests. *Journal of Central European Agriculture.*

Sahayaraj, K. and Paulraj, M.G., 2003. Insect pests and beneficial arthropods of groundnut in relation to wind velocity. *Asian Journal of Microbiology, Biotechnology and Environmental Science*, 5(2): 233–235.

Sahayaraj, K., 2004. Reduviids in biological control. In: *Indian Insect Predators in Biological Control*, (Ed.) Sahayaraj, K. Daya Publishing House, New Delhi, India, p. 134–176.

Singh, O.H., Devjani, P. and Singh, T.K., 2002. Life table of *Cotesia glomeratus* (Linnaeus), a larval parasitoid of *Pieris canidia* (Sparrman). *Journal of Biological Control*, 16(1): 11–13.

Southwood, T.R.E., 1978. *Ecological Methods with Particular Reference to the Study of Insect Population*. Chapman and Hall, London, pp. 524.

Stewart, L.A., Hemiptinne, J.L. and Dixon, A.F.G., 1991. Reproductive tactics of ladybird beetles: Relationships between egg size, ovariole number and developmental time. *Functional Ecology*, 5: 380–385.

Strohmeyer, H.H., Stamp, N.E., Jarzomski, C.M. and Bowers, M.D., 1998. Prey species and prey diet affect growth of invertebrate predators. *Ecological Entomology*, 16: 68–79.

Thompson, W.R., 1951. The specificity of host relations in predaceous insects. *Canadian Entomology*, 83: 262–280.

Thompson, S.N. and Hagen, K.S., 1999. Nutrition of entomophagous insects and other arthropods. In: *Handbook of Biological Control*, (Eds.) T.S. Bellows and T.W. Fisher. Academic Press, New York, p. 594–638.

Venkatesan, S., Seenivasagan, R. and Karuppasamy, G., 1997. Influence of prey species on feeding response, development and reproduction of the reduviid *Cydnocoris gilvus* Burm. (Reduviidae : Heteroptera). *Entomon.*, 22(1): 21–27.

Wiebe, A.P. and Obrycki, J.J., 2002. Prey suitability of *Galerucella pusilla* eggs for two generalist predators, *Coleomegilla maculata* and *Chrysoperla carnea*. *Biological Control*, 23: 143–148.

Wigglesworth, V.B. 1934. Physiology of ecdysis in *Rodnius prolixus* (Hemiptera) II. Factors controlling moulting and metamorphosis. *Journal of Microscience*, 77: 19–222.

Chapter 22

Impact of Mercury on Hypothalamic Nuclear Center (NPO) in a Freshwater Teleost *Danio equipinatus* with Special Reference to Reproductive Physiology

Binay Kumar Singh, Kaushalendra Kumar, Md. Shamsul Haque and Ritesh Mandal
Department of Zoology, Government P.G. College, Shahdol – 484 001 (M.P.) India

ABSTRACT

Neurosecretory cells have vital role in the regulation of reproductive physiology of teleosts. Cells of nucleus preopticus (NPO) shows significant deposition of neurosecretory material during spawning period. It has been observed that due to mercury intoxication in the teleost, *Danio equipinatus* inhabitant of river Sone the deposition of neurosecretory material becomes low resulting in reduction in the reproductive physiology.

Keywords: *Nucleus preopticus, Neurosecretory material, Danio equipinatus.*

Introduction

In fishes, the various processes of reproductive physiology run under the influence of neuroendocrine factors. The Cells of Nucleus

Preopticus (NPO) is one of the most predominant factors that control the overall phenomena of reproductive physiology by releasing neurosecretory substances. A large number of workers (Scharrer, 1932; Bargmann, 1962; Dodd and Kerr, 1963; Sathyanesan, 1965; Saxena, 1980; Joshi and Joshi, 1989; Pahari and Pahari, 1979 and Tischenko *et al.*, 1976) have established a co-relationship between cells of nucleus preopticus the process of reproduction and the impact of heavy metals upon it. But nevertheless, there are few more hidden facts yet to be explored.

Present work deals with the impact of mercury on the hypothalamic cells of nucleus preopticus in a teleost fish, *Danio equipinatus* (Ham) inhabiting river Sone. River Sone receives salts of mercury as effluents discharge from Orient Paper Mills, Amlai that affect adversely causing damage to the nucleus preopticus and thereby, declining reproductive capabilities.

Materials and Methods

Female freshwater teleost, *Danio equipinatus* were collected separately from river Sone and from other water resource (River Mudna) in the first week of each month for a complete reproductive cycle. Fishes of both the groups were brought to laboratory and sacrified. Their brain and ovary were dissected out, fixed and stained for recording observations.

Results and Discussion

In *Danio equipinatus* collected from other water resource free from mercury intoxication, the cells of nucleus preopticus were well developed, paired and inverted 'L' shaped structure with rich neurosecretory granules concentrated around the nucleus. The ovaries were full-grown and remain packed with all stages of oocytes including mature ovum also during spawning phase (Figure 22.1).

In the fishes collected from river Sone it was observed that the cells of nucleus preopticus were hypertrophied with lesser concentration of neurosecretory granules and the ovary was also loosely packed with fewer oocytes and corpora atretica (Figure 22.2).

By observing the above facts in the fish *Danio*, the decline in reproductive capabilities probably might be due to hypertrophied cells of nucleus preopticus. Such a hypertrophy in the cells of nucleus preopticus is due to mercury intoxication in the water of river Sone

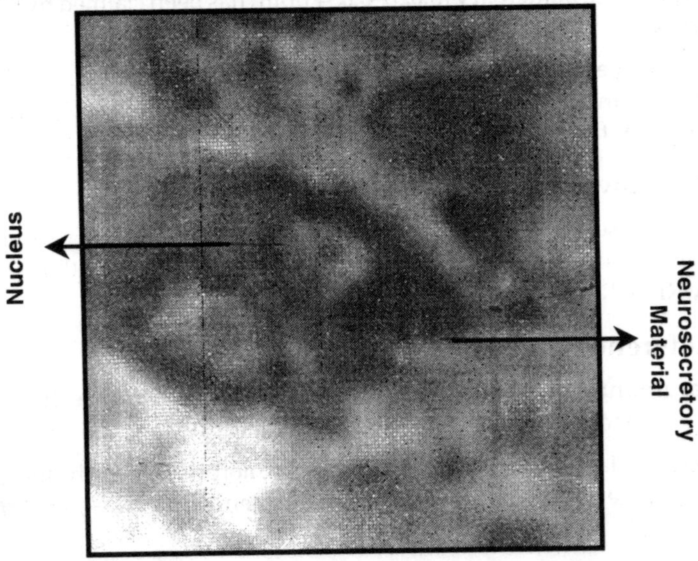

Figure 22.1: Showing Neurosecretory Materials Concentrated Around the Nucleus

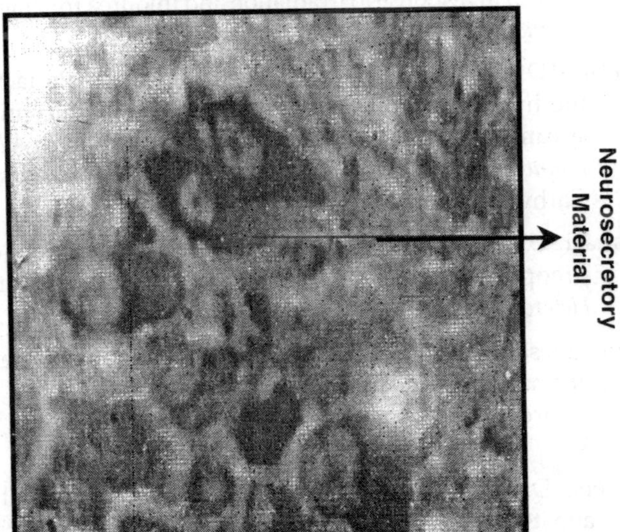

Figure 22.2: Showing Lesser Concentration of Neurosecretory Materials Around the Nucleus

as presence of mercury (0.028–0.054 ppm) has been claimed by CSE (2004).

The earlier observations of Haider and Sathyanesan (1973), Pahari and Pahari (1979), Saxena (1976), Upadhyay (1984) in different fish species also support this opinion.

Acknowledgement

The senior author is grateful to UGC, CRO, Bhopal for providing the financial assistance vide sanction letter No. 4-001 (03) 2004 (MRP/CRO), Dated 27–03–2004.

References

Bargmann, W., 1962. Neurosecretorische nervenfaserh und adenohypophyse. *Anat. Anz.*, 109: 206–261.

Dodd, J.M. and Kerr, T., 1963. Comparative morphology and histology of the hypothalamo-hypophyseal system. *Symp. Zool.*, 13: 909–927.

Haider, S. and Sathyanesan, A.G., 1973. Changes in the neurosecretory system of freshwater teleost, *Esomus danrica* (Ham.) in response to irradiation and thiourea treatment, *Strah. Lentherapie*, 145(4): 436–442.

Joshi, K.D. and Joshi, P.C., 1989. Correlation of climatic factors with the hypothalamo-hypophyseal-gonadal activity in the hill stream teleost, *Puntius dukai* (Day). In: *Nat. Symp. on Advances in Limnology and Conservation of Endangered Fish Species*, H.N.B. Garhwal University, Srinagar, pp. 35.

Pahari, U. and Pahari, B., 1979. Effect of inorganic merou on preoptico-neurohypophyseal system of freshwater fish, *Heteropneustes fossilis*. In: *Proc. 66th Ind. Sci. Cong. Asso.*, pp. 71.

Sathyanesan, A.G., 1965, Hypothalamo-hypophyseal system in the normal and hypophysectomized teleost, *Porichthys notatus*, Girard and its response to continuous light. *J. Morph.*, 117: 25–48.

Saxena, D.N., 1976. The hypothalamo-neurohypophyseal system and its physiological co-relationship with the reproductive cycle in Indian freshwater goby, *Glossogobius giuris* (Ham). *Zool. Jehrb. Anat.*, 96: 45–73.

Scharrer, E., 1932. Secretory cells in the mid brain of the European minnow, *Phoxinus laevis* (L.). *J. Comp. Neurol.*, 55: 573–576.

Upadhyay, M., 1984. Studies on hypothalamo-neurohypophyseal system in freshwater fish, *Esomus danrica* with special reference to reproductive cycle. *Ph.D. Thesis*, A.P.S. Univ., Rewa.

Chapter 23

Annual Sexual Rhythm in Relation to Environmental Cues in the Freshwater Turtle, *Kachuga tecta* (Gray) and *Kachuga smithi* (Gray) in Jammu Shiwaliks, J&K State

Anil K. Verma* and V.K. Gupta**

**Head, Department of Zoology, Government P.G. College, Udhampur, J&K State*

***Division of Pharmacology, Regional Research Laboratory (CSIR), Canal Road, Jammu – 180 001*

ABSTRACT

Impact of environmental factors like temperature, photoperiod, humidity and rainfall on the testicular cycle of the two sympatric freshwater turtle species *viz., Kachuga tecta* (Gray) and *Kachuga smithi* (Gray) inhabiting Chenab river-fed distributory Gho-Manhasan Khul in Jammu has been investigated. The study indicates that the reproductive cycle in the male individuals of these two species like the other reptilian forms, is influenced not only by the complex interactions of the environmental cues but the internal rhythm also. The interspecific differences in the reproductive cycles and the differential nature of responses exhibited by these forms in response to the similar set of factors reflected the differences in the relative control of annual sexual rhythm by the internal cues in concert with the Environmental monitoring and not by the latter (extrinsic factors) alone. The interspecific variations in the reproductive cycles of the two turtle species reflects the differences in the extent to which these factors are involved and the relative role played by them in regulating/monitoring different phases of testicular cycle under prevailing agro-environmental conditions in these two sympatric turtle species in northern India.

Keywords: *Kachuga tecta, Kachuga smithi; Testes; GSI; Environmental factors.*

Introduction

Freshwater turtles are known to exhibit marked changes in the intensity and periodicity of feeding, reproductive and behavioural activities in relation to the seasonal shift in environmental condition (Bury, 1979). The inactivity usually occurs in winter period but during warm spells, the turtles resume their activities (Cagle, 1954). Although variations in the seasonal periodicity of these activities in various turtle species in response to environmental factors have been recorded world over (Bury, 1979; Moll, 1979; Litch, 1972). Nevertheless, a fruitful attention about study of the reproductive aspects in relation to environmental factors, has been found scant in case of Indian turtles so much that the hard-shelled freshwater turtle belonging to family: Emydidae have remained least attended to by the biologists save for Duda *et al.* (1993, 1999); Yadav (1982); Verma and Sahi (1996, 1997); Verma and Duda (1998); Verma *et al.* (1999); and Vyas (1997).

Therefore, this study was envisaged to generate data pertinent to turtle reproduction and add further information to the existing knowledge of the impact of environmental cues on the reproductive cycles of *K. tecta* and *K. smithi* inhabiting river Chenab fed distributaries in Jammu Shiwaliks. These waterbodies face hazards in their water-level and vegetation at times, in different months of the year alongwith the variation of the temperature regimen and rainfall which ultimately affect the annual activity and the reproductive strategies adopted by the inhabiting turtle forms.

Material and Methods

Turtle specimens collected during this study were all mature individuals, maturity assessed by the method of Verma and Sahi (1996). The detailed description of the study area characteristics, various collecting techniques employed; seasonal gravimetric changes in the testes further processing of the gonads etc. have been explained elsewhere in details (Verma, 1992).

Observations and Discussion

The seasonal changes in the temperature, humidity, rainfall and photoperiod recorded at the collection site during present study have been depicted in Figure 23.1 and seasonal changes in the testes weight and gonado-somatic indices (GSI) of the male *K. tecta* and *K. smithi* have been respectively depicted in Figures 23.2–23.4 and Tables 23.1–23.4.

Table 23.1: Showing Size-pattern of the Testes and Body Weight Recorded for the Adult Males of *K. tecta* during the Years 1989–91 (Range in Parentheses)

Months	*Number of Observations (n = 36)*	*Body Weight Mean ± S.D. (in gms)*	*Testes Weight Mean ± S.D. (in mgs)*	*Percentage Change (+/–)*
March	6	117.66 ± 12.85 (104–130)	202 ± 22.01 (180–230)	– 37.6%
April	5	116.80 ± 9.09 (104–127)	235 ± 92.94 (165–370)	+ 16%
May	4	118.50 ± 13.43 (107–136)	292 ± 64.56 (218–360)	+ 24%
June	3	126.00 ± 7.21 (120–134)	326 ± 86.12 (272–425)	+ 11.6%
July	4	124.25 ± 12.47 (108–135)	868 ± 108.64 (762–1015)	+ 166%
August	5	132.40 ± 8.20 (123–140)	954.2 ± 351.23 (395–1373)	+ 9.9%
September	4	123.50 ± 25.30 (105–160)	148 ± 28.31 (110–174)	– 84.4%
October	3	141.00 ± 53.40 (96–200)	141 ± 45.88 (108–194)	– 4.7%
November	2	110.50 ± 4.95 (107–14)	324 ± 66.47 (277–371)	+ 129%

Table 23.2: Showing the Month-wise Variations of GSI in the Adult Male Turtle *K. tecta* Investigated During the Years 1989–1991, at Jammu

Month	*No. of Observations (n = 36)*	*Gonado Somatic Index (GSI)*
March	6	0.17
April	5	0.20
May	4	0.24
June	3	0.25
July	4	0.70
August	5	0.72
September	4	0.12
October	3	0.10
November	2	0.29

Table 23.3: Showing Size-pattern of the Testes and Body Weight Recorded for the Adult Males of *K. smithi* Sacrificed during the Years 1989–91 (Range in Parentheses)

Months	*Number of Observations (n = 34)*	*Body Weight Mean ± S.D. (in gms)*	*Testes Weight Mean ± S.D. (in mgs)*	*Percentage Change (+/–)*
March	4	127.75 ± 29.00 (95–131)	349 ± 184.84 (182–533)	– 10.01%
April	4	123.25 ± 23.92 (90–145)	332 ± 141.36 (158–497)	– 4.87%
May	5	127.20 ± 24.79 (105–168)	495 ± 115.2 (320–620)	+ 49.09%
June	4	128.75 ± 11.47 (120–141)	527 ± 139.41 (610–720)	+ 6.46%
July	4	118.50 ± 25.63 (91–145)	750 ± 164.19 (596–906)	+ 42.3%
August	5	127.40 ± 15.42 (110–146)	1357 ± 438.41 (908–2003)	+ 80.93%
September	4	121.25 ± 12.28 (109–135)	939 ± 65.21 (838–1020)	– 30.8%
October	4	118.00 ± 17.83 (98–139)	443 ± 182.1 (198–633)	– 52.8%

Table 23.4: Showing the Month-wise Variation of GSI in the Adult Male Turtle *K. smithi* Investigated During the Years 1989–1991, at Jammu

Month	*No. of Observations (n = 34)*	*Gonado Somatic Index (GSI)*
March	4	0.30
April	4	0.26
May	5	0.38
June	4	0.40
July	4	0.63
August	5	1.03
September	4	0.84
October	4	0.34

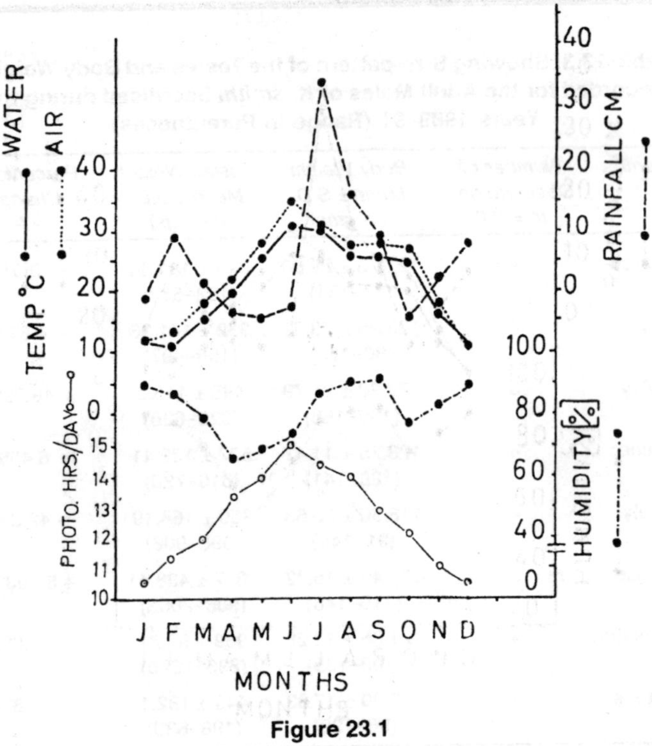

Figure 23.1

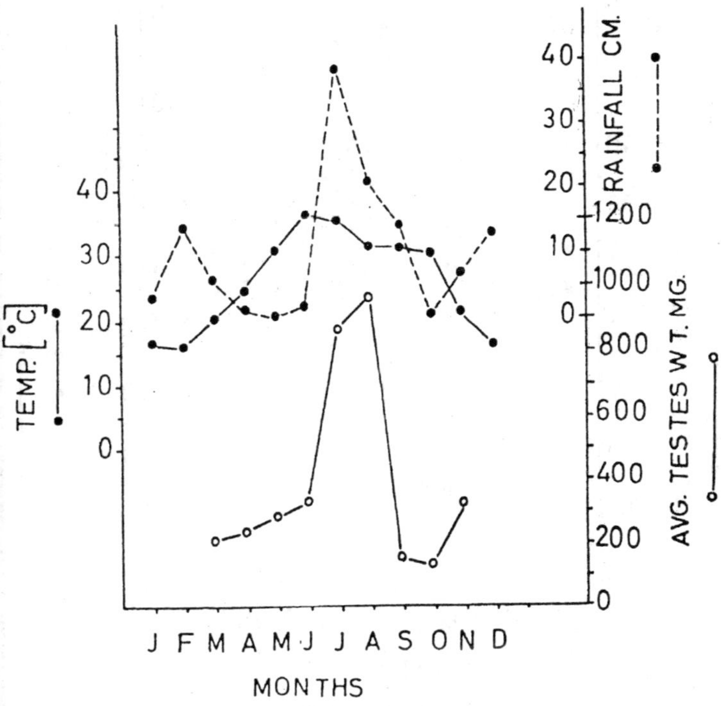

Figure 23.2

The turtles emerge from hibernation in March with relatively small testes of comparatively low weight. These increased progressively and steadily April onwards till June whereafter, the testes gained enormous weights during late July until maximum weight was obtained in the month of August in both the turtle species. August onwards a sudden rather rapid reduction in testes weights until October was recorded. The turtles entered hibernation in November with their testes already regressed. The spermatogenic activity in both turtles peaked in August and coincided with peak GSI (Figure 23.4).

Sexual periodicity of many species is influenced by many extrinsic and intrinsic factors; and of the extrinsic factors that influence reproduction, light, temperature, humidity, rainfall and food supply have been established as the important (Baker, 1938;

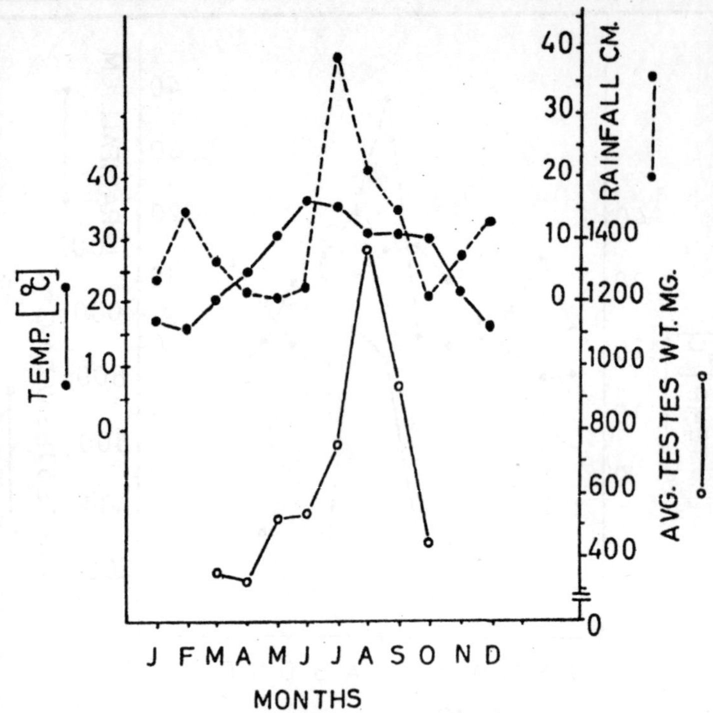

Figure 23.3

Lack, 1954) cues. The events during the testicular cycle of these turtles reveals that the temperature, light, and to some extent food supply, emerge as the important factors to influence, the sexual cycle, a finding consistent with that of Licht (1972) who has found that temperature was the most important timing cue for breeding cycles in reptiles and that photoperiod at best had a secondary influence. It has been found that the spermatogenic cycle of the testes in both the turtles (*K. tecta* and *K. smithi*) closely followed the changes in ambient temperature in that, the spermatogenic activity started rapidly in May when ambient temperature was on the increase and the activity underwent a cessation, late September onwards.

The testicular recrudescence was in progress in April/May (when the average monthly temperature was around 30°C) and it extended for the two successive months (June–July) with the rising

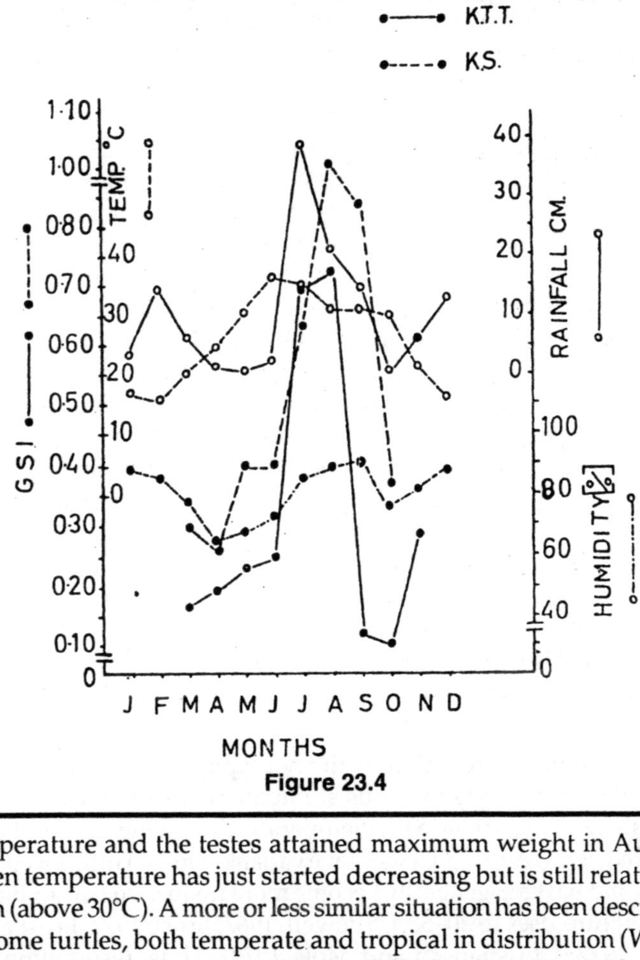

Figure 23.4

temperature and the testes attained maximum weight in August when temperature has just started decreasing but is still relatively high (above 30°C). A more or less similar situation has been described in some turtles, both temperate and tropical in distribution (White and Murphy, 1973; Lofts and Tsui, 1977; Singh, 1977; McPherson and Marion 1981; Yadav, 1982; Ganzhorn and Licht, 1983; Gasith and Sidis, 1985; Mitchell, 1985; Rao and Shaad, 1985).

While the atmospheric temperature maybe increasingly on the higher side (Figure 23.1) during these months to be deterimental for spermatogenesis, it is suggested that the adverse thermal effect may be offset by the increased rainfall which not only expectedly lowers the overall thermal condition of the water body in which these lives but also provides an effective thermal buffer to maintain the temperature of the stream a good few degree below the ambient temperature, to help these turtles to embark upon final active testicular maturation phase, as evidenced by rising GSI in both turtles in June/July to peak in August. Warm temperature have been demonstrated as important for testicular maturation in some snakes (Aldridge, 1975; Hawley and Aleksiuk, 1976). Sudden rise in the weight of the, testes (Figures 23.2–23.3) during the months of July–August/September thus stands explained, the progressive decline in testes weight thereafter, leading to a testicular regression and onset of the refractory phase, at a time when the average monthly temperature is well below 30°C (in October/November) demands explanation. The increasing winter temperature in April and onwards and concomitant onset of spermatogenesis could therefore, be physiologically related; a view consistent with the findings of, earlier workers (Licht, 1972; Spaet, 1973; McPherson and Marion, 1981; Duvall, Guillette and Jones, 1982; Marion, 1982; Gasith and Sidis, 1985; Mitchell, 1985). It could, therefore, be generalized that elevated body temperatures appear to be a requirement for testicular recrudescence, although factors other than temperature may also intervene in some components of the cycle. Since present study was not designed to explore photoperiodism in detail, the possibility of photoperiodic dependence for recrudescence, cannot be, however, ruled out.

Although the present study did not explore in detail the precise thermal requirements for testis stimulation, it seems reasonable, nevertheless, to conclude that temperatures intermediate to those recorded here (*i.e.* 16–36°C) will have variable effects. Turtle's feeding habits and basking generally become more frequent when ambient temperature rises to 20–30°C (Powell, 1967; Ernst, 1972; Ernst and Ernst, 1972; Christiansen and Moll, 1973). Thus, resumption of feeding after a winter hibernal spell and occasional basking would facilitate testicular recrudescence in spring and early summer, and pattern of testes recrudescence would, presumably represent a combination of rising average water temperatures and opportunity for basking.

However, testicular regression may not be regulated by the same proximate cues as recrudescence, although the temperature regimen during the August/September and September/October are more or less the same as in March/April/May. It is, therefore, safe to presume that the testis regression in the turtles (*K. tecta* and *K. smithi*) like some other turtles and lizards (Licht, 1973; Fischer, 1974; Ganzhorn and Licht, 1983), may at least be, partially due to the onset of refractoriness (transient condition) in testes in late summer, and/or in late autumn and by the time this insensitive condition terminated, environmental (ambient) temperatures are not likely to be high enough for a second increase in testis activity until sometime in the next summer. Contrarily, McPherson and Marion (1981) have attributed testicular regression in (*Sternotherus odoratus*) to the fall/ decline in water temperature and photoperiod, which does not parallel the events in the emydid turtles described.

The peak spermiogenesis in both the turtles *i.e. K. tecta* and *K. smithi* registered in the month of August, is not inconsistent with Moll's (1979) generalization that "a peak sperm production in majority of temperate zone turtles is reached in August", but differs to the extent that *K. tecta* and *K. smithi* do not fall within the temperate zone limits. Similar reports of August peak of spermatogenesis even in tropical turtles are quite frequent in literature (Singh, 1977; Moll, 1979; Yadav, 1982; Gasith and Sidis, 1985; Mitchell, 1985; Rao and Shaad, 1985) and, therefore, contrasts with the present finding. It may therefore, be concluded that diverse cycle may be met in the turtles independent of the geographical or climatic zonation.

The seasonal variations in the gonado somatic index (GSI) in relation to the environmental factors, in the two turtles (*K. tecta* and *K. smithi*) suggest that the temperature and duration of sunshine (photoperiod) may be the main factors affecting the gonado somatic index. The GSI in both the turtles rises with rising temperature; increasing duration of sunshine (Figure 23.4) and it's decrease in June (in *K. smithi*) when the temperature and sunshine are still on the increase suggests the intervention of some other factor(s) possibly the intrinsic factors. Rainfall does not seem to be related to the gonado somatic index in these turtles. The curve plotted for the GSI in both the species follows more or less similar contour (Figure 23.4) excepting that there is an insignificant rise in GSI during November in case of *K. tecta*, the period when both the temperature and photoperiod has considerably decreased and the pre-hibernal GSI slightly higher than in the month following its emergence, unlike in the other turtle *K. smithi*, where animal enters hibernation with a reduced GSI which does not increase in October (pre-hibernal month) as would be expected if these climatic factors were the only ones of importance affecting testis growth. This clearly suggests that some other factor or factors must be operating the *K. smithi* to inhibit testes growth (recrudescence) during September and October, but may be less effective in the *K. tecta* at this season. Likewise, temperature and photoperiod are known to control/regulate the timing of spermatogenic events in other turtles (Burger, 1937; McPherson and Marion, 1981; Ganzhorn and Licht, 1983; Mitchell, 1985) but their mode of action in relation to other environmental factors is still not precisely known.

Perusal of the Figure 23.4 indicates that GSI in *K. tecta* touched the lowest a month before the turtle entered the hibernation in late November, but not before a mild spermatogenic activity was set in the month of October and in *K. smithi*, on the other hand, GSI reached nadir a month after the turtle emerged in March from hibernation. This finding makes an interesting and an important observation on the refractory phase of male cycle in these two emydid turtles.

While the refractory phase of the testicular cycle in *K. tecta* obviously is a short duration phase, extending from September to October, in *K. smithi* on the other hand it turns out to be a more extended phase spreading from September to April. It is clear from these observations that the conditions of temperature and/or photoperiod obtaining in the month of October are the optimal conditions for termination of refractory phase in one turtle (*K. tecta*) while entirely different set of conditions of temperature and photoperiod like those obtaining in the month of April, are effective in terminating the refractory phase in *K. smithi*. All other conditions of food availability and water quality remaining the same for both the turtles which infact are strict sympatriats.

A logical deduction to follow from what has been said above is that differential responses of these two sympatric turtles in respect of the different timings of the termination of refractory phase in them, following simultaneous attainment of the peak in spermatogenic activity in August, would strongly favour implication of internal rhythms which are apt to differ with species in regulating the sexual cycle in these turtles as has been also evidenced in other reptiles (Licht, 1972, 1984; Ganzhorn and Licht, 1983; Mitchell, 1985).

Secondly, the finding that despite different timings of termination of refractory phase, whence different timings of the onset of testicular activity (recrudescence and early spermatogenesis) in these two sympatic turtles, the peak in spermatogenic activity is reached temporally almost in the same months (July/August in *K. tecta* and August/September in *K. smithi*) would only point to markedly different rates at which spermatogenic activity takes place in these turtles, it being much faster in *K. smithi* than in *K. tecta*. The different rates of spermatogenic activity in the two strictly sympatric emydid species would further support the assumption made above that involvement of intrinsic factors like internal rhythms in regulation of testicular activity/cycle can not be easily dismissed.

References

Aldridge, R.D., 1975. Environmental control of spermatogenesis in the rattle snake *Crotalus vinidis*. *Copeia*, p. 493–496.

Baker, J.R., 1938. The evolution of breeding seasons. In: *Evolution*, (Ed.) G. de Bee. Oxford, Clarendon, pp. 161–177.

Burger, J.W., 1937. Experimental sexual photoperiodicity in the male turtle *Pseudemys elegans* (Wild). *Am. Nat.*, 71: 481–487.

Bury, R.B., 1979. Population ecology of freshwater turtles. In: *Turtles: Perspectives and Research*, (Eds.) M. Harless and H. Morlock. John Willey and Sons. New York, p. 571–602.

Cagle, F.R., 1954. Observations on the life-cycles of painted turtles (genus *Chrysemys*). *Am. Midl. Nat.*, 52: 225–235.

Christiansen, J.L. and Moll, E.O., 1973. Latitudinal reproductive variation within a single subspecies of painted turtle, *Chrysemys picta belli*. *Herpetologica*, 29(2): 152–163.

Duda, P.L.; Verma, A.K. and Sahi, D.N., 1993. Sex-ratios in freshwater turtles from Jammu, India. *Hamadryad*, 18: 10–12.

Duda, P.L., Verma, A.K. and Sahi, D.N., 1999. Ecological notes on the freshwater hard-shelled Turtles (Family: Emydidae) of Jammu and Kashmir State, India. In: *Advances in Fish and Wildlife Ecology and Biology*, Vol. II (Ed.) B.L. Kaul. Daya Publishing House, New Delhi, p. 181–185.

Duvall, D.L., Guillette, J.T. and Jones, R.E., 1982. Environmental control of reptilian reproductive cycles. In: *Biology of the Reptilia*, (Eds.) Carl Gans. and F.H. Pough. Academic Press, London, 14: 209–231.

Ernst, C.H., 1972. Temperature activity relationship in the painted turtle, *Chrysemys pictaa*. *Copeia*, p. 217–222.

Ernst, C.H. and Ernst B.M., 1972. Biology of *Chrysemys picta belli* in south western Minnesota. In: *Proc. Min. Acad. Sci.*, 38: 77–80.

Fischer, K., 1974. Die Stererung der Fortpflan zungszyklen bei mann lichen reptilien, Forschretterder. *Zoologia*, 22: 362–390.

Ganzhorn, D. and Licht, P., 1983. Regulation of seasonal gonadal cycles by temperature in the painted turtle *Chrysemys picta*. *Copeia*, p. 343–358.

Gasith, A. and Sidis, I., 1985. Sexual activity in the Terrapin *Mauremys caspica rivulata* in Israel, in relation to the testicular cycle and climatic factors. *J. of Herpeto.*, p. 255–260.

Hawley, A.W.L. and Aleksiuk, M., 1976. The influence of photoperiod and temperature on seasonal testicular recrudescence in the red-sided garter snake (*Thamnophis sirtalis parietalis*). *Comp. Biochem. Physiol.*, 53: 215–221.

Lack, D., 1954. *The Natural Regulation of Animal Numbers*. Claredon Press, Oxford.

Licht, P., 1972. Problems in the experimentation of annual physiological cycles in reptiles. In: *Hibernation– Hypothermia IV Symposium.* Elsevier Publ. Co., Amesterdam, p. 681–712.

Licht, P., 1973. Thermal and photic influences on reptilian reproduction. *Excerpta Med. Int. Congr. Series*, 273: 185–190.

Lofts, B., and Tsui, H., 1977. Histological and histochemical changes in the gonads epididymes of the male soft-shelled turtle, *Trionyx sinensis. J. Zool., Lond.*, 1181: 57–88.

Marion, K.R., 1982. Reproductive cues for gonadal development in reptiles: Temperature and photoperiod effects on the testicular cycle of the lizard *Sceloporus undulatus. Herpetologica*, 38: 26–39.

McPherson, R.J. and Marion, K.R., 1981. Seasonal testicular cycle of the stinkpot turtle (*Stemotherus odoratus*) in central Albama. *Herpetologica*, 37: 33–40.

Mitchell, J.C., 1985. Variation in the male reproductive cycle in a Population of stinkpot turtles, *Sternotherus odoratus* from Virginia. *Copeia*, (1): 56–58.

Moll. E.O., 1979. Reproductive cycles and adaptations In: *Turtles: Perspectives and Research*, (Eds.) M. Harless and H. Morlock, John Wiley and Sons, New York, p. 305–331.

Powel, C., 1967. Female sexual cycle of *Chrysemys picta* and *Clemmys inscuplta* in Nova Scotia. *Can. Field Nat.*, 81: 134–140.

Rao, R.J. and Shaad, F.U., 1985. Sexual cycle of male freshwater turtle *Trionyx gangeticus* (Cuvier). *Herpetologica*, 41(4): 437–442.

Singh, D.P., 1977. Annual sexual rhythm in relation to environmental factors in a tropical pond turtle, *Lissemys punctata granosa*. *Herpetologica* 33: 190–194.

Spaet, R.H., 1973. Seasonal variation in the tubular and interstitial areas of the testes in *Sternotherus odoratus* (Latreille). *Master's Thesis*. Eastern Illinois University.

Verma, A.K., 1992. Studies on the male reproductive cycles of some emydid turtles of Jammu. *Ph.D. Thesis*, University of Jammu, Jammu, pp. 169.

Verma, A.K., and Sahi, D.N., 1996. On the size at maturity in the male freshwater turtles, *Kachuga tecta* and *Kachuga smithi* in Jammu, India. *Zoo' Sprint*, 11(7): 3 and 5.

Verma, A.K. and Sahi, D.N., 1997. Sexual dimorphism in *Kachuga tecta* and *Kachuga smithi* (Testudines : Bataguridae) in Jammu, India. *Hamadryad*, 22(2): 125–128.

Verma, A.K. and Duda, P.L., 1998. Testicular weight cycle of the freshwater turtle, *Kachuga tecta* in Jammu, J&K state. *J. of Ecobiol.*, 10(2): 117–122.

Verma, A.K., Sahi, D.N. and Duda, P.L., 1999. Morphology of the male reproductive organs in the Indian Sawback turtle *Kachuga tecta* and Brown roofed turtle, *Kachuga smithi* from J&K state. In: Fish and Wildlife Ecology and Biology, (Ed.) B.L. Kaul. Daya Publishing New Delhi, p. 160–165.

Vyas, R., 1997. Notes on growth and maturity in the Indian roofed terrapin (*Kachuga tecta*). *J. Bombay Nat. Hist. Soc.*, 94: 160–162.

White, J.B. and Murphy, G.G., 1973. The reproductive cycle and sexual dimorphism of the common snapping turtle. *Chelydra serpentina serpentina*. *Herpetologica*, 29: 240–246.

Yadav, M.R., 1982. Breeding biology of Indian tropical pond turtle, *Lissemys punctata granosa* (Schoepff). *Ph.D. Thesis*, Banaras Hindu University, Banaras.

Chapter 24

Reproductive Biology of Olive Ridley Sea Turtle (*Lepidochelys olivacea*): Insights from Research in Orissa Coast of India

Basudev Tripathy

Department of Environmental Sciences,
Andhra University, Visakhapatnam – 530 003 (Andhra Pradesh)
E-mail: tripathy@yahoo.co.uk

ABSTRACT

Reproductive biology forms a critical part in the life cycle of the living organisms. The reproductive biology and behavior of sea turtle is complex and interesting. While courtship and mating takes place in the water, nesting takes place on land. The various processes in reproduction include oviposition, non-attended incubation, hatching, hatchlings emergence and return to the sea. The hatchlings thereafter remain in the sea until they reach maturity and females return to their natal beaches to nest. Together, all these phases form an ecologically successful reproductive strategy. Orissa, on the east coast of India, is well known for the seasonal mass aggregation of migratory endangered olive ridley sea turtles for reproduction. Along

* Corresponding Address: C/o KV Bolarum, Q. No. A/2, Allenby Lines, J.J. Nagar P.O., Secunderabad – 500 087 (Andhra Pradesh); Phone: 040–27860650

the Orissa coast, the reproductive biology and behaviour of olive ridley sea turtles have been studied for the last two decades. This paper provides a comprehensive account of the offshore and onshore reproductive biology and behaviour of olive ridley sea turtles with a special focus on the studies conducted along the Orissa coast of India.

Keywords: *Sea turtle, Lepidochelys olivacea, Reproductive patch, Arribada, Hatchlings, Threats, Orissa.*

Introduction

Currently there is a surge of interest in life history strategies of sea turtles. Much of this interest revolves around an increased awareness that some populations are endangered and some species are ecologically and economically valuable. Although some aspects of a sea turtle life history such as growth rate and longevity are inadequately unknown, other aspects like fecundity and nesting behaviour are well documented. Reproduction in sea turtle occur within three general constraints; copulation, nesting and hatching (Georges *et al.*, 1993). Within these constraints, sea turtle share a number of general reproductive characteristics such as iteroperous breeding, stereotyped nesting laying relatively large number of eggs on multiple occasion and strong site fidelity to the nesting ground. Together these constraints and characteristics form the context of an ecologically successful reproductive strategy for the turtle. Following its reproductive effort, the female returns to the foraging areas which is sometime thousands of kilometers apart from the breeding ground and accumulate the energy to support her next reproductive effort (Miller, 1997). The production of the next generation of turtles results from a synergism of the efforts of the ecological conditions in the foraging areas on the energetic of the female parent and of beach environmental conditions for development of the embryos. To be successful, reproduction must occur when environmental conditions support adult activity. Further, environmental condition of the nesting beach must favour embryonic development and survival. Additionally, hatchlings must emerge on onshore and the offshore conditions enhance their chances of survival.

As a consequence of their food and habitat needs, the various species of sea turtles are unevenly distributed throughout the tropical and sub tropical seas of the world (Pritchard, 1997). Among the seven species of sea turtles occurring worldwide, the flat back turtle (*Natator depressus*) is endemic to Australian continent and the Kemp's ridley turtle (*Lepidochelys kempi*) is restricted to the Gulf of Mexico (Limpus, 1971; Márquez, 1994). The other five species *i.e.* Leatherback turtle (*Dermochelys coriacea*), Hawksbill turtle (*Eretmochelys imbricata*), Loggerhead turtle (*Caretta caretta*), Green turtle (*Chelonia mydas*) and Olive ridley turtle (*Lepidochelys olivacea*) have wide range of distribution from cold temperate waters to warm temperate and tropical areas (Pritchard, 1971; Hirth, 1971; Witzell, 1983; Dodd, 1988; Márquez, 1990).

The olive ridley turtle (*Lepidochelys olivacea*) is considered to be the most abundant turtle having circum-global distribution (Pritchard, 1997). This is also the smallest of all the sea turtle species. Male and female olive ridleys migrate from their foraging grounds and aggregated in large numbers at selected sites for some part of the year for their reproduction (mating, nesting and hatching). The reproductive cycle of olive ridley turtle is annual (Márquez, 1990). After a successful copulation in the water, female ascent onto the beach for laying eggs. Nesting takes place either in a solitary way or in great simultaneous aggregations. The olive ridley sea turtle is well known for its synchronous mass nesting behaviour also known as *arribada* (Spanish for "arrival") in which several hundred thousands of female olive ridleys ascent to the beach in couple of nights and nest *en-masse.* Although the fact of mass nesting is well known, the phenomenon has so far been poorly understood by scientists. Most author credit the mass nesting behaviour with a survival value, seeing it as a strategy to swamp the predators with a very temporary over-abundance of food (Pritchard, 1969). After a prolonged (~ eight weeks) incubation period, the eggs hatch inside the nest and hatchlings then emerged out from the nest and crawl back to the sea. They remain there till attending their adulthood and return to breeding areas after maturity. However, not much is known on the detail reproductive biology of this species. This article attempted at reviewing some of the important aspects of the reproductive biology and nesting behaviour of the olive ridley sea turtle and with a special focus on the research conducted along the Orissa coast of India.

Reproduction in Sea Turtle

The reproductive biology in sea turtle seems to be very interesting and complex. The reproduction process is more or less similar among all the species (Hirth, 1980; Hendrickson, 1982; Miller, 1997). All species of turtle migrate, at least short distances between foraging and breeding areas, although the exact reason for migration of turtles is not known. Suitable sandy beaches with easy access for egg laying area are sometimes far from feeding areas and separated by thousands of kilometers and this may be one of the reason, turtles requires to travel long distances between their foraging and nesting habitat. Nevertheless, it is now well established that sea turtles migrate between their foraging and reproductive area with a high degree of accuracy of returning back to the same locality year after year (Limpus *et al.*, 1984; Bjorndal, 1982; Limpus *et al.*, 1992). Tag returns from many research studies across the world have established important aspects of the reproductive biology and movement of sea turtles (Balazs, 1999). Although turtles may not necessarily return to the beach of their birth, tagging and genetic studies have revealed that they return to the region of their birth for reproduction, which include courtship and mating in the near shore waters of the nesting ground and laying of eggs on the beach in relatively close proximity (0 to 5 km) during subsequent nesting attempts (Meylan, 1982; Limpus *et al.*, 1992).

Reproduction in turtles is not an annual phenomenon except for kemp's ridley turtle and possibly for olive ridley sea turtle (National Research Council, 1990; Márquez, 1994; Pandav and Choudhury, 2000; Tripathy, 2005). The remigration interval for females is from one to nine years; which varies among species (Van Buskirk and Crowder, 1994; Limpus *et al.*, 1984; Miller, 1997). Similarly, the inter-nesting period is also different among the species. *Chelonidae* with the exception of the ridleys, have an average interesting period of 12 to 16 days (Ehrhart, 1982; Miller, 1997). Information relating to the courtship and mating of sea turtles in general is not adequately known except for anecdotal accounts (Booth and Peters, 1972; Ehrhart, 1982; Dash and Kar, 1990). The range of species specific signals that contribute to receptiveness remains undefined. Mounting pair of turtles is often seen at the surface of the water but copulation occurs anywhere within the water column (Hughes, 1974; Limpus, 1995). During mating, male mount on the female on her back, with the claws of fore flippers and hind flippers, hooking the carapace of the female. Female receives bites on her flippers, neck and head that leave open sores requiring several weeks to heal. Several males are known to mate with several females and for several times during the breeding season. During the mating season, mating takes place almost continuously. The hemipenis of male extends much beyond the tail and enter into the cloaca of the female in a semicircular position. Although a pair may remain coupled for hours, the actual transfer of sperm may not require the entire period. Also there is the quick disposal of copulating pairs after mating is over or anytime during the copulation. However, mating pairs are less disturbed by human related activities such as fishing close to the mating place and water displacement on the mating pair due to boat movement (Tripathy, *pers. obser.*). Also males show fidelity to particular courtship areas in successive reproductive migrations (Limpus, 1993).

In general, sea turtles come to the seashore for nesting during night with possible exception of *Lepidochelys kempi* which is known for day nesting (Márquez, 1994). Sea turtle show strong fidelity to their nesting beach. However, it is not fully known as to why a population or an individual sea turtle uses a particular beach for nesting. Nevertheless, the general requirements for nesting beaches identified are suitable accessibility, away from any chance of inundation of eggs and good grain size to facilitate nest construction and gas exchange (Mortimer, 1990). The nesting behavior and nesting process in sea turtle complete through a series of successive stages. At the outset, the female swim parallel to the coast before moving onto the beach for egg laying. During this movement, they continuously raise their head out of the water may be to take note of the topography and profile of the beach to which they approach later in the night. Thereafter, they tend to crawl onto the shore beach. The gravid female are usually pushed onto the shore by the incoming tide and as soon as the tide recede they come in contact with the wet sand of the intertidal zone. At this stage, turtle can easily sense the disturbance by light, movement of human beings or disturbances by any other animals, which enable them to go back to the sea to escape the disturbances otherwise they crawl on to the shores slowly till they reach the selective place of egg laying. Soon after reaching the nesting site, the turtle began to excavate the body pit and clear the nesting site. These two processes occur almost simultaneously. It then starts rotating its body by moving either to the left or to the right by the movement of all the four flippers working together. During the body pit formation, the hind flippers push sands backward and also lift the sand from the pit. Soon after the completion of the process of pit excavation, it starts digging the egg chamber with the help of the two hind flippers alternately. During this process the posterior part oscillate left and right side of the cavity. As the egg chamber deepened, the turtle raise its anterior portion upward. After positioning itself over the completed egg chambers, the egg extrusion began. It lay egg singly or in-group of two or three at a time. During the oviposition, tears rolled down from both the eyes, perhaps as to help removing the sand particles from the eyes. After egg laying is completed, the turtle cover its egg chamber. The process of covering and packing the egg chamber is entirely done by the rear flippers alternately as in digging. After some amount of sand spread over the nest by a complete sequence of limb movements, it thumbs the nest supporting the full weight of its body to compact the sands. As soon as covering of the egg chamber is over, the female then returns back into the sea. Thus, there is absolutely non-attended incubation of the nest by the female.

Temperature is considered to be the most crucial factor influencing the incubation of the eggs, be they are in natural condition or artificially planted. Temperature also determines the sex of the turtle. The surrounding temperature of the sand and the ambient environment determine the sex ratio of the turtle during the embryonic development. The temperature dependent sex determination, along with the inaccessibility of males, makes determining the natural sex ratios and the population size difficult. The pivotal incubation temperature for turtle eggs varies between different species and population but tends to be around 26–32°C (Ackerman, 1997). Generality is that cooler temperatures produce males and warmer temperatures produce females and this hold for all species.

As soon as the embryo developed into a juvenile, the process of hatching starts with the young ones emerge out by rupturing the egg shell with the help a horny protuberance on the nasal tip commonly known as "egg tooth" which towards the end of the incubation period disappear, after few days of the juvenile emerged out from the nest (Miller, 1997). The process of breaking of the eggshell is also supplemented by the newly developed soft claws of fore flippers of the juvenile. As soon as the juvenile comes out of the egg chamber, it tends to crawl towards the sea (Lohmann *et al.*, 1997).

Reproductive Biology in Olive Ridley Sea Turtle

The olive ridley turtle (*Lepidochelys olivacea*) is widely distributed throughout the tropics, with the exception of the Gulf of Mexico, and appearing in the highest densities in the northern Indian Ocean and eastern Pacific Ocean (Márquez, 1990; Reichart, 1993; Ernst *et al.*, 1994). While the kemp's ridley (*Lepidochelys kempii*) has a single nesting population, olive ridley turtles are considered the most abundant of sea turtles (Pritchard, 1997). The annual migrations of both male and female olive ridleys to the same breeding area where fishing intensity is high, puts them at risk of getting entangled and drown in fishing nets. Similarly, the seasonal and ontogenic shifts in habitat of olive ridleys present challenges to biologists in assessing the population in a given period of time. The large aggregations at sea and at nesting beaches have also led to debates over whether sea turtles migrate in socially structured groups or independently (Plotkin *et al.*, 1995). Therefore, the spatio-temporal distribution of turtles during migration and breeding is not adequately known.

Courtship and Mating

Olive ridley sea turtle occupy the neritic zone during the breeding season. Males and females migrate to the breeding ground and aggregate in the coastal waters close to the shore near the mass nesting beaches. Kalb *et al.*, (1992) defined this assemblage of olive ridleys as, '*Reproductive Patch*'. According to them, this reproductive patch is composed of mature reproductive males and females engaged in courtship and mating. However, some amount of mating does take place far from the shore also (Pitman, 1990; Kopitsky *et al.*, 2000). Some males appear to remain in oceanic waters, are non-aggregated, and mate opportunistically as they intercept females en-route to near-shore breeding grounds and nesting beaches (Plotkin *et al.*, 1995; 1996; Kopitsky *et al.*, 2000). The reproductive patches in the area are dynamic throughout the season, varying in surface area and occasionally in location.

Along the Orissa coast the courtship and mating of olive ridley have been observed at least since early 1980s. Review of literature

(Dash and Kar, 1990; Pandav and Choudhury, 2000) indicates that male and female olive ridleys migrate to the coastal waters of Orissa traversing the Tamil Nadu and Andhra Pradesh coast along the east coast of India in October–November and remain in the coastal waters of Orissa till May–June (Dash and Kar, 1990; Shanker *et al.*, 2003). Mating takes place during early winter *i.e.* October–January. December is considered to be the peak period for mating of olive ridleys along the coast of Orissa and on the east coast of India (Dash and Kar, 1990; Pandav, 2000; Tripathy, 2005). Although, olive ridley turtles mating reported to occur all along the coastal waters of Orissa, major reproductive congregations are known to take place near the rookeries (Dash and Kar, 1990; Pandav, 2000; Ram, 2000; Tripathy, 2004). While the dynamics and size of reproductive patches along Orissa coast is not studied in details, the recent work of Ram (2000) at Gahirmatha and result of offshore congregations study by Tripathy (2004) at Rushikulya rookery indicates that the reproductive patch varies between 65 sq. km to 75 sq. km in size and keep on changing with time and space. Reproductively active male olive ridleys are also found to be aggressive during the mating period and try to mount on many objects. The studies by Pandav (2000) and Tripathy (2004) along the Orissa coast have observed mounting of three males over a female and male over male. The gestation period of olive ridley turtle in Orissa is not adequately known. However, laparoscopy study elsewhere and tag recaptures from breeding grounds in Orissa (Kalb, 1999; Pandav, 2000) suggest that the time lag between mating and egg laying for olive ridley turtle varies between 15 and 45 days.

Nesting Activities

The nesting of olive ridley sea turtles takes place both as *arribada* and in solitary ways. Solitary nesting emergence of olive ridley turtle is known to take place along the Orissa coast almost every month (Dash and Kar, 1990). However, the numbers of solitary nesting turtles along the Orissa coast are more during the January to May, which indicates that this is the main nesting season of olive ridley sea turtle along the Orissa coast (Dash and Kar, 1990).

The major nesting aggregations for olive ridley turtles (*arribada*) occur in few beaches of Costa Rica and Mexico in the east Pacific (Pritchard, 1997), and along the Orissa coast of India (Bustard 1976; Kar and Bhaskar, 1982; Pandav *et al.*, 1994). There are currently five olive ridley *arribada* beaches along the Pacific coast of the Americas, two in northern Costa Rica, two in southern Nicaragua, and one in southern Mexico. The Playa Escobilla in Oaxaca, Mexico currently supports the largest population of *arribada* nests in the western hemisphere (Kalb, 1999). Mexico has mass nesting populations of both species of the ridleys. The Pacific beaches of Oaxaca are known for olive ridley mass nesting, while the Atlantic coast of Tamaulipas is the only nesting ground of the kemp's ridley turtle. The Oaxaca beach (La Escobilla) supports a single *arribada* with an average of approximately 100,000 annual nestings/year. During 1996, the nesting population crossed the figure of 8,00,000 (Fuentes *et al.*, 1998). Since the total ban on adult olive ridley harvest in 1990, the number of annual nesting on this beach has increased from 1,50,000 in 1990 to 8,35,000 in 1996 and these turtles are now considered a recovering population (Ballestero *et al.*, 1998).

A significant portion of the world's olive ridley sea turtle population nest *en-masse* along the Orissa coast of India. The Orissa coast harbours three major *arribada* sites (Gahirmatha, Devi and Rushikulya rookery). The Gahirmatha rookery is known to be the largest reported *arribada* nesting ground for olive ridley sea turtles in the world (Bustard, 1976). Gahirmatha is a Marine Sanctuary and the mass nesting beach is part of Bhitarkanika National Park. It is located at the mouth of the River Maipura, between the Dharma River and the Paradeep port, along the northern Orissa coast. The mass nesting of olive ridley sea turtles at Gahirmatha was first reported by Dr. H.R. Bustard, an FAO/UNDP consultant during his all India crocodile survey in 1974 (Bustard, 1976). Estimates of mass nesting at Gahirmatha have ranged between 100,000 to 800,000 in different years [Patnaik *et al.*, 2001; Orissa Forest Department, 2004 (http://www.wildlifeorissa.org.in)]. Since its discovery, the mass nesting of olive ridley sea turtles at Gahirmatha was known to takes place along the 15 km mainland beach near the Maipura river mouth, from Ekakulanasi to Habalikhati, but this beach got fragmented in two, during a cyclonic storm of 1989 after which mass nesting was restricted to a four km long beach, renamed as the Ekakulanasi rookery (Pandav *et al.*, 1994). The 1999 'super-cyclone' that hit the Orissa coast, once again split Ekakulanasi into few smaller islands. Now mass nesting takes place on these small islands, each less than one km in length (Tripathy, *pers. obser.*). The mass nesting of olive ridley turtles at Gahirmatha takes place between December to March and the first *arribada* at Gahirmatha is normally followed by a second one of much lower intensity after a gap of 45–60 days (Dash and Kar, 1990). However, the recent trends in mass nesting at Gahirmatha indicate a failure of the second *arribada* (Bivash Pandav, unpublished information).

While more than a hundred thousand turtles are believed to nest annually at Gahirmatha, tens of thousands nest at other rookeries, mainly Devi and the Rushikulya river mouth (Kar, 1982; Pandav, 2000). The Devi rookery is located 100 km south of Gahirmatha, in Puri district of Orissa. This rookery was discovered in 1981 as the second mass nesting rookery along the Orissa coast but due to poor attention and no monitoring by any agencies, later the mass nesting beach was invaded with intensive *Casuarina* plantation (Pandav *et al.*, 1994). This probably has resulted in major shifting in nesting at this rookery by the turtles in an island located well inside the sea. At present major nesting activities takes place at the island only. Also there is no mass nesting monitoring at this rookery, hence the annual number of annual *arribada* nesters at Devi is not known except a study by Pandav (2000) during 1995 and 1997.

The third and the recently discovered mass nesting rookery for olive ridley turtles along the Orissa coast is located at the river mouth of Rushikulya, situated in the southern most Ganjam district of Orissa (Pandav *et al.*, 1994). This rookery is 300 km south of Gahirmatha and recent study indicate that mass nesting of turtles at this rookery is a regular event at least since 1994, although the number of turtles estimated annually fluctuate in great numbers (Pandav, 2000; Tripathy, 2005). Unlike the Gahirmatha and the Devi rookery (Tripathy, *pers. obser*) where sporadic nesting is almost continuous for the entire season (> 10 turtles/night), at the Rushikulya rookery, the trend of sporadic nesting is irregular with nesting intensity (intermediate nesting) increasing only before the commencement of the *arribada* and during the rest of the nights, there is either no nesting or low sporadic nesting (< 5 turtles/night).

Although the factor that synchronizes the huge aggregation of female turtles at selected sites is still unclear, Pritchard (1969) suggested that *arribadas* evolved as a form of predator swamping. This predator-swamping theory was further supported by Eckrich and Owens (1995) when they found that nests predation during the first 24 hours was significantly lower for *arribada* nests than solitary. The reason for selection of few selected beaches also attributes to the behaviour of *arribada* nesters and strong site fidelity (Plotkin *et al.*, 1995; Kalb, 1999). The study along Orissa coast by Pandav (2000) and Tripathy (2005) documented a significant low predation of nests laid during *arribadas* and supports the above theory. However, the level of predation to sporadic nests were quite high and is a matter of concern as hatching rates of solitary nests are higher than those of *arribada* nesters and hence play an important role in survival of the turtle. The nest predation pressure at the Rushikulya and the Devi rookery of Orissa is high due to human settlement (fishing villages) close to the mass nesting beach.

The occurrence of the *arribada* can be correlated with the phase of moon as has been reported by Márquez *et al.* (1982), Márquez and Van Dissel (1982), Kar and Dash (1984) and Dash and Kar (1990). In particular, *arribadas* at Gahirmatha along the Orissa coast have traditionally shown a greater regularity and closer affinity to the near tidal days of the lunar cycle. During the first and third quarter moon, tidal fluctuations are minimum. Much smaller neap tides occur during seventh day after full moon or new moon. Moreover, tides always start rising after sun down, which allows the turtles an early emergence to the beach during this period. This may be the reason, turtles ascend the beach in larger numbers during the neap phase of the tidal cycle.

Although there have been several attempts to define the qualities of a nesting beach, it is not obvious why a population of sea turtles or an individual sea turtle uses a particular beach (Mortimer, 1990). The temperature, weather condition, physiography of nesting beaches and its adjacent sea, conditions of tide, temperature and surface current circulation may play an important role to induce the females for selection of the site (Dash and Kar, 1990). These factors probably stimulate the olive ridleys for choosing the selected areas along the Orissa coast to deposit their eggs. However, detailed study on the environmental parameters responsible for *arribada* of olive ridley sea turtles along the Orissa coast is required.

Earlier tagging studies on olive ridleys have recorded the movement between their nesting beaches with a range of 85 to 160 km (Schulz, 1971; Meylan, 1995). Compared to these studies, the movement of olive ridleys between nesting beaches in Orissa (inter-rookery movement of female turtles; > 300 km) seems to be the longest for the species recorded so far (Pandav, 2000; Tripathy, 2005). Eckert *et al.* (1989) recorded movements between nesting beaches for leatherback sea turtle with a range of 30 to 110 km. Bjorndal *et al.*, (1983) reported that the distance between the intra-seasonal re-nesting attempts of loggerheads ranged up to 290 km. Other records of intra-seasonal nesting movements suggest that sea turtles are capable of moving long distances, but the proportion of individuals doing so is low (Limpus and Reed, 1985). Movement of olive ridleys between nesting beaches has also been observed between the two mass nesting sites in Costa Rica; Nancite and Ostional beaches (Cornelius and Robinson, 1985). This dynamic movement between beaches is remarkable and may represent an exception to the natal homing hypothesis (Bowen *et al.*, 1992). It is possible that beach exchange is part of a complex phenomenon that olive ridley turtles use to colonize or even move to another beach altogether.

So far, in Orissa there are six instances where olive ridleys have shifted their nesting beach (Dash and Kar, 1990; Pandav and Choudhury, 2000; Tripathy, 2005). Although the shifting of beach phenomenon by olive ridleys of Orissa is poorly understood, there could be more than one reason for this. The change in the river mouth and thereby beach topography and offshore disturbances may be playing important role in beach selection by turtles. Continuous monitoring at these nesting beaches in Orissa is essential to determine the degree of such movements between nesting beaches. While inter-rookery movement by olive ridley sea turtles in Orissa is known recently (Pandav and Choudhury, 1998; Pandav and Choudhury, 2000) and turtles tagged in Orissa were subsequently recorded in Sri Lanka as well, study by Tripathy (2005a) reveal the onward tag returns of olive ridley turtles tagged abroad (a Philippines tagged turtles was recaptured at the Rushikulya rookery during mass nesting of 2004). This indicates that olive ridley may be migrating to Indian coastal waters from many parts of the Indian Ocean and Pacific Ocean areas. More significantly, the genetic studies (Shanker *et al.*, 2004a) also reveal the distinctiveness of the population on the east coast of India and suggested that they may be ancestral to populations in the Atlantic and Pacific oceans.

The two species of *Lepidochelys* exhibit the longest inter-nesting periods (20 to 28 days for *L. kempi* and 17 to 30 days for *L. olivacea;*

Chavez, 1969 and Pritchard, 1969). The study by Pandav (2000) and Tripathy (2005) along the Orissa coast found the mean inter-nesting period (IP) of olive ridleys to be 22 days, (range: 20 to 25 days). The typical IP (internesting period) among other genera of *Chelonia* is 10 to 15 days (Ehrhart, 1982). On beaches where *arribadas* occur, the IP is usually the time between successive *arribadas* (Cornelius and Robinson, 1985). Depending on the beach, *arribadas* typically occur at two to four week intervals (Cornelius and Robinson, 1986; Márquez *et al.*, 1976; Pritchard, 1969) but delays can lengthen the IP to over two months. The difference between *Lepidochelys* and the other five species of turtle appears to be the ability to retain eggs for a longer time while awaiting the next *arribadas* (Kalb, 1999).

The amount of time required to complete the nesting process varies among the species (Hirth, 1980). Olive ridley turtle take nearly one hour to complete the nesting activities on the beach, from emerging from the surf to covering of egg chamber till going back to the sea. During oviposition the hind flippers are positioned to cover the nest chamber. In olive ridley turtle, both hind flippers are extended outwards on the sand behind the turtle so that the opening to the nest is not covered by the flippers but is covered by the posterior margin of the carapace. During egg laying, turtles are relatively tolerant of a modest level of external disturbance and the level of tolerance increases as the turtle lays more eggs. All marine turtles lay white spherical cleidoic eggs with flexible calcareous shells (Miller, 1985). However, the size of eggs varies between individual and also between species. The average size of olive ridley turtle egg is similar to that of kemp's ridley and hawksbill but much smaller than loggerhead turtle and is also much smaller in size as compared to green and leatherback turtle. Studies conducted along the Orissa coast suggest that the egg size of olive ridley turtle ranges between 3.6 to 3.9 cm (Biswas, 1982; Dash and Kar, 1990; Tripathy, 2005). Also, in relation to other sea turtle species and their own body size, olive ridley turtle were found to lay large clutches of small sized eggs (Van Buskirk and Crowder, 1994). The average number of eggs laid by a female per clutch ranges from several dozens to more than 155 and the egg diameter of olive ridleys ranges from 32.1 to 48.1 mm (Marquez, 1990; Deraniyagala, 1939; Schulz, 1975; Márquez *et al.*, 1976; Dash and Kar, 1990; Tripathy, 2005). However, significant variation occurs between different geographic locations.

Incubation and Hatching of Eggs

The incubation period of olive ridleys usually extends between 45 to 65 days (Márquez, 1990; Dash and Kar, 1990; Hewavisenthi, 1994; Kalb, 1999; Pandav, 2000; Tripathy, 2005). Usually it takes longer incubation period in late winter then the summer months. Therefore, temperature is considered as a vital factor influencing the incubation period of the nests, be they are natural or artificial and also determine the sex of the turtle. Dash and Kar (1990), Pandav and Choudhury (2000) and Tripathy (2005) have documented the duration of incubation period for olive ridley sea turtles along the Orissa to range between 48 to 55 days. The fertility of marine turtle eggs is high, typically greater than 80 per cent (Miller, 1985). The fertility rate of olive ridley eggs in Orissa beach is greater compared to other beaches (> 90 per cent; Dash and Kar, 1990; Tripathy *et al.*, 2003). However, substantial differences occur in hatching and emergence success of the eggs due to the peculiarity of the geomorphology at the nesting sites (Pandav, 2000; Tripathy, 2005).

The eggs in olive ridley turtle hatch after 6–8 weeks of incubation, depending on the temperature (Dash and Kar, 1990; Miller, 1997; Tripathy *et al.*, 2003). The emergence of sea turtle hatchlings from their nest has been recorded mostly during night or cool hours of the day (Bustard, 1967; Fowler 1979). During the night, the beach surface temperature is low and is known to favour emergence of hatchlings, and their movement towards the sea is guided by natural light in the horizon (Bustard, 1967; Witherington and Bjorndal, 1991; O'hara, 1980). During diurnal emergence, the heat generated by beach sand lowers their activity level, leaving them vulnerable to predation (Mrosovsky, 1978). Besides, hatchlings get desiccated and die due to high temperature during the day. Predation is much greater on hatchlings that emerge during the day than those that emerge at night, because most predators are diurnal and hatchlings are very visible on the sand (Stancyk, 1982; Brown and McDonald, 1995).

Information is lacking regarding trends in clutch sizes over the course of a nesting season for olive ridley sea turtles. There are some reports available on the size of the olive ridley turtle eggs from Indian coastline (Biswas, 1977; Dash and Kar, 1990). Clutches of sea turtle eggs typically have high hatching success (80 per cent or more) unless external factors such as predation, environmental changes microbial infections etc. intrude (Miller, 1997; Hitchins *et al.*, 2004).

The emergence success of olive ridley turtle hatchlings has been extensively studied in last two decades (Dash and Kat, 1990; Pandav, 2000; Tripathy *et al.*, 2003; Tripathy, 2005). Hatching success is considered to be influenced by a number of variable factors such as egg fertility, egg death by dehydration or predation, nest destruction by tides, extreme weather condition and poaching (Hitchin *et al.*, 2004). Hatching success of olive ridley sea turtle is very high provided the environmental conditions are conducive and the nests are not exposed to poaching either by man or other animals (Dash and Kar, 1990). Studies conducted on olive ridley turtle also reveal high variable rates of hatching success on the same beach depending on environmental conditions during the different nesting years. Dash and Kar's 11 years study at the Gahirmatha beach (1990) showed a variation of 85.27 to 90.81 per cent hatching success in different years. Similar study by Pandav (2000) and Tripathy (2005) indicate a lower, hatching success of 63.6 per cent at the Gahirmatha and high hatching success at the Rushikulya rookery of Orissa (95.3 per cent).

Threats to Olive Ridley Turtles, their Eggs and Hatchlings

There are natural causes of sea turtle mortality, which includes killing by other animals, and abiotic factors that are unfavourable both in the sea and on the land such as hurricane, erosion, rain etc. Many species from crabs to sharks are aquatic predators for sea turtle hatchlings and adults. However, the major cause for population decline of sea turtle globally is due to intense exploitation over many decades for meat, shell and eggs by humans. Today, many die as incidental catch in fisheries, and others are exposed to consequences of development such as habitat loss.

Due to the various threats and its endangered status, the olive ridley sea turtle is included in the schedule I of the Indian Wild Life (Protection) Act, 1972 and is prohibited from any kind of exploitation of the animal and its habitat. However, the population along the Orissa coast and rest of India is continuing to decline (Shanker and Choudhury, 2005). The major cause of mortality of adult turtle is intentional and accidental killing during fishing operation in the high seas. An increase in mortality of olive ridley sea turtles along the Orissa coast was documented from a few thousands in the early

1990s to more than 10,000 per year by 1999 (Pandav, 2000). A review of data suggest that this population may be on the verge of a decline, based on evidence from the failure of *arribadas* in recent years along many important beaches of Orissa and decline in adult sizes and high fishery-related mortality (Shanker *et al.*, 2004).

Lighting is a major problem recognized for sea turtle hatchlings and adult turtles (Witherington, 1992). Under natural conditions, sea turtle hatchlings move directly towards the sea after they emerge out from the nest. However, when any kind of artificial source of light is present near the nesting beach, they tend to move towards the source of that light as it disrupts the sea finding behaviour of the hatchlings (Miller, 1997). When sea turtle hatchlings emerge on beaches with artificial light, they have been previously recorded to become disoriented and move away from sea, towards the source of light (Frick, 1976; Witherington, 1992). The presence of artificial light near the nesting beach often causes mortality in hatchlings by directing them away from the sea (Peters and Verheijen, 1994). Artificial illumination near the beach is a major huddle for olive ridley sea turtle hatchlings all along the Orissa coast (Pandav and Choudhury, 2000; Tripathy *et al.*, 2003). In particular, the Rushikulya rookery beach is comparatively brightly illuminated than the other two rookeries in Orissa due to the presence of various artificial light sources such as the townships, roads and factory. As a result of this, the disoriented hatchlings get entrapped in vegetation and during daytime, large numbers of hatchlings found stranded behind the nesting beach. Because of this reason, high juvenile mortality takes place right on the nesting beach (Tripathy, 2005).

The river mouths are dynamics along the Orissa coast. As a result of this, substantial egg loss occur on the nesting beach. In almost every nesting colony, some nests are lost due to erosion-accretion and tidal inundation. The extent of egg mortality varies widely among beaches and between different years. Nests deposited on shifting beaches are more susceptible to destruction from erosion. Dash and Kar (1990), Pandav and Choudhury (2000), Mortimer (1981) and Cornelius and Robinson (1986) suggests that heavy loss of the post ovipositional eggs may occur when beach erosion takes place on mass nesting beaches. More than 25 per cent of eggs get exposed every year due to beach erosion at Gahirmatha and Rushikulya rookery of Orissa. Although erosion is a natural process, it can be greatly influenced by human action. Erosion along the

Orissa coast particularly at Gahirmatha beach is taking place mostly after the 1970's when *Casuarina* was extensively planted along the coast (Pandav and Choudhury, 2000). However, in the absence of dense *Casuarina* near the Rushikulya rookery, the beach erosion at this place could be attributed to the changes in offshore currents due to continuous change in the direction of flow of river. Nevertheless, detail studies are required to understand the actual cause of erosion so that fate of nesting beach at Rushikulya can be predicted.

The other factors that affect turtle eggs, juveniles and the hatchlings are the biotic components present on the beach. During nesting, a large number of nests get predated by various predators like feral dog, hyenas, jackals etc. immediately after nesting is over. The benthic sediment fauna chiefly the ghost crabs also predate on the eggs. The predation of eggs is greater in Devi and Rushikulya rookery. While the main reason at Devi is due to temporary human settlement in the island, at the Rushikulya rookery of Orissa, historically the beach was separated from mainland connection by a channel, which runs parallel to the shore for eight-kilometers before it joins the northern end of Chilka. However, over the time, this channel has become relatively shallow and because of human habitation near the beach, there are bridges constructed at many places for access to the beach. This allows mammalian predators to access the beach easily and thereby predations by mammals are more at this rookery.

Although it controls illuminating the beach, *Casuarina* plantations provides shelter to many predators including feral dogs. As a result of which, the predation pressure on turtle nest increases on beaches with *Casuarina* plantation. *Casuarina* plantation near the beach is also detrimental to the nesting sea turtle population in more than one way. Dense *Casuarina* causes excessive shading on the nesting beach. Studies in Florida suggest that nests laid in the shaded areas are subject to lower incubation temperature and which can alter the sex ratio of the population (Schmelz and Mezich, 1988). Davis and Whiting (1977) reported declines in nesting activities of loggerhead in Everglades National Park where dense stands of *Casuarina* took over the native beach vegetation. *Casuarina* has been extensively planted all along the coast of Orissa and which is harmful for adult turtles and hatchlings in many ways (Pandav, 2000).

Conclusion

Great deals of studies have been made on many aspects of reproductive biology of sea turtles and in particular the olive ridley sea turtle is an extensively studied animal along the Orissa coast of India. However, there is much more research required on numerous other biological and environmental parameters, which are key to successful reproduction in olive sea turtle. In particular, our knowledge on the offshore migration and behaviour during courtship and mating is poor. Similarly, lack of studies on the behaviour in foraging areas and energy utilization leave a vast gap in understanding the physiology of the animal. The future studies should also focus on the behaviour of reproductively active males and females in their breeding ground. Continuous monitoring of the area will help in getting a better understanding of various aspects of the sea turtle biology such as the degree of inter-rookery movement, rate of mortality, probability of tag loss and estimation of male and female sea turtles visiting the coast. Study of changes in coastal geo-morphology using satellite imageries and monitoring the route followed by the turtles through satellite telemetry will provide crucial information on many important aspects of the sea turtle biology and behaviour. Research in support of conservation such as continuous monitoring of nesting beach, turtle-fisheries interaction and conservation genetic study is urgently required for the conservation and management of olive ridley sea turtles of Orissa, India and the Indian subcontinent.

Acknowledgements

I am thankful to Dr. Kartik Shanker for his comments and suggestions on the abstract of this manuscript. Bivash Pandav, Kartik Shanker and B.C. Choudhury have contributed greatly to my research work on olive ridley sea turtles in Orissa. The author was supported by the Coastal and Marine Conservation Programme of Ashoka Trust for Research in Ecology and the Environment, Bangalore during preparation of the manuscript.

References

Ackerman, R.A., 1997. The nest environment and the embryonic development of sea turtles. pp. 51–82, In: *The Biology of Sea Turtles,* (Eds.) P. Lutz and J. Musick. CRC Press, Boca Raton, 432p.

Balazs, G.H., 1999. Factors to consider in the tagging of sea turtles. pp. 101–109, In: *Research and Management Techniques for Conservation of Sea Turtles,* (Eds.) K.L. Eckert, K.A. Bjorndal, F.A. Abrell-Grobois and M. M. Donnelly. IUCN Marine Turtle Specialist Group Publication No. 4, CGC, Blanchard, Pennsylvania, USA, 235 p.

Ballestero, J. Randall, M.A. and R. Rojas, 1998. Management, conservation and sustainable use of olive ridley sea turtle eggs (*Lepidochelys olivacea*) in the Ostional Wildlife Refuge, Costa Rica: An 11 year review. pp. 4–5, In: *Proceedings of the 18*th *International Sea Turtle Symposium,* (Eds.) F.A. Abreu-Grobois, R. Briseno-Duenas, R. Márquez-Millan, L. Sarti (Compilers). US Department of Commerce, NOAA Technical Memorandum, MFS-SEFSC-436.

Biswas, S.L.. Acharyo and B.C. Mahapatra, 1977. Observation on incubating eggs of *Lepidochelys olivacea* (Eschscoltz) from Orissa coast in natural and artificial condition. *Science and Culture,* 143: 43–45.

Bjorndal, K.A., 1982. In: *Biology and Conservation of Sea Turtles.* Smithsonian Institution Press, Washington, D. C., 615 p.

Bjorndal, K.A., A.B. Meyland and B.J. Turner, 1983. Sea turtle nesting at Melbourne beach, Florida, I. Size, growth and reproductive biology. *Biological Conservation,* 26: 65–76.

Booth, J. and J. Peters, 1912. Behavioural studies on the green turtle (*Chelonia mydas*) in the sea. *Animal Behaviour,* 20: 808.

Bowen, B.W., A.B. Meylan, J.P. Ross, C.J. Limpus, G.H. Balazs and J.C. Avise, 1992. Global population structure and natural history of the green turtle (*Chelonia mydas*) in terms of matriarchal phylogeny. *Evolution,* 46: 865–881.

Brown, L. and W. MacDonald, 1995. Predation on green Turtle *Chelonia mydas* nests by wild canids at Akyatan Beach, *Biological Conservation,* 71: 55–60.

Bustard, H.R., 1967. Mechanism of nocturnal emergence from the nest in green turtle hatchlings. *Nature,* p. 214–317.

Bustard, H.R., 1976. World's largest sea turtle rookery ? *Tiger Paper,* 3: 3.

Cornelius, S.E. and D. Robinson, 1986. Post nesting movements of female olive ridley turtles tagged in Costa Rica. *Vida Silvestre Neotropical*, 1(1): 12–23.

Cornelius, S.E. and D.C. Robinson, 1985. Abundance, distribution and movements of olive ridely sea turtles in Costa Rica. Report to USFWS, Albuquerque, New Mexico.

Dash, M.C. and C.S. Kar, 1990. *The Turtle Paradise: Gahirmatha*. Interprint, New Delhi, p. 295.

Davis, G.E. and M.C. Whiting, 1997. Loggerhead sea turtle nesting in Everlades National Park, Florida, USA. *Herpetologica*, 33: 18–28.

Deraniyagala, P.E.P., 1939. *The Tetrapod Reptiles of Ceylon, Vol 1: Testudinates and Crocodilians*. National Museum, Colombo, Ceylon, p. 412.

Dodd, C.K. Jr., 1988. Synopsis of the biological data on the loggerhead sea turtle *Caretta caretta* (Linnaeus, 1758). U.S. Fish and Wildlife Service, Biological Report, 88: 1–110.

Eckert, K.L., S.A. Eckert, T.W. Adams, and A.D. Tucker, 1989. Internesting migrations by leatherback sea turtles (*Dermochelys coriacea*) in the West Indies. *Herpetologica*, 45: 190–194.

Eckrich, C.E. and D.W. Owens, 1995. Solitary versus *arribada* nesting in olive ridley sea turtles (*Lepidochelys olivacea*): A test of the predator-satiation hypothesis. *Herpetologica*, 51(3): 349–354.

Ehrhart, L.M., 1982. A review of sea turtle reproduction. pp. 29–38, In: *Biology and Conservation of Sea Turtles*, (Ed.) K.A. Bjorndal. Smithsonian Institution Press, Washington, D.C., 615 p.

Ernst, C.H., J.E. Lovich and R.W. Barbourr, 1994. Turtles of the United States and Canada, Smithsonian Institution Press, Washington, D.C., 578 p.

Fowler, L., 1979. Hatching success and predation in the green sea turtle (*Chelonia mydas*) at Tortuguero, Costa Rica. *Ecology*, 60: 946–955.

Frick, J., 1976. Orientation and behavior of hatchlings green turtle *Chelonia mydas* in the sea. *Animal Behaviour*, 24: 849–857.

Fuentes, P.A., L. Sarti and P. Ulloa, 1998. Conservation and management of sea turtles in Mexico. pp. 6–7, In: *Proceedings of*

18th International Sea Turtle Symposium, (Eds.) F.A. Abreu-Grobois, R. Briseno-Duenas, R. Márquez-Millan, L. Sarti (Compilers). US Dept of Commerce, NOAA Technical Memorandum MFS-SEFSC-436.

Georges, A.C., J. Limpus and C.J. Parmenter, 1993. Natural history of the Chelonia. In: *Fauna of Australia, Amphibia and Reptilia*, (Eds.) Glasby, C.J., G.J.B. Ross and P.L. Beesley. Australian Government Publishing Service, Canberra, 2A: 120.

Hendrickson, J.R., 1982. Nesting behaviour of sea turtles with emphasis on physical and behavioural determinants of nesting success or failure. pp. 53–57, In: *Biology and Conservation of Sea Turtles*, (Ed.) K.A. Bjorndal. Smithsonian Institution Press, Washington D.C., 583 p.

Hewavisenthi, S., 1994. The embryo and hatchling mortality of the green turtle (*Chelonia mydas*) and olive ridley (*Lepidochelys olivacea*) turtles in relation to clutch size. *Herpetological Journal*, 4: 73–76.

Hirth, H.F., 1971. Synopsis of the Biological data on the green turtle *Chelonia mydas* (Linnaeus) 1758, FAO, Rome.

Hirth, H.F., 1980. Nesting behavior and reproductive biology of sea turtles. *American Zoologist*, 20: 507–523.

Hitchins, P.M., O. Bourquin and S. Hitchins, 2004. Nesting success of hawksbill turtles (*Eretmochelys imbricata*) on Cousine Island, Seychelles. *Journal of Zoology*, 264: 383–389.

Hughes, G.R., 1974. The sea turtle of South East Africa. I. Status, morphology and distributions. *S. Afr. Assoc. Mar. Biol. Res. Ocenogr. Res. Inst. Invest. Rep.*, No. 35.

Kalb, H.J., 1999. Behavior and physiology of solitary and *arribada* olive ridley sea turtle (*Lepidochelys olivacea*) during the internesting period. *Ph.D. Thesis*, Texas A&M University (Unpublished).

Kalb, H.J., R. Valverde and D.W. Owens, 1992. What is the reproductive patch of the olive ridley sea turtle? pp. 57–61, In: *Proceedings of the 12th Annual Workshop on Sea Turtle Conservation and Biology*. Jekyll Island, Georgia.

Kar, C.S. and S. Bhaskar, 1982. The status of sea turtle in the eastern Indian Ocean. pp. 365–372, In: *Biology and Conservation of Sea*

Turtles, (Ed.) K.A. Bjorndal Smithsonian Institution Press, Washington D.C., 615 p.

Kar, C.S., 1982. Discovery of second mass nesting ground for Pacific ridley sea turtles in Orissa, India. *Marine Turtle Newsletter*, 23: 3.

Kar, C.S. and M.C. Dash, 1984. Mass nesting beaches of the olive ridley *Lepidochelys olivacea* (Escholtz, 1829) in Orissa and the behavior during an *arribada.* pp. 36–48, In: *Proceedings of the Workshop on Sea Turtle Conservation,* (Ed.) E.G. Silas. CMFRI Special Bulletin, 18: 119.

Kopitsky, K., R.L. Pitman, and P.D. Dutton, 2000. *Reproductive ecology of olive ridley sea turtles in the open ocean in the eastern tropical pacific,* pp. 90–91, In: *Proceedings of the 20th Annual Symposium on Sea Turtle Biology and Conservation,* (Eds.) A. Mosier, A. Foley and B. Brost (Compilers). U.S. Department of Commerce, NOAA Technical Memorandum, NMFS-SEFSC-477.

Limpus, C.J., 1971. The flatback turtle, *Chelonia depressa* Garman, in southeast Queensland, Australia. *Herpetologica*, 27: 431.

Limpus, C.J., 1985. A study of the loggerhead turtle, *Caretta caretta* in Queensland. *Ph.D. Thesis,* University of Queensland, Brisbane.

Limpus, C.J., 1993. The green turtle, *Chelonia mydas,* in Queensland: Breeding males in the southern Great Barrier Reef. *Wildlife Research,* 20: 513.

Limpus, C.J. A. Fleay and V. Baker, 1984. The flatback turtle, *Chelonia depressa,* in Queensland: Reproductive periodicity, philopatry and recruitment. *Australian Wildlife Research,* 11: 579.

Limpus, C.J. and P. Reed, 1985. The green turtle, *Chelonia mydas,* in Queensland: Population structure in a coral reef feeding ground. In: *Biology of Australasian Frogs and Reptiles,* (Eds.) Grigg, G.R. Shine and H. Ehrmann. Surrey Beaty and Sons, Sydney, pp. 47.

Limpus, C.J., J.D. Miller, C.J. Parmenter, D. Reimer, N. McLachlan and R. Webb, 1992. Migration of green (*Chelonia mydas*) and loggerhead (*Caretta caretta*) turtles to and from eastern Australian rookeries. *Wildlife Research,* 19: 347.

Limpus, C.J., A. Fleay, and M. Guinea, 1984. Sea turtles of the Capricornia section, Great Barrier Reef. pp. 61–78, In: *The Capricornia Section of the Great Barrier Reef: Past, Present and Future,*

(Eds.) W.T. Ward and P. Saenger. Royal Society of Queensland and Australian Coral Reef Society.

Lohman, K.J., Witherington, B.E. Lohmann, C.M.F, and M. Salmon, 1997. Orientation, navigation, and natal beach homing in sea turtles. pp. 107–135, In: *The Biology of Sea Turtles*, (Eds.) P. Lutz and J. Musick. CRC Press, Boca Raton, 432 p.

Lutz, P.L. and J.A. Musick, 1997. In: *The Biology of Sea Turtles*. Boca Raton, Florida, CRC Press, 432 p.

Márquez, M.R., 1994. Synopsis of biological data on the Kemp's ridley turtle *Lepidochelys kempi* (Garman, 1880). NOAA Technical Memorandum. NMFS-SEFSC-343, Miami, Florida, 91 p.

Márquez, M.R., O.A. Villaneuva and C. Penaflores, 1976. Sinopsis de datos biologicos sobrc la tortuga golfina *Lepidochelys olivacea* (Eschscholtz, 1829). Inp Sinopsis Pesca No. 2. Sicspinp, Mexico.

Márquez, M.R., O.A. Villaneuva and P.M. Snachez, 1982. The population of Kemp's ridley sea turtle in the Gulf of Mexico. pp. 159–164, In: *Biology and Conservation of Sea Turtles*, (Ed.) K.A. Bjorndal. Smithsonian Institute Press, Washington, D.C. 583 p.

Márquez, R., 1990. FAO Species Catalogue: *Sea Turtles of the World*: An Annotated and Illustrated Catalogue of the Sea Turtle Species Known to Date in FAO Fisheries Synopsis, 125(11): 81.

Márquez, R. and H.G. Van Dissel, 1982. A method for evaluating the number of massed nesting olive ridley sea turtle, *Lepidochelys olivacea* during an arribazon, with comments on arribazon behaviour. *Netherlands Journal of Zoology*, 32(3): 419–425.

Meylan, A., 1995. Sea turtle migration: Evidence from tag returns. pp. 91–100. In: *Biology and Conservation of Sea Turtles*, (Ed.) K.A. Bjorndal. Smithsonian Institution Press, Washington, D.C., 615 p.

Miller, J.D., 1985. Embryology of marine turtles. In: *Biology of the Reptilia*. (Eds.) C. Gams, F. Billett and P.F.A. Maderson. Wiley-Interscience, New York, 14A: 305.

Miller, J.D., 1997. Reproduction in Sea Turtles. pp. 51–82, In: *The Biology of Sea Turtles*, (Eds.) P. Lutz and J. Musick. CRC Press, Boca Raton, 432 p.

Mortimer, J.A., 1990. The influence of beach sand characteristics on the nesting behaviour and clutch survival of green turtle (*Chelonia mydas*). *Copeia*, p. 802.

Mortimer, J.A., 1981. Reproductive ecology of the green turtle, *Chelonia mydas* at Ascension Island. *Ph.D. Dissertation*, University of Florida, Gainesville, 162 p. (Unpublished).

Mrosovsky, N., 1978. Editorial. *Marine Turtle Newsletter*, 9: 1–2.

National Research Council, 1990. *The Decline of the Sea Turtles*. National Academy of Science Press, Washington, D.C., 259 p.

O'Hara, J., 1980. Thermal influences on the swimming speed of loggerhead turtle hatchlings. *Copeia*, 4: 773–780.

Pandav, B., 2000. Conservation and management of olive ridley sea turtle (*Lepidochelys olivacea*) along the Orissa coast. *Ph.D. Thesis*, Utkal University, Orissa, (Unpublished).

Pandav, B. and B.C. Choudhury, 1998. Olive ridley tagged in Orissa recovered in the coastal waters of eastern Sri Lanka. *Marine Turtle Newsletter* 82: 9–10.

Pandav, B. and B.C. Choudhury, 2000. Conservation and management of olive ridley sea turtle (*Lepidochelys olivacea*) in Orissa. Project Final Report, Wildlife Institute of India, 77 p.

Pandav, B. and B.C. Choudhury and C.S. Kar, 1994. A status survey of olive ridley sea turtle (*Lepidochelys olivacea*) and its nesting habitats along the Orissa coast, India. Wildlife Institute of India, 48 p.

Patnaik, S.K., C.S. Kar and S.K. Kar, 2001. A quarter century of sea turtle conservation in Orissa. Wildlife Wing, Forest Department, Government of Orissa, Bhubaneswar, 34 p.

Peters, A. and K.J.F. Verheijen, 1994. Impact of artificial lightings on the seaward orientation of hatchling loggerhead turtles. *Journal of Herpetology*, 28(1): 112–114.

Pitman, R.L., 1990. Pelagic distribution and biology of sea turtles in the eastern tropical Pacific, pp. 143–148. In: *Proceedings of the 10th Annual Workshop on Sea Turtle Biology and Conservation*, Miami, Florida, (Eds.) T.H. Richardson, J.I. Richardson and M. Donnelly. NOAA Tech. Memo. NMFS-SEFC-278.

Plotkin, P.T., Byles, R.A., Rostal, D.C. and Owens, D.W., 1995. Independent versus socially facilitated oceanic migrations of the olive ridley, *Lepidochelys olivacea*. *Marine Biology*, 122: 137–143.

Plotkin, P.T., D.W. Owens, R.A. Byles, R. Patterson, 1996. Departure of male olive ridley turtles, *Lepidochelys olivacea* from a near shore breeding ground. *Herpetologica*, 52(1): 1–7.

Pritchard, P.C.H., 1971. The leatherback or leathery turtle, *Dermochelys coriacea*. International Union for Conservation of Nature Monograph 1: 1–39. Morges, Switzerland.

Pritchard, P.C.H., 1969. Studies of the systematics and reproductive cycles of the genus *Lepidochelys*. *Ph.D. Thesis*, University of Florida, Gainesville, Florida (Unpublished).

Pritchard, P.C.H., 1997. Evolution, phylogeny and current status. pp. 1–28. In: *The Biology of Sea Turtles*, (Eds.) P.L. Lutz and J.A. Musick. Boca Raton, Florida, CRC Press, 432 p.

Ram, K., 2000. Behavioral ecology of the olive ridley sea turtle *Lepidochelys olivacea* (Eschscholtz, 1827) during the breeding period. *Masters Dissertation*, Salim Ali School of Ecology, Pondicherry University, India, (Unpublished).

Reichart, H.A., 1993: Synopsis of biological data on the olive ridley sea turtle *Lepidochelys olivacea* (Eschscholtz 1829) in the western Atlantic. U.S. Department of Commerce, NOAA Technical Memorandum NMFS-SEFSC-336, 78–84 p.

Schmelz, G.W. and R.R. Mezich, 1988. A preliminary investigation of the potential impact of Australian pines on the nesting activities of the loggerhead turtle. pp. 63–66. In: *Proceedings of the 8th Annual Workshop on Sea Turtle Conservation and Biology*, (Ed.) B.A. Schroeder. NOAA Technical Memorandum, NMFS-SEFC-214.

Sea Turtle Conservation, 2004. Orissa Forest and Wildlife Department, Government of Orissa (http://www.wildlifeorissa.org.in) accessed on 1st August 2005.

Shanker, K. and B.C. Choudhury (Eds.), 2005. *Marine Turtles of Indian Sub-continent: Status, Threats and Conservation*. Universities Press, Hyderabad.

Shanker, K., B. Pandav and B.C. Choudhury, 2004. An assessment of the olive ridley turtle (*Lepidochelys olivacea*) nesting population in Orissa, India. *Biological Conservation,* 115: 149–160.

Shanker, K., B.C. Choudhury, B. Panda, B. Tripathy, C.S. Kar, S.K. Kar, N.K. Gupta and J.G. Frazier, 2003 Tracking olive ridley turtles from Orissa. pp. 50–51, In: *Proceedings of the 22nd Annual Symposium on Sea Turtle Biology and Conservation,* Miami, Florida, USA, (Ed.) J. Seminoff. NOAA Technical Memorandum NMFS-SEFSC-503.

Shanker, K.R., K. Aggarwal, J. Rama Devi, B.C. Choudhury, and L. Singh, 2004a. Phylogeography of olive ridley turtles (*Lepidochelys olivacea*) on the east coast of India: Implications for conservation theory. *Molecular Ecology,* 13: 1899–1909.

Stancyk, S.E., 1982. Non-human predators of sea turtles and their control. pp. 139–152. In: *Biology and Conservation of Sea Turtles,* (Ed.) K.A. Bjorndal. Smithsonian Institution Press, Washington, D.C. 615 p.

Tripathy, B., 2004. A study of the offshore distribution of olive ridley turtles (*Lepidochelys olivacea*) in the coastal waters of Rushikulya rookery along the Orissa coast, India. Report of Wildlife Conservation Society, India Programme, CWS, Bangalore, 34 p.

Tripathy, B., 2005. A study on the ecology and conservation of olive ridley sea turtle (*Lepidochelys olivacea*) at the Rushikulya rookery of Orissa coast, India. *Ph.D. Thesis,* Andhra University, Visakhapatnam, India (Unpublished).

Tripathy, B., 2005. Unidentified tag of olive ridley *Lepidochelys olivacea* found in Rushikulya rookery of India. *Hamadryad* (In Press).

Tripathy, B., B. Pandav and R.C. Panigrahy, 2003. Hatching success and orientation of *Lepidochelys olivacea* (Eschscholtz) at Rushikulya rookery, Orissa, India. *Hamadryad,* 27(2): 185–192.

Van Buskirk, J. and L.B. Crowder, 1994. Life-history variation in marine turtles. *Copeia,* p. 66–81.

Witherington, B.E., 1992. Behavioural responses of nesting sea turtles to artificial lighting. *Herpetologica,* 48: 31–39.

Witherington, B.E. and K.A. Bjorndal, 1991. Influences of artificial lighting on the seaward orientation of hatchling loggerhead turtles *Caretta caretta*. *Biological Conservation*, 55: 139–149.

Witzell, W.N., 1983. Synopsis of biological data on the hawksbill turtle, *Eretmochelys imbricata* (Linneaus, 1766). FAO, *Fisheries Synopsis*, 137: 78.

Chapter 25

Endocrine Disruptors and Reproductive Health: A Critical Assessment

Sunil Kumar* and Y. Verma

Division of Reproductive Toxicology and Histochemistry,
National Institute of Occupational Health (ICMR),
Meghani Nagar, Ahmedabad – 380 016, India

ABSTRACT

There is a growing concern about chemicals in the environment interfering with hormone system. There is speculation that these chemicals cause adverse effects on reproduction and development. However, on the basis of changes observed in wild life and aquatic fauna, increasing rates of testicular cancer, other diseases of male reproductive organs, and with an increasing evidence of decline in semen quality, one can postulate that hazardous substances in the environment especially environmental endocrine disruptors, adversely affect reproductive health. The available data on various endocrine disrupters and risk factors (malnutrition and infections which could exaggerate the risk) suggest greater vulnerability of population of developing or underdeveloped

* Corresponding author: Deputy Director (Sr. Grade), National Institute of Occupational Health (ICMR), Meghani Nagar, Ahmedabad – 380 016, Phone: (079) 22868351, 352; Fax: 22866110; E-mail: sunilnioh@yahoo.com

countries to reproductive hazards from exposure to these exogenous environmental chemicals. However, still debate is going on about the adverse effects of EDCs and reproduction. Thus, there is a need to constantly monitor both exposure and affect parameters such as congenital malformations, subnormal growth and development, testicular cancer, hypospadias, cryptorchidism, breast cancer, etc. The effects of exposure must be studied in the children of adults, who have been exposed to suspected hormonal disrupters. The issue of these environmental hormone disruptive chemicals is of concern to all of us as there could be far reaching effects on reproduction and development.

Keywords: *Persistent chemicals; Endocrine disruptors; Hormone; Semen; Pesticides; Metals; Dioxin; Reproductive health.*

Introduction

Over the past two-three decade, there has been a growing concern regarding the adverse effects of synthetic chemicals released into the environment on the development and/or function of the reproductive, endocrine, immune and nervous systems of various animal species including wildlife fauna. The synthetic chemicals and their metabolites have entered in the environment either to increase agricultural yields or to manufacture industrial products. Thus, thousand tones of synthetic chemicals, most of which did not exist 50 to 60 years back, are produced annually worldwide. They are used in the manufacture of consumer goods and industrial products including pesticides, fertilizers, plastics, etc. Many of them persist in the environment, which may have hormone-disrupting effects. However, only a few synthetic chemicals have been screened so far for their hormone disruptive activity. Persistent nature of these chemicals in the environment and their bioaccumulation in the body raises the possibility of long-term deleterious health effects to aquatic fauna, wildlife as well as other animals including human beings.

These exogenous chemicals may bind with hormone receptors in the body and inhibit the activity of natural hormones, elicit hormone like effects or circulate in the body. Thus these chemicals may interfere in various physiological functions of the body. Hormones are essential for the regulation of many biological and physiological processes and act as messengers to communicate information to the cells. Natural hormones produced by the endocrine glands travel through the blood to distant places in the body and bind to specific cell sites called hormonal receptors, where they exert their effects on growth, development, reproduction, behavior and various other functions of the body. Some of the environmental chemicals bind or block the receptors and produce abnormal signals that might lead to various impairments. Crisp *et al.* (1998) broadly defined the endocrine disruptors as exogenous agents that interfere with the production, release, transport, metabolism, binding action or elimination of natural hormones that are responsible in the body for the maintenance of homeostasis and the regulation of the developmental process and/or behavior (Crisp *et al.*, 1998). Such chemicals are reported to have the potential to cause adverse effects on wildlife and might also affect human health. Evidence regarding in *utero* exposure to certain synthetic estrogens such as diethylstilbestrol (DES) has an adverse reproductive effect in men and carcinogenic effect in the daughters of women treated during pregnancy is accumulating (USEPA, 1997). Some endocrine-disrupting effects, previously thought to be estrogenic, are in fact having anti-androgenic effects causing great concern particularly for male reproduction. Substances, which cross the placenta and disrupt the hormone-regulating endocrine system by mimicking the female hormones are known as xenoestrogens. Kumar *et al.* (1999) emphasized the need to conduct systematic studies on the effects of endocrine disruptors on the health of animals as well as humans and to screen environmentally persistent chemicals for their endocrine disruptive activity. International Programme on Chemical Safety (IPCS) concluded that one of the major issues that need to be addressed when evaluating the impact of endocrine disrupting chemicals (EDCs) on human health and the environment is whether the effects reported in the literature represent isolated cases or more global responses (IPCS, 2002).

Endocrine disruptors can be defined as exogenous substances that alter function of the endocrine system, which in turn cause adverse health affects to living beings or their progeny. Compounds that affect the hormone system are frequently referred to as "endocrine-disrupting chemicals" or "endocrine disruptors" (IATP, 2001). It is a known fact that interference of hormone function during development is dangerous to the developing embryo and newborn and later manifest into adulthood and very small doses can produce profound adverse effects on the young ones exposed in *utero*.

Types of Endocrine Disruptors

Several structurally different chemicals have been reported to have endocrine disruptive effects either *in vivo* or *in vitro*. Thus these chemicals cannot be divided on the basis of their chemical structure. These chemicals can be divided mainly into three types on the basis of their action in the cells, organs or body as follows;

1. The exogenous chemicals that mimic like natural hormones such as estrogen and androgen.
2. Chemicals that block hormone receptors in the cell, preventing natural hormones from entering and performing their function and can alter the number of hormone receptors in the foetuses if exposed during critical period of development.
3. Exogenous chemicals when circulate in the body that may lead to the disruption in the synthesis and homeostasis of natural hormones.

Sources of Endocrine–Disrupting Chemicals

A wide range of chemicals with diverse chemical structures derived from plants and environmental origins and synthetic chemicals or their metabolites are reported to have hormonal activity. Hormone disrupting chemicals have been found in common day-to-day items such as plastic containers, plastic wrappers, detergents, soap, cosmetics, pesticides, and also in the municipal sewage. They are also produced from burning of chlorine containing items. Endocrine-Disrupting Chemicals (EDSs) such as nonyl phenol, alkyl phenol ethoxylates (APEs) and phthalates are often found in common household items such as detergents, cosmetics, and household cleaners and even in plastic food containers (IATP, 2001). Some food items such as vegetables, fruits and cereals, etc. carry residues of persistent pesticides and some of them may have hormone-disrupting effects. Several pesticides are known or suspected to be endocrine disruptors that enter in our body through food and water. Endocrine hormone disrupting chemicals can be found in measurable quantities in the fat of every living being, including people in the remote areas of the world or living in high altitude where they have never used such chemicals.

Compounds of plant origin having hormonal activity are generally called phyto-estrogens. They are found in edible plants and/or grasses. However, we proposed that these compounds should be named phyto-endocrine disrupting chemicals (PEDCs) as some of the compounds of plant origin might have anti androgenic activity. There is evidence of biological effects of these chemicals if consumed in large quantities.

Living beings including humans can also be exposed to heavy metals or their oxides such as lead, mercury, and cadmium through environmental and industrial pollution. Some of the metals are considered to have adverse endocrine disruptive or reproductive effects. The growing list of hormone-disrupting chemicals includes a number of other persistent chemical compounds including polychlorinated biphenyls (PCBs), which are widely used in electrical transformers, cosmetics, varnishes, inks, carbonless copy paper, pesticides and for fire resistant coatings to wood and plastic (WWF, 1999a). Dioxin, an endocrine disruptor and human carcinogen, is an unwanted by-product of a number of human activities, including incineration, pesticide production, and manufacturing of polyvinyl chloride (PVC) plastic. U.S. Environmental Protection Agency (USEPA) identified emissions from medical waste incinerators as one of the major sources of dioxin pollution (USEPA, 1997). Further, there is also a report that sewage effluents are also having estrogenic activity (Purdom *et al.*, 1994). This may be due to a large number of chemicals discharged into the sewage and some of them have already been reported to have hormonal activity.

Another group of compound, which might have hormone disruptive activity, is phthalate compounds (some of them reported to have endocrine and reproductive disruptive effects). Phthalates are used as plasticizers in the production of polymeric materials such as poly vinyl chloride (PVC), imparting flexibility and workability both during the manufacturing process and to the end product (Harris *et al.* 1997). When used in this way, they are not chemically bound to the product (Autin, 1973) and may therefore leach into the surrounding medium (Jaeger and Rubin, 1972). These compounds are the most abundant man-made chemicals in the environment. Phthalates are found in food and in the environment because plastics are used so commonly in day to day activities all over the world (Huber *et al.* 1996). They are non-degradable for a very long time and when the plastic burns a number of harmful compounds are produced including the known endocrine disruptor, dioxin.

Endocrine Disruptors and Reproductive Health

Hormones play an important role in the proper development of the growing foetus as well as neonates, and play vital role in reproduction, pregnancy and lactation. The developing offspring is the most sensitive target for endocrine disruption since much of the development of neural, reproductive, and immune systems takes place in the womb up to early childhood (Bern, 1992). The foetus is particularly vulnerable even to the minutest concentration of chemicals. These chemicals are passed from mother to offspring, through the womb and breast milk in mammals including humans and via the egg in reptiles, amphibians, fishes and birds, which may affect the offspring's normal growth and development. The timing of exposure to hormone-disrupting chemicals during pregnancy is as important as the dose exposure. Exposure to very small amounts of a hormone disrupting chemical at an important period of development of the foetus or infant can be more harmful than greater exposures at other times or to adults. A number of laboratory studies have demonstrated that exposure of foetus to endocrine disrupting chemicals can profoundly disturb organ differentiation and development (Gray, 1992; Peterson *et al.*, 1993). It is known that some organs appear to be particular at risk due to maternal exposure to these chemicals, especially those organs with receptors for gonadal hormones for developmental abnormalities in offspring. In female foetuses this includes the mammary glands, fallopian tubes, uterus, cervix and vagina, and in male foetuses it includes the prostate, seminal vesicles, epididymis and testes. In both sexes the external genitalia, brain, skeleton, thyroid, liver, kidney and immune system are also targets for endocrine disrupting chemicals (Colbey, 1980; Leatherland and Sonstegard, 1982; Colborn *et al.*, 1993). A number of animal species have shown signs of abnormal development and growth, such as birth defects, abnormal sexual development, aberrant behaviour, etc. which has been linked to hormone disrupting chemicals. These chemicals interfere with chemical signals that guide development before birth and during early life. They also prevent individuals from growing up according to their genetic character. The effects of hormone disruption can be unpredictable, subtle, and not readily visible until adulthood. If a person was exposed *in utero* to low doses of these chemicals affect may be seen during adult life in the form of reduced fertility. This can be detected only when exposed individuals go for pregnancy or their out come.

Exposure to low doses of endocrine-disrupting chemicals may have little effect on the exposed adult organism, but the offspring of that organism may suffer drastic repercussions. Fox (1992) reported that newly hatched herring gull chicks collected in the lake Ontario area in USA; a region highly contaminated with DDT, had altered reproductive system. Male chicks had oviducts and gonads resembling ovaries and oviduct system of female birds developed abnormally (Fox, 1992). Reeder *et al.*, (2005) has studied the intersexuality and cricket frog decline. Compared with the pre-organochlorine era studied (1852–1929), the percentage of inter sex cricket frogs increased during the period of industrial growth and initial uses of poly chlorinated biphenyls (1930–1945), was highest during the greatest manufacture and use of DDT and PCBs (1946–1959), began declining with the increase in public concern and environmental regulation and then prevented sales of DDT in USA (1960–1979) and continued to decline through the period of gradual reduction in the environmental residues of pesticides and PCBs in the mid western USA (1980–2001) (Reeder, 2005). Further, reduction in penis size and serum testosterone levels in juvenile alligator population of lake Apopka in Florida, USA, has also been attributed to the contamination of the lake by DDT-metabolite p, p–DDE and p, p DDD (Guillette *et al.*, 1996). Later the same group of investigators also draw attention to two Nematocides that are established reproductive toxins in humans, di bromochloropropane (DBCP) and ethylene dibromide (EDB), which could also have played a role in the reproductive failure observed in alligators from lake Apopka in the early 1980s (Semenza *et al.*, 1997). Further, there has been concern that exposure to endocrine disruptors may contribute to the sex ratio changes in the offsprings of exposed populations. Very recently Tiido *èt al.* (2005) reported that exposure to persistent organochlorine pollutants may increase the proportion of ejaculated Y-bearing spermatozoa (Tildo *et al.*, 2005). This study add to the growing body of evidence that exposure to POPs may alter sex ratio. DDE has also been identified as the principal pollutant causing eggshell thinning and reproductive failure in predatory birds. Even though DDT has been banned in many countries there are instances where DDE, levels in eggs and prey species are still high enough to cause reproductive failure (WWF, 1999b). Further, another group of pesticide *i.e.* pyrethroid compounds which are considered to be safe till recently as compared to organochlorine and organophosphate compounds have also been reported to be able to disrupt estrogen function *in vitro* (sumithrin, fenvalerate and d-transallethrin) reported by Go Vera *et al.*, 1999. However, very little is known about the estrogenic potential of these compounds *in vivo* and need further study. Cypermethrin, a pyrethroid pesticide has also been reported to have a significant sub-lethal effect upon the pheromonal mediated endocrine system in mature male *Salmon parr* (Morre and Waring, 2001). They suggested that low levels of cypermethrin in the aquatic environment might have a significant effect on Atlantic salmon populations through disruption of reproductive functions. Thus, there is a need to evaluate synthetic pyrethroids for their endocrine disruptive activity.

Reports on abnormal sexual development in reptiles (Guillette *et al.*, 1994; 1995) and birds (Fry, 1995) as well as feminized response in male fish (Jobling and Sumpter, 1993) have also suggested an association between these affects and environmental chemicals functioning as estrogens. Davis *et al.* (1993) and Colborn *et al.* (1993) have also reported that many chemicals exhibit biological activities associated with female sex hormone estrogen (Davis *et al.*, 1993; Murature *et al.*, 1987). Murature *et al.* (1987) suggested that the concentration of phthalate compounds (particularly DBP) in the cellular fraction of sperm from adult men is negatively correlated with either sperm density or the total number of sperm (Mylchreest *et al.*, 1998). However, little is known about the estrogenic potential of phthalate compounds *in vivo*, and there is a need for further study. Mylchreest *et al.* (1998) studied the effect of DBP on the prenatal and neonatal development of reproductive tract and reported that DBP specifically impaired the androgen dependent development of the male reproductive tract, suggesting that DBP may not be estrogenic but anti-androgenic in rats at high doses (Jobling *et al., 1995*). However, this needs confirmation as DBP has been reported to have estrogenic effects *in vitro* (Jobling *et al.*, 1995), and low level of estrogenic potential of DBP may not be reflected in the study of Mylchreest *et al.* (1998). Recent studies of Colon *et al.* (2000) suggested a possible association between plasticizers with known estrogenic and anti-androgenic activity and the cause of premature breast development in a human female population. They have carried out an investigation in Puerto Rico where highest known incidence of premature thelarche has been reported. Significantly higher levels of phthalate compounds [dimethyl, diethyl, dibutyl and di-(2-ethyl hexyl)] and its major metabolite, mono-(2-ethyl hexyl) phthalate were reported in 28 (68 per cent) samples from thelarche subjects as compared to only one in control samples (Korach *et al.*, 1987)

A number of experimental studies reinforce the concept that environmental xenobiotics can have deleterious effects on the reproductive health of the animals (Korach *et al.*, 1987; Bulger and Kuffer, 1985). Further circumstantial evidence is accumulating that environmental xenobiotics may be disrupting reproductive processes in human males (Chilvers *et al.*, 1984; John Rad Cliffe Hospital Group, 1992; Giwereman and Skakkebaek, 1992). It has been realized that some of the endocrine disruptive effects, previously thought to be estrogenic in fact have anti-androgenic effects. Several environmental chemicals including metabolites of the fungicide vinclozolin and the pesticide such as DDT disrupt male reproductive development and function by inhibiting androgen receptor mediated events reported by Kelce *et al.* (1998). They also reported that these chemicals commonly bind androgen receptor with moderate affinity and act as antagonists by inhibiting transcription of androgen dependent genes. A study conducted by Rupa *et al.* (1991) on reproductive performance in population exposed to pesticides in cotton field in Andhra Pradesh, India, indicated a significant decrease in fertile males and a significant increase in abortions among the wives of the exposed males as compared to control groups (Yu *et al.*, 2000). However, they did not correlate these abnormalities with pesticide residue levels in biological materials of the exposed workers. Recently Saiyed *et al.* (2003) reported that endosulfan exposure in male children may delay sexual maturity and interfere with sex hormone synthesis (Saiyed *et al.*, 2003). Yu *et al.* (2000) published the information on the people poisoned by cooking oil contaminated heat-degraded polychlorinated biphenyls (PCBs). They reported that these chemicals persist in the victim's body and some are toxic to female reproductive system in experimental animals. They identified the exposed women interviewed for reproductive dysfunction. These findings suggest that high level of PCB/ Polychlorinated dibenzofuran (PCDF) exposure has some effects on female endocrine and reproductive function (Yu *et al.*, 2000). Weidner *et al.* (1999) reported a significant increased risk of cryptorchidism, but not for hypospadias found in sons of women working in gardening. However, risks were not increased in sons of men working in farming, or gardening. The increased risk of cryptorchidism among sons of female gardeners could suggest an association with prenatal exposure to occupationally related chemicals (Weidner, 1998). Very recently Yoshida *et al.* (2005) mentioned an association of cryptorchidism with homozygosity for the specific estrogen receptor alpha (ESR 1) haplotype which suggests the relevance of genetic susceptibility to environmental endocrine disruptors in the development of cryptorchidism (Yoshida, 2005). Most of the available studies from different parts of the world did not correlate reproductive impairments with the doses or residues of chemicals in the affected subjects.

Adlercreutz (1998) reviewed the available information on phytoestrogens and reported that phytoestrogens bind to estrogen receptors and show significant effects in animals, humans, and in cell cultures (Adlercreutz, 1998). The most well known estrogenic effect of phytoestrogens is the "Clover Disease" in Australian sheep leading to infertility (Price and Fenwick, 1985). Adlercreutz, (1998) also suggested that populations with higher consumption of phytoestrogens have lower cancer mortality and/ or incidence than Western populations with low intake of these compounds. Further, he reported that there seems to be no evidence that intake of phytoestrogens from soya or other leguminosae or from sprouts may cause infertility in adult human and small amounts of estrogens or phytoestrogens are not deleterious to the human foetus but, may even be beneficial for growth (Adlercreutz, 1998). Recently Strom *et al.* (2001) reported that exposure to Soy formula (contains high level of phytoestrogens) does not appear to lead to different general health or reproductive outcomes than exposure to cow milk formula (Strom *et al.*, 2001). IPCS (2002) suggested that phytoestrogens are of particular interest because they are diverse in structure, undergo complex metabolic processes, and are ubiquitous in the environment (IPCS, 2002).

Dioxin may be produced when medical waste incinerators burned items such as chlorine-bleached paper products, PVC blood bags and tubing, and other chlorine-containing waste. Burning of chlorine-contaminated waste create dioxins that are most toxic chemicals ever identified. Dioxin from incinerator smokes can travels through the air on dust or water particles and eventually falls to the ground. Sometimes it travels as far as hundreds or thousand miles away from the incinerator that release it. Dioxin may also land on grasslands (IATP, 2001). As the animals eat grasses, they ingest small amounts of dioxin, which bio-accumulates in the fat. When people eat meat or dairy products, they also consume some dioxin. USEPA (1997) estimates that 90 per cent of human exposure to dioxin comes from food, and that up to three-fourths of that comes from beef and dairy products (IATP, 2001). Based on animals experiment Yonemoto (2000) reviewed the effects of dioxin on reproduction and development reported that exposure to dioxin during pregnancy and lactation induces various functional effects on offspring at very low doses. In humans, even if there is no exposure to dioxin after birth, there might be effects on thyroid function in infants exposed to dioxin from breast milk (Yonemoto, 2000).

Semen Quality and Persistent Chemicals

Although controversial, there have been several recent reports of decline in human sperm counts over the last fifty years. Wildlife has been reported with deformed genitalia, aberrant mating behavior, sterility, infertility, and other physical and behavioral anomalies. Some scientists believe that hormone-disrupting chemicals have contributed to various health and reproductive problems in men and women, including a rise in incidence of breast, prostate and testicular cancers, decline in sperm count, an increasing incidence of endometriosis (a disease of women where uterine endometrial tissue grows outside the uterus), and increase in genital defects. Several workers have reported apparent drop in semen quality from different parts of the world particularly from western countries (Bromwich *et al.*, 1994; Carlson *et al.*, 1992; Suominen and Vierula, 1993). A Meta analysis of 61 studies, including 14947 normal men, was undertaken by Carlsen *et al.* (1992). They showed a significant decrease in sperm concentration (113 million/ml to 66 million/ml) and sperm volume (from 3.40 ml to 2.75 ml) over the period 1938–1990. These findings have sensitized the scientific community and have stimulated further research on this aspect. However, no such changes in semen quality have been reported by the other investigators (Fisch *et al.*, 1996; Paulsen *et al.*, 1996). Some of the studies criticized the pro-decrease paper on statistical point of view (Olsen *et al.*, 1995). Recently, Swan *et al.* (1998) reported that there was no evidence of a decline in semen quality in non-western countries while they observed a significant decline in Europe and United States (Swan, 1998). Heinze, (1998) reported that recent studies from Europe and the United States indicated large interregional differences in sperm density. He noted that interregional differences in the United States (New York city vs Los Angeles) were as large as the reported differences in mean sperm density in 1938 versus 1990 (Heinze, 1998). Still an intense debate is going on among the scientific community about the real picture regarding the decline of semen quality. Mehta and Anand Kumar (1997) from Bangalore, India also reported a decline in semen quality. They correlated these changes with various air pollution indices such as suspended particulate matter, sulphur dioxide, and lead (Mehta and Anandkumar, 1997). Gopalkrishnan (1997) from Bombay, India mentioned a definitive trend of decline in semen quality. She reported that there is a decrease in the mean percentage of normal morphology with less than 30 per cent normal forms (Gopalkrishnan, 1997). Orejuela *et al.* (1998) emphasized that generalization of a worldwide decline trend in semen quality is still risky, and highlight the need for innovative studies with good quality data to address this important issue related to human reproduction (Orejuela, 1998). Kumar (2004) mentioned that there is a need for well planned epidemiological studies to constantly monitor both exposure and affected, reproductive parameters including trend of semen quality after controlling various confounding factors (Kumar, 2004).

Scientists are slowly realizing the fact that there may be some adverse effects of environmental chemicals particularly estrogenic or anti-androgenic compounds on the reproductive function of human as well as animals. Sharpe and Skakkeback (1993) postulated that an apparent drop in sperm count may be due to developmental exposure to estrogenic xenobiotics (Sharpe and Skakkeback, 1993). McLachlan *et al.* (1998) provided experimental support to this hypothesis. They reported that prenatal exposure to synthetic estrogen diethyl stilbestrol (DES) is associated with poor semen quality, prostatic disease, cryptorchidism and testicular neoplasia in mice (McLachlan *et al.*, 1998). Recently Behrens *et al.* (2000) reported that even a short *in utero* exposure of rats to DES during a critical period for gonadal development results in cannibalism and reduced testis. Further, DES tended to reduce the number of sertoli cells (Behrens *et al.*, 2000). The continuing accumulation of evidence from oestrogen receptor knockout mouse animal models demonstrated that oestrogen plays an essential and direct role in the development and maintenance of male fertility (Jones and Simpson, 2000). Guo *et al.* (2000) published a report on reproductive function of boys prenatally exposed to polychlorinated biphenyls and dibenzofurans. They found that sperms of these exposed boys had greater abnormal morphology, reduced motility, and reduced capacity to penetrate hamster's oocytes (Guo *et al.*, 2000). Recently, Skakkeback *et al.* (2001) suggested that poor semen quality, testicular cancer, undescended testis, and hypospadias are symptoms of one underlying entity, the testicular dysgenesis syndrom (TDS), which may be increasing due to adverse environmental influences. Further, they suggested that TDS is the result of disrupting embryonic programming and gonadal development during foetal life (Skakkeback *et al.*, 2001).

There are several lacunas in the available data on health impairment and exposure to endocrine disruptors. The most important are non-availability of exposure data both environmental as well as biological, mixed exposure, dose of the chemical, and timing and duration of exposure.

Possible pathway of endocrine disruptive effects of the endocrine disruptors is shown in Figure 25.1. However, there may be several other mechanisms of action, which may be responsible for the endocrine and reproductive effects induced by these chemicals. Peterson *et al.* (1997) divided endocrine disruptors in to two major classes *i.e.* reproductive and anti-thyroid, which are capable of altering male reproductive system development and function (Peterson *et al.*, 1997) They further, divided these in to eleven sub classes on the basis of mechanism or site of action *i.e.* estrogens, antiestrogens, androgens, antiandrogens, progestins, anti-progestins, aryl hydrocarbon receptors (AHR) agonist (chemicals that affects by an AHR mechanism), toxicants that affects reproduction by acting primarily by the central nervous system (CNS), toxicants that are toxic to endocrine cells, toxicants that affects germ cells, chemicals that alter circulating thyroid hormone

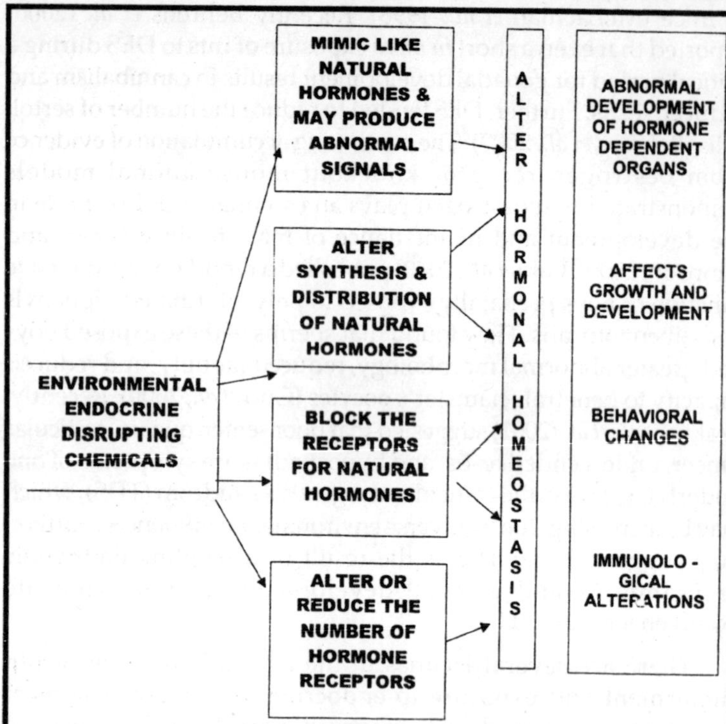

Figure 25.1: Possible Pathway of Endocrine Disruption and Reproductive Health

concentrations. Thus these persistent chemicals act at various levels of development with different mechanism of action. Studies in humans correlating reproductive dysfunction with the exposure level are scanty and need comprehensive epidemiological study.

Acknowledgements

Authors are thankful to the Dr. H.N. Saiyed, Director NIOH for encouragements.

References

Adlercreutz, H., 1998. Human health and phytoestrogens. In: *Reproductive and Developmental Toxicology,* (Ed.) Kenneth S. Korach. Marcel Dekker Inc. New York, pp. 299–371.

Autin, J., 1973. Toxicity and health threats of phthalate esters: Review of the literature. *Environ. Health Perspect.*, 4: 3–26.

Behrens, G.H., Torjesen, P., Petersen, P.M., Tretli, S., Grotmol, T., Haugen, M.T.B., Sorensen, D.R., 2000. Reproductive function in male after brief *in utero* exposure to diethylstilboestrol. *Int. J. of Andrology*, 23: 366–371.

Bern, H., 1992. The fragile fetus. In: *Chemically Induced Alterations in Sexual and Functional Development: The Wildlife and Human Connection*, (Eds.) Colborn, T. and C. Clement. Princeton Scientific Publishing, Princeton, New Jersey, pp. 9–15.

Bromwich, P.J., Cohen, J., Stewart, A. Walker, 1994. Decline in sperm counts an artifact of changed reference of "normal"? *Br. Med. J.*, 309: 19–22.

Bulger, W.H. and Kuffer, D., 1985. Estrogenic activity of pesticides and other xenobiotics on the uterus and male reproductive tract. In: *Endocrine Toxicology*, (Eds.) J.A. Thomas, K.S. Korach, J.A. McLachlan. Ravan Press, New York, pp. 1–33.

Carlsen, E.A., Giwercman, N., Keiding, and Skakkeback, N.E., 1992. Evidence for decreasing quality of semen during past 50 years. *Br. Med. J.*, 305: 609–613.

Chilvers, C., Pike, M.C., Fogelman, K. and Woodsworth, M.E.J., Apparent doubling of frequency of undescended testes in England and Wales in 1962–1981. *Lancet*, 2: 330–332.

Colborn, T., Vamsaal, F.S. and Sato, A.M., 1993. Developmental effects of endocrine chemicals in wildlife and humans. *Environ. Health. Perspect.*, 101: 378–384.

Colby, H.D., 1980. Regulation of hepatic and steroid metabolism by androgens and estrogens. In: *Advances in Sex Hormones Research*, (Eds.) J.A. Thomas, R.I. Singhal and M.D. Baltimore. Urban and Schwarzenberg, pp. 27–71.

Colon, D., Caro, Carles Bourding, J. and Rasario, O., 2000. Identification of phthalate esters in the serum of young Puerto Ricen girls with premature breast development. *Environ. Health. Perspect.*, 108: 895–900.

Crisp, T.M., Clegg, E.D., Copper, R.L., Wood, W.P., Anderson, D.G., Baetcke, K.P., Hoffmann, J.L., Morrow, M.S., Rodier, J.E., Schaeffer,

J.E., Tourt, L.W., Mourice, G.Z. and Patel, Y.M., 1998. Environmental endocrine disruption: An effects assessment and analysis. *Environ. Health Perspct.*, 106 (Suppl. 1): 11–53.

Davis, D.L., Bradlow, H.L., Wolff, M., Woodruff, T., Hoel, D.G. and Anton-Culver, H., 1993. Medical hypothesis: Xenoestrogens as preventable cause of breast cancer. *Environ. Health Perspect.*, 101: 372–377.

Fisch, H., Goluboff, E.T., Olsen, J.H., Fieldshauh, J., Broder, S.J. and Brand, D.H., 1996. Semen analysis in 1283 men from the united state over a 25 year period: No decline infertility. *Fertil. Steril.*, 65: 1009–1014.

Fox, G., 1992. Epidemiological and pathological evidence of contaminant-induced alterations in sexual development in free living wildlife. In: *Chemically Induced Alterations in Sexual and Functional Development: The Wildlife and Human Connection*, (Eds.) Colborn, T. and C. Clement. Princeton Scientific Publishing, Princeton, New Jersey, pp. 147–158.

Fry, D.M., 1995. Reproductive effects in birds exposed to pesticides and industrial chemicals. *Environ. Health Perspect.*, 103: 165–171.

Giwercman, A. and Skakkebaek, N.E., 1992. The human testis: An organ at risk. *Int. J. Androl.*, 15: 373–375.

Go Vera, J., Wolff Garey and Pogo, B.G.T., 1999. Estrogenic potential of certain pyrethroid compounds in the MCF-7 human breast carcinoma cell lines. *Environ. Health Perspect.*, 107: 173–178.

Gopalkrishnan, K., 1997. Decreasing sperm counts fact or fiction. *ICMR Bull.*, 27: 77–82.

Gray, L.E., 1992. Chemical induced alterations of sexual differentiation: A review of effects in human and rodents. In: *Chemically Induced Alterations in Sexual and Functional Development: The Wildlife and Human Connection*, (Eds.) Colborn, T. and C. Clement. Princeton Scientific Publishing, Princeton, New Jersey, pp. 203–230.

Guillette, L.J. Jr., Gross, T.M., Masson, G.R., Matter, J.M., Percival, H.F. and Woodward, A.R., 1994. Developmental abnormalities of the gonad and abnormal sex hormone concentrations in juvenile alligators from contaminated and control lakes in Florida. *Environ. Health Perspect.*, 102: 680–688.

Guillette, L.J. Jr., Gross, T.S., Gross, D., Rooney, A.A. and Percival, H.F., 1995. Gonadal steroidogenesis *in vitro* from juvenile aligators obtained from contaminated or control lakes. *Environ. Health Persp.*, 103: 31–36.

Guillette, L.J. Jr., Pickford, D.B., Crain, D.A., Roony, A.A. and Percival, H.F., 1996. Reduction in penis size and plasma testosterone concentration in juvenile aligators living in contaminated environment. *Gen. Camp. Endocrinol.*, 101: 32–34.

Guo, Y.L., Hsu, P.C., Hsu, C.C. and Lambert, G.H., 2000. Semen quality after prenatal exposure to polychlorinated biphenyls and dibenzofurans. *Lancet*, 356: 1240–1241.

Harris, C.A, Henttu, P., Parker, M.G. and Sumpter, J.P., 1997. The estrogenic activity of Phthalate esters *in vitro. Environ. Hlth. Pres.*, 105: 802–811.

Heinze, J., 1998. Interregional differences undermine sperm trend conclusion. *Environ. Health Perspect.*, 106A: 369–370.

Huber, W.W., Grasl-kraupp, B., and Schulte-Herman, R., 1996. Hepatocarcinogenic potential of 2 Di (2-ethyl hexyl phthalate) in rodents and its implications on human risk. *Crit. Rev. Toxicol.*, 26: 365–481.

IATP, 2001. Hormone-disrupting chemicals (EDCs). Keeping industrial pollution out of the food system. The Institute for Agriculture and Trade Policy (IATP websites), Minneapolis, USA.

IPCS, 2002. Global assessment of the state-of-the-Science of Endocrine disruptors. (Eds.) Terri Damstra, Sue Barlow, Aake Bergman, Robert Kavlock glen Van Der Kraak. WHO.

Jaeger, R.J. and Rubin, R.J., 1972. Migration of phthalate esters plasticizers from polyvinyl chloride blood begs in to stored human blood and its localization in human tissues. *New Eng. J. Med.*, 287: 1114–118.

Jobling, S. and Sumpter, J.P., 1993. Detergent compounds in sewage effluents are weakly estrogenic: An *in vitro* study using rainbow trout (*Oncorhynchus mykiss*) hepatocytes. *Aquatic Toxicol.*, 27: 361–372.

Jobling, S., Raynolds, T., White, R., Parker, M.G. and Sumpter, J.P., 1995. A variety of environmentally persistent chemicals including some phthalate plasticizers are weakly estrogenic. *Environ. Hlth. Pres.*, 103: 582–587.

John Rad Cliffe Hospital Study Group, 1992. Cryptorchidism: An apparent substantial increase since 1960. *Br. Med. J.*, 305: 609–613.

Jones, M.E. and Simpson, E.R., 2000. Oestrogens in male reproduction. Best Practice and Research Clinical Endocrinology and Metabolism: 14: 505–516.

Kelce, W.R., Gray, L.E. and Wilson, E.M., 1998. Antiandrogens as environmental endocrine disruptors. *Reprod. Fertil. Deve.*, 10: 105–111.

Korach, K.S., Sarver, P., Chae, K. and McLachlan, J.A., 1987. Estrogens receptors-binding activity of polychlorinated hydroxy biphenyls; conformationally restricted structural probes. *Mol. Pharmacol.*, 33: 120–126.

Kumar, S., 2004. Semen quality and persistent environmental chemicals: Current status. *ART and Science*, 4: 8–11.

Kumar, S., Zaidi, S.S.A., Diwan, A. and Saiyed, H.N., 1999. Environmental estrogenic compounds and male reproductive health. *Indian J. Occ. and Environ. Med.*, 3: 177–180.

Leatherland, J.F. and Sonstegard, R.A.B., 1982. Thyroid responses in cats fed diets formulated with great lakes Coho salmon. *Bull. Environ. Contam. Toxicol.*, 29: 341–346.

McLachlan, J.A., Newbold, R.R., Li, S.F. and Nigishi, M., Are estrogens carcinogenic during development of testis? *AMPIS*, 106: 240–242.

Mehta, R.H. and Anandkumar, T.C., Declining semen quality in Bangaloreans: A preliminary report. *Current Science*, 2: 621–623.

Morre, A. and Waring, C.P., 2001. The effects of synthetic pyrethroid pesticide on some aspects of reproduction in Atlantic salmon (*Salmo salar* L.). *Aquatic Toxicology*, 52(1): 1–12.

Murature, D.A., Tang, S.Y., Steinhardt, G. and Dougherty, R.C., 1987. Phthalate Esters and semen quality parameters. *Biomed. Environ. Mass Spectrom* , 13: 473– 477.

Mylchreest, E., Cattley, R.C. and Foster, P.M.O., 1998. Male reproductive tract malformations in rats following gestational and lactational exposure to Di (n–butyl) phthalate: An anti-androgenic mechanism. *Toxicol. Sciences,* 43: 47–60.

Olsen, G.W., Bonder, K.M., Ramlow, J.M., Ross, C.E. and Lipshultz, L.T., 1995. Have sperm counts been reduced 50 per cent in 50 years. *Fertil. Steril.,* 63: 887–893.

Orejuela, F., Lipshultz, L.T. and Lamb, D.J., 1998. Debate about sperm count decline. *Environ. Hlth. Perspect.,* 106A: 370.

Paulson, C.A., Berman, N.G. and Wang, C., 1996. Data from men in greater seattle area reveals no downward trend in semen quality further evidence that deterioration of semen quality is not geographically uniform. *Fertil, Steril.,* 65: 1015–1020.

Petersen, R.E., Theobold, H.M. and Kimmel, G.L., 1993. Developmental and reproductive toxicity of dioxins and related compounds cross species comparison. *Crit. Rev. Toxicol.,* 23: 283–335.

Peterson, R.E., Cooke, P.S., Kelce, W.R. and Gray, E.L., 1997. Environmental endocrine disruptors. In: *Comprehensive Toxicology: Reproductive and Endocrine Toxicology,* Vol. 10, (Eds.) I. Glenn Sipes, A. Charlene, A. McQueen Jay Gandolfi. Pergamon Press, pp. 181–191.

Price, K.R. and Fenwick, G.R., 1985. *Food. Add. Contamin.,* 2: 73.*

Purdom, C.E., Hardiman, P.H., Bye, V.J., Eno, N.C., Taylor, C.R. and Sumpter, J.P., 1994. Estrogenic effects of effluents from sewage treatment works. *Chem. Ecol.,* 8: 275–285.

Reeder, A.L., Ruiz, M.O., Pessier, A., Brown, L.E., Levengood, J.M., Phillips, C.A., Wheeler, M.B., Warner, Reand Beasley, V.R., 2005. *Environ. Health. Perspect.,* 113(3): 261–265.

Rupa, D.S., Reddy, P.P. and Reddy, D.S., 1991. Reproductive performance in, population exposed to pesticides in cotton fields in India. *Environ. Res.,* 55: 123–128.

Saiyed, H.N., Dewan, A., Bhatnagar, V., Shenoy, U., Shenoy, R., Rajmohan, H., Patel, K., Kashyap, R., Kulkarni, P., Rajan, B. and Lakkad, B., 2003. Effect of endosulfan on male reproductive development. *Environ. Health Pers.,* 111: 1958–1962.

Semenza, J.C., Guillette, L.J Jr., Tolbert, P.E., Jackson, R.J., Rubin, C.H., 1997. Reproductive toxins and alligators abnormalities at lake Apopka. *Environ. Health Perspectives,* 105: 1030–1032.

Sharpe, R. and Skakkeback, N.E., 1993. Are estrogens involved in falling sperm counts and disorders of male reproductive tract? *Lancet,* 341: 1392–1395.

Skakkebaek, N.E., Rajpert, DeMeyts, E. and Main, K.M., 2000. Testicular digenesis syndrom: An increasing common developmental disorders with environmental aspects. *APMIS,* 109(103 Suppl.): 522–528.

Strom, B.L., Macones, G.A., Schinnar, R., Stallings, V.A., Ziegler, E.E., Drulis, J.M., Barnhart, K., Nelson, S.E., Sammel, M.D., Hanson, S.A., 2001. Exposure to Soy based formula in infancy and endocrinological and reproductive outcomes in young adulthood. *JAMA,* 286(7): 807–814.

Suominen, J. and Vierula, M., 1993. Semen quality of finish men. *Br. Med. J.,* 306: 1579.

Swan, S.H., Elkin, E., Fenster, L., 1998. Response: A reanalysis of sperm density of sperm data. *Environ. Health Perspect.,* 106A: 370–371.

Tiido, T., Rignell-Hydbom, A., Jonsson, B., Giwecrman, Y.L., Rylander, L., Hagmar, L., and Giwercman, A., 2005. Exposure to persistent organochlorine pollutants associates with human sperm Y : X chromosome ratio. *Human Reprod.* 28 (In Press).

US EPA, 1997. Back ground on Endocrine disruptors (EPA web site Home/Search/comments) September 24.

Weidner, I.S., Jensen, T.K., Moiler, H., Skakkebaek, N.E., 1998. Cryptorchidism and hypospadias in sons of gardeners and farmers. *Environmental Health Perspectives,* 106: 793–796.

WWF, 1999a. Background information on pollutants considered to be hormone disruptors. World Wildlife Fund Canada.

WWF, 1999b. Hazards and exposure associated with DDT and Synthetic. Pyrethroids used for Vector control, pp. 1–59.

Yonemoto, J., 2000. The effects of dioxin on reproduction and development. *Industrial Health,* 38: 259–268.

Yoshida, R., Fukami, M., Sasagawa, I., Hasegawa, T., Kamatani, N. and Ogata, T., 2005. Association of cryptorchidism with a specific halotype of the estrogen receptor alpha gene: Implication for the susceptibility to estrogenic environmental endocrine disrupters. *Clinical Endocrinol. Metab.*, 17.

Yu, M.L., Hsu, C.C., Guo, Y.L. and Rogan, W.J., 2000. Menstruation and reproduction in women with polychlorinated biphenyls (PCBs) poisoning: Long term follow up interviews of the women from the Taiwan Yucheng cohort. *Int. J. of Epidemiology*, 29: 672–677.

Yu, M.L., Hsu, C.C., Guo, Y.L., Rogan, W.J., 2000. Menstruction and reproduction in women with polychlorinated biphenyls (PCBs) poisoning: Long term follow up interviews of the women from the Taiwan Yucheng cohort. *Int. J Epidemiology*, 29: 672–677.

Chapter 26

Alleviation of Diclofenac Induced Testicular Toxicity by Vitamin E Supplementation in Albino Rats

G. Vanithakumari and K. Sakthi Shree***

**Head, Department of Zoology, Bharathiar University, Coimbatore – 641 046, Tamil Nadu*

***Department of Zoology, Government Arts College (Autonomous), Coimbatore – 641 018, Tamil Nadu*

Introduction

The mammalian reproduction process is long, complex and uniquely susceptible to disturbances by external agents. The study of reproductive system and its functioning, forms on integral part of evaluation of infertility and impotency, which seems to be prevalent widely in recent years, among humans. The amount of research into endocrine disruption has exploded over the past decade, and reevaluation of the state of research in this area is timely. The phrase "endocrine disrupter" is anything that can cause an imbalance within the endocrine system. The endocrine system is not only concerned with the reproductive tract, but also with the entire hormone producing organs/glands, which maintain bodily homeostasis (*e.g.*, thyroid, parathyroid, anterior and posterior pituitary, pancreas, adrenals, pineal and the gonads (IPCS, 2002). There are debates about whether endocrine reproductive health is really declining, and whether endocrine disrupting chemicals play any role in the perceived decline.

The endocrine system can be regarded as a number of interconnecting and interacting axes. The major regulator of the mature reproductive system is the hypothalamo-pituitary-gonadal (HPG) axis that functions as a classical negative feedback system. This consists of the gonadotropin-releasing hormone (GnRH) neurons in the hypothalamus (brain) that release GnRH into the portal blood supply. GnRH stimulates the release of gonadotropin-Luteinizing hormone (LH) and Follicle-stimulating hormone (FSH) from the gonadotroph cells in the pituitary gland. LH and FSH travel through the systemic circulation and act on the endocrine active cells of gonad (testis or ovary). Within the testis, the Leydig cells secrete testosterone and oestradiol, and the Sertoli cells secrete inhibin B. These gonadal signals feedback to the pituitary and hypothalamus, to regulate the release of GnRH and gonadotropins (IPCS, 2002). There is the potential for endocrine disrupting chemicals (EDCS) to act at any level of the HPG axis, but there is general support for the view that the development and programming of the axis during foetal life could be the most sensitive window to permanently alter the homeostatic mechanisms of the endocrine system (IPCS, 2002).

The intake of over the counter drugs has increased with a blatant disregard for the continuous propaganda against self-medication. Their short-term relief from discomforts due to ailments usually paves way for the development of complications in the reproductive system, which may be expressed directly or indirectly. Carlsen *et al.*, (1992) (Carlsen *et al.*, 1992) published a provocative study, which suggested that human semen quality has declined by 50 per cent from 1930 to 1991. Many subsequent studies criticized or re-analyzed the original meta-analysis, but the trends and conclusions have been reaffirmed (Skakkeback, 2001). Hence, an identification of toxic chemicals and the effects produced on their intake is a necessary method to obtain proof for advocating against their usage. The prevalent anti-inflammatory agent, which has gained popularity in recent times, diclofenac sodium (DS), which is used widely for relief from aches and pain, has been taken for study to elaborate its mechanism of action on male reproductive system, especially testes, and expose its advantageous and disadvantageous effects.

Activity of Diclofenac Sodium

Diclofenac sodium (DS) is a potent NSAID. Being a phenyl acetic derivative, it is a powerful inhibitor of cyclooxygenase (COX) and at higher concentration also reduces the formation of products of lipoxygenase (LOX) pathway (Deleuran, 2001). Toxic NSAIDS including DS depleted cellular ATP before the enzyme leakage. They are uncouplers of mitochondrial oxidative phosphorylation, which result in the impairment of ATP synthesis (Masubhuchi *et al.*, 1998).

The metabolism of DS in humans partitions between acyl glucuronidation and phenyl hydroxylation, with the former reaction catalyzed by uridine-5-diphosphoglucunosyl transferase 2B7, while the latter is catalyzed by cytochrome P450 (CYP) 2C9 and 3A4 (Boelsterli *et al.*, 1995). Formation of protein adducts are evident following oxidative metabolism of diclofenac catalyzed by CYP enzymes. Therefore, it is conceivable that acyl glucuronide and benzoquinone imines derived from diclofenac modify protein covalently, thereby, producing toxicity in-patient via either direct disruption of critical cellular function or elicitation of immunological responses (Tang, 2003).

Inhibition of PG synthesis is considered to be fundamental to the mechanism of action of DS (Kirchner *et al.*, 1997). PGs play an important role in causing inflammation, pain and fever (Data Sheet, 2002). PGs are secreted by the male accessory reproductive structures in large quantities and are very much needed for normal male reproductive function. They are produced via action of COX enzymes, which exist in two isoforms-COX-1 and COX-2 (Deleuran, 2001). Isoenzyme COX-1 catalyses synthesis of prostanoids and regulate normal physiological processes, while isoenzyme COX-2 catalyses generation of PG that mediate inflammation, pain and fever (Goldstein, 2000). COX-2 dependent production of prostanoids *in vivo* is powerfully regulated by the supply of arachidonic acid from either exogenous or endogenous sources. Therefore, the supply of arachidonic acid is the key regulator of COX-2 activity, which in turn influences the potencies of NSAID drugs (Hamilton, 1975).

Cytochromes P450 are the key enzymes for activating and inactivating many drugs. Therefore, individual expression levels of cytochromes P450 may play a crucial role in drug safety and drug efficacy. Over expression of cytochrome P450 may yield rapid turnover and elimination of drugs before the target site was reached and any pharmacological effect is observed. Therefore, it may be vital to know the individual cytochrome P450 status in order to select the appropriate drug before drug resistance occurs. Expression levels and activity of cytochromes P450 depend on many different factors. These factors include tissue and organ specific expression, sex- and age-dependent expression, genetic differences yielding polymorphic forms, competitive inhibition or induction of cytochromes P450 due to multiple drug interaction, nutrition and diet (Doehmer *et al.*, 1973).

Diclofenac is considered to be potentially teratogenic in humans (Chan *et al.*, 2001), as recommended oral dosage itself is a teratogenic concentration becoming accumulated in foetal tissue. Significant morphological changes were observed in liver of offspring, whose parents were treated with DS (Gokcimen *et al.*, 2001). DS embryotoxicity may be related to free oxygen radical damage (Chan *et al.*, 2001). During, pregnancy, the systemic blood pressure is decreased and DS enhances uterine blood velocity in rats.

A dose dependent significant incidence of abnormal sperms, expose the mutagenicity of DS. However, Diclofenac has also been reported to transiently retrieve partial epididymal obstruction formed by inflammation as a result of infection. Obstruction results in oligospermia or azoospermia results, and this is rectified to a certain extent by the drug's anti-inflammatory action (Martin *et al.*, 1997). Selvaraj (2002) has reported about the increase in number and size of clear cells and vacuoles in the epithelium of caput, corpus and caudal segments of epididymis on diclofenac treatment in rats. The regression of principal cells is seen and this may be due to impaired basal cell secretion of PGE_2 (Dun *et al.*, 2000). A decrease in duct diameter, epithelial cell height was also observed. This resembles the castration effects reported by Rajalakshmi *et al.*, (1985).

Though abundant literature on diclofenac toxicity of other organ systems is present, studies on the effect of DS on male reproductive system is very limited. Dose related inhibition of androgen and 16-androstene biosynthesis, when co-incubated with pregnenolone has been reported by Kwan *et al.*, (1993). A selective significant and sustained decrease in plasma level of prolactin through PG inhibition has been reported by Joris *et al.* (1998). Diclofenac has been reported to transiently retrieve partial epididymal obstruction formed by inflammation as a result of infection (Martin *et al.*, 1997). Therefore, an attempt to disclose the effects of diclofenac on testes was undertaken in the present investigation.

Role of Oxidative Stress

A free radical is any species containing one or more unpaired electrons and capable of independent existence. The free radicals in biological systems are derivatives of oxygen and are referred to as reactive oxygen species (ROS). These include hydroperoxyl radical (HO_2*), superoxide radical (O_2*), hydrogen peroxide (H_2O_2) and hydroxyl radical (OH*). The production of free radicals in living cells may be due to accidental release as by-products of metabolism or deliberate during phagocytosis, or under oxidative like high oxygen concentration (Rosen, 1995). Oxidative phosphorylation results in the formation of ATP, on which living organisms are dependent for their various internal activities. In addition, ROS are also inevitably produced during the metabolic processes, which in turn modify various cellular components by oxidation, thus impairing their physiologic function. Therefore, oxygen is generally considered to be a double-edged sword. ROS production is enhanced under pathological conditions such as inflammation and ischemic reperfusion, and involved in various diseases and aging (Halliwell and Gutteridge, 1999).

The most deleterious effect caused by ROS is the peroxidation of membranes of lipids. Lipid peroxidation can directly damage membrane structure and indirectly damage other cell components by the production of reactive aldehydes. Enzymatic system such as oxygenases and oxidases produce ROS, utilizing oxygen molecule as substrate. Xanthine oxidase, an enzyme involved in purine metabolism generates ROS under ischemic condition in the cardiovascular system (Harrison, 2002). Cyclooxygenase, which catalyses the initial oxidation step of arachidonate to prostaglandins, and induced under inflammatory conditions, also generates ROS, in addition to the mitochondrial electron transport system (Baud and Aradaillou, 1986). Immune and inflammatory responses cause tissue damage in rheumatic diseases, as this involves cartilage and bone destruction by free oxygen radicals.

Antioxidant is any substance that when present at low concentrations compared with those of an oxidizable substrate significantly delays or prevents oxidation of that substrate. In animal systems, agents functioning as antioxidants catalytically remove free radicals and other reactive species. Enzymes like catalase, superoxide dismutase, peroxidase and proteins like transferrin, heptoglobins, haemopexin and low molecular mass antioxidants like glutathione, α-tocopherol, bilirubin and uric acid are antioxidants that protect biomolecules against oxidative damage (Halliwell and Gutteridge, 1999).

The breakage of free radical chain reaction by antioxidants is due to their ability to transfer phenolic hydrogen to a peroxy free radical of peroxidised polyunsaturated fatty acid. The phenoxy free radical formed, reacts with vitamin C to regenerate the tocopherol, or react with peroxy free radical to form a non-free radical product. The oxidation product is conjugated with glucuronic acid and secreted in bile (Mayes, 1996). In this way, the antioxidants help in controlling the level of free radicals, thus protecting cells from damage.

The generation of ROS in the male reproductive tract is potentially dangerous due to their toxic effects at high levels on sperm quality and function. The intracellular antioxidant enzymes cannot protect the plasma membrane that surrounds the acrosome and tail forcing spermatozoa to supplement their limited intrinsic antioxidant defences (Agarwal *et al.*, 2003). The lipids of spermatozoa are rich in PUFA side chain. This high PUFA content gives the sperm plasma membrane, considerable fluidity, needed for it to participate in the membrane fusion events associated with fertilization. In peroxidised sperm, fluidity decrease and capacity to fertilize is diminished. Oxidative stress also causes rise in intracellular free Ca^2 and depletion of ATP, interfering with sperm motility (Aitken and Fisher, 1994).

Use of Vitamin E Supplementation

Vitamin E is a fat-soluble vitamin having a known anti-sterility activity. It prevents reabsorption of foetuses in pregnant female rats and testicular degeneration in males. Its chemical name 'tocopherol' is derived from the Greek word meaning 'to bear offspring'. It has been identified as a dietary factor essential for reproduction in rats and is being recommended for a variety of defects like acne, artherosclerosis, cystic fibrosis, diabetus, infertility, muscular dystrophy, peptic ulcer and rheumatic fever (Fitch and Johnson, 1977). α-tocopherol acts as an antioxidant preventing the development of chain reactions and undergoes auto-oxidation readily in unsaturated oil. This reaction is reversible, resulting in the formation of tocopheroxy radical, which can be reduced chemically back to tocopherol (Boguth, 1970).

The first line of defense against peroxidation of polyunsaturated fatty acids of cellular and sub cellular membrane phospholipids is through vitamin E, as they possess affinities for α-tocopherol. Therefore, vitamin E concentrates at these sites (Mayes, 1996). Peroxide formation involves the formation of a free radical (R*) by a polyunsaturated fatty acid (RH) followed by the addition of oxygen to form peroxide (ROO*). This is followed by chain propagation by union with another polyunsaturated fatty acid to produce a free radical. Replacing one of the substrate can inhibit this chain reaction. This replacement is done by vitamin E as an antioxidant by the method of hydrogen abstraction.

Vitamin E has no effect on structural abnormalities in spermatozoa. It plays an important role in maintaining the viability of spermatid population, making it possible for the epithelial cells to acquire their fully differentiated structural appearance (Bensoussan *et al.*, 1998). Though testis weight was not affected, the histological changes in testis of vitamin-E deficient diet fed rats for four months, were observed by Minakata *et al.* (1987). Suleiman *et al.* (1996) have showed the improvement of sperm motility by decreasing MDA concentrations in spermatozoa in asthenospermia patients, by oral vitamin E treatment. The supplementation of diet with vitamin E improved the serum quality, reducing the production of free radicals in male rabbits (Yousef *et al.*, 2003).

Vitamin E protects the organism against the reactive oxygen species (ROS) mediated damage of spermatozoa. Oral administration of vitamin E significantly improved the *in vitro* functioning of human spermatozoa, assessed by the Zona binding test (Kessopoulou *et al.*, 1995). The supplementation of Vitamin E significantly increased the number of spermatozoa per centimeter of ejaculate. The protective role of vitamin E and growth stimulating hormone in boar semen against its fatty acid peroxidation, has been evidenced by

Brezezinska-Slebodzinska *et al.* (1995). Long-term feeding effect of gossypol on Holstein bulls caused some affected reproductive traits, which were countered and improved by Vitamin-E supplementation. The supplementation of this vitamin is of clinical importance for prolonged storage of sperms whenever needed (Verma, and Kanwar, 1999).

Diclofenac is a powerful non-steroidal anti-inflammatory agent, which inhibits PG synthesis. As POs play an important role in males, affecting the secretion of androgen indirectly, thus present investigation has been undertaken, to determine the adverse effects caused by drug intake. As altered androgen levels are responsible for disturbances in fertility, the determination of LH and FSH levels were undertaken to evaluate the hypothalamo-hypophyseal-testicular axis. In addition, to the testosterone levels, testicular steroidogenic pathway enzymes 3β-hydroxysteroid dehydrogenase and 17β-hydroxysteroid dehydrogenase were also studied. This measurement of androgen and gonadotropin titres in serum of drug treated animals will also enlighten the direct or indirect effect of the drug on the reproductive system. In addition, the activities of functionally related testicular enzymes involved pyruvate-malate shuttle (NADP-isocitrate dehydrogenase, malic enzymes malate dehydrogenase and ATP-citrate lyase) as well as pentose phosphate pathway enzyme activities (Glucose-6-phosphate dehydrogenase and 6-phospho gluconate dehydrogenase) which are essential for the generation of NADPH and acetyl CoA for steroidogenesis and lipogenesis, were also included in the study. The antioxidant effect of vitamin E in drug treated animals, in this study will enable the therapeutic use of this vitamin as supplementation in toxic conditions, thus increasing its importance in reproductive studies.

Observations and Inference

A reduction in food intake is observed with a slight decreasing trend in body weight on DS treatment. This may be due to gastrointestinal intolerance and resultant imbalance in protein metabolism, as seen by a slight increase in serum protein on DS treatment. This in turn affects food intake, thus, bringing down the body weight (Veena, 1995). The co-administration of vitamin E with DS induced a partial restoration of body weight. This may be due to the anti-oxidant properties and consequent protective effects of vitamin E (Brigelius-Flohe, 1999) (Table 26.1).

Table 26.1: Effect of Diclofenac Sodium and Vitamin E Supplementation on Body Weight and Sex Organ Weights of Albino Rats

Organ/Group	*Control*	*Diclofenac Treated*	*Vitamin E Supplemented*
Body weight (gm/100 gm BW)	218.905±6.999	205.470±3.589*	215.00±2.050
Testis (mg/100 gm BW)	1.393±0.041	1.142±0.044**	1.193±0.049*
Vas deferens (mg/100 gm BW)	79.057±0.819	67.014±0.757**	68.900±0.973**
Caput Epididymis (mg/100 gm BW)	160.217±0.633	140.434±0.627**	145.808±0.882**
Corpus Epididymis (mg/100 gm BW)	31.213±0.686	19.529±0.231**	22.782±0.123**
Cauda Epididymis (mg/100 gm BW)	165.46±1.188	145.986±1.171**	145.656±1.126**
Seminal Vesicle (mg/100 gm BW)	161.898±0.959	122.070±1.37**	126.978±0.875**
Coagulating Gland (mg/100 gm BW)	82.250±1.279	66.204+0.426**	69.100±0.364**
Dorsolateral Prostate (mg/100 gm BW)	87.827±1.124	60.406±0.620**	63.204±0.487**
Ventral Prostate (mg/100 gm BW)	99.134±0.657	80.268±1.332**	82.408±1.192**

Descriptive statistics given are mean ± SE.

*: Significant at 5 per cent level by Dunnett's test.
**: Significant at 1 per cent level by Dunnett's test.

The low Testosterone levels due to the reduced Leydig cell number was observed histologically by Anbalagan (2004) on ciprofloxacin administration on testis. Similar decrease in Leydig cell number has been observed by Vijaya (1996), on acetaminophen and caffeine administration. The reduced testosterone levels in the present study may be due to reduction of the same, being similar to the report about the inhibitory effect of DS by Selvaraj (2002).

The decrease in serum LH and FSH titres was observed in this study suggesting an underproduction from pituitary in response to decreased GnRH production from hypothalamus. The hypothalamo-pituitary-gonadal system forms a closed loop feedback control mechanism. Testosterone and oestradiol exhibit a negative feedback control on both LH and FSH secretion at hypothalamus, thus suppressing the frequency of LH release (Plant *et al.*, 1978). Pituitary is also subjected to such a negative feedback control, which in turn reduces its responses to GnRH (Tillbrook *et al.*, 1991). When serum testosterone level is low, serum gonadotropins may be expected to be high by a negative feedback control. The lowering of testosterone as well as gonadotropin titres in the present study, suggest the change of negative feedback control and the action of the drug directly on the hypothalamo-hypophyseal-testicular axis, decreasing all the three hormone levels. Probably the adverse effect of DS is mediated first at the level of hypothalamus or pituitary and decrease in testosterone level is only a secondary change (Table 26.2).

Table 26.2: Effect of Diclofenac Sodium and Vitamin E Supplementation on Serum Gonadotropins and Testosterone of Albino Rats

Hormone/Group	*Control*	*Diclofenac Treated*	*Vitamin E Supplemented*
FSH (mIU/ml)	15.323±0.541	9.248±0.399**	10.040±0.417**
LH (mIU/ml)	2.697±0.107	1.394±0.048**	1.612±0.035**
Testosterone (ng/ml)	1.167±0.036	0.906±0.043**	0.962±0.036**

Descriptive statistics given are mean ± SE.

*: Significant at 5 per cent level by Dunnett's test.

**: Significant at 1 per cent level by Dunnett's test.

The potent action of vitamin E in hormone production in pituitary-gonadal axis in humans and rats has been reported by Umeda *et al.*, (1982). Vitamin E supplementation to DS treated animals was able to restore the hormone levels partially mainly in large duration group, suggesting the insufficiency of vitamin E dose used. Probably a longer duration intake of a larger dose might help in the restoration to normal hormone levels. As LH, FSH and testosterone are the major regulators of testicular structure and function (Fritz, 1978); the observed low titres of serum FSH and testosterone may be responsible for the decrease in testicular weight. The testis is highly susceptible to change when exposed to adverse drugs leading to a decrease in normal weight. Seminiferous tubules make up to 90 per cent of the wet weight of the normal testes of rat (Christensen and Mason, 1965).

Many workers have reported a reduction in testicular weight due to spermatogenic arrest on administration of drugs, chemicals and pesticides (Mann and Lutwak-Mann, 1981). A similar reduction in serum titres of FSH, LH and testosterone, with corresponding reduction in testicular weight on DS treatment has also been reported by Achiraman (1997) and Selvaraj (2002). Supplementation of drug treated animals with vitamin E was beneficial only to a certain extent similar to the effect reported by Selvaraj (2002). In the present study, DS treatment caused a significant reduction in the weight of epididymis as well as other accessory sex organs. This may be due to the reduction of circulating testosterone, which is the main hormone responsible for maintaining the accessory sex organ structure and function. Vitamin E supplementation was not effective in preventing the reduction in accessory sex organ weights. Similar decrease in the weight of epididymis, coagulating gland, dorsolateral and central prostatic lobes and seminal vesicle, due to low testosterone titres, on methotrexate treatment, has been reported by Badri Narayanan (1995).

The testis and accessory sex glands are known to be adversely affected by many pharmacological agents. Cicero *et al.* (1976) have reported about the fall in blood serum LH and testosterone levels followed by involution of accessory organs on methadone treatment in rats. Testicular atrophy and a high incidence of sperm abnormalities are frequently seen in chronically alcoholic men (Van Thiel *et al.*, 1979a). DS brought about an extensive degenerative change in majority of the tubules, with karyotype and pycnotic changes, and, on further administration, necrosis of the tubules indicating the adverse effect of the drug. A similar impairment of spermatogenesis has been observed by Ezzat and Gohary (1994), on caffeine administration. The absence of secondary spermatocytes, spermatids and spermatozoa indicated the impairment of spermatogenesis at the late spermatocyte stage–this may be due to the reduced testosterone levels caused by DS.

The degeneration of most tubules with reduction in Leydig cell number was also observed under DS influence. The reduction in cell population and their exfoliation had caused extraction of Sertoli cell specifically at the transition from stages VII–XIII of the seminiferous epithelium (Cameron *et al.*, 1993; O'Donnell *et al.*, 1994). Adjacent Leydig cells may be involved in this as has been suggested by morphometric observations (Bergh, 1983). As FSH is needed for spermatogonial proliferation and meiotic onset, lowered FSH levels observed in the present study must be responsible for the abrupt arrest of spermatogenesis and the decreased number of spermatogonial population as reported by Huckins *et al.* (1973). The reduction in the number of Leydig cells seen in the present study correlates with the drastic decrease in serum testosterone levels and altered steroidogenesis. Therefore, in accordance with the observation of Siva Kumar *et al.* (2004), the damage in Leydig cell steroidogenesis might have been due to defective expression of LH receptors. The lowered serum FSH levels suggests the lowering of ABP production in Sertoli cells which are needed for transport of androgens from Leydig cells to seminiferous tubules as well as in conversion of DHT (dihydrotestosterone) within the Sertoli cells (Bustos-Obregon, 1994). Supplementation of DS treated rats with vitamin E caused a similar structural revival with no progress in the functional capacity, with the presence of most tubules empty of spermatids (Figure 26.1).

The 43–60 per cent reduction in the activity of 3β-HSD as well as 17β-HSD enzymes in the testis, observed after DS administration, are very significant. 3β HSD participates in the conversion of pregnenolone to progesterone with an accumulation of pregnenolone (Ewing and Brown, 1977). In this study decrease in LH levels is in correlation with the decrease in steroidogenic enzymes. This drastic reduction indicates poor conversion of pregnenolone and its resultant accumulation. The $CytP_{450}$ enzyme activity is required for the interaction of conversion of progesterone to androstenedione, and subsequent testosterone production. Vitamin E supplementation with DS caused an increase in concentration of both 3β-HSD and 17β-HSD. This suggests the working of the antioxidant mechanism (Hsu *et al.*, 1998), but the concentration is not sufficient for the recovery to normal control levels. (Table 26.3).

In tissues like testis, which are involved in active fatty acid and steroid synthesis, tricarboxylic acid (TCA) cycle occurs at a slow rate and most of the citrate formed is exchanged in the cytosol by tricarboxylic transporter (Lehninger *et al.*, 1993). ICDH is one of the

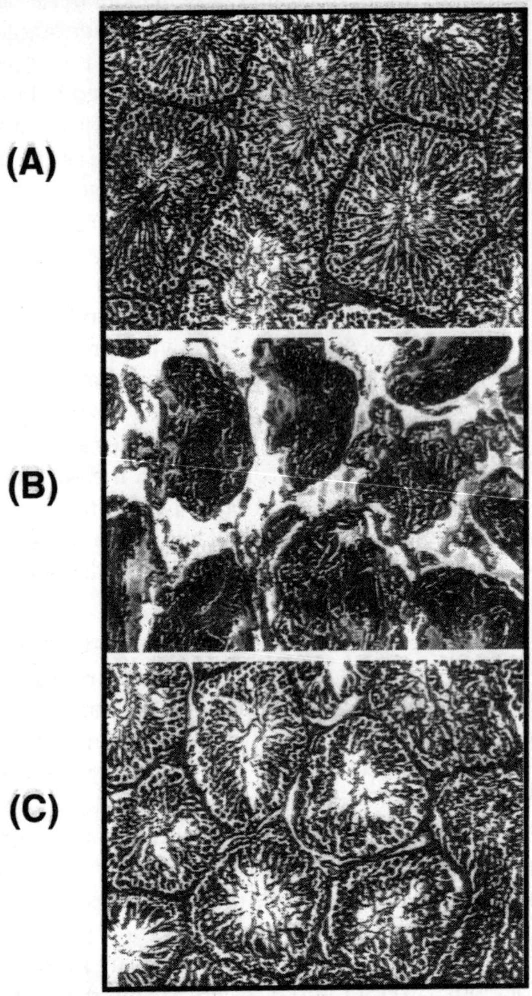

Figure 26.1: Histoarchitecture of Control, Diclofenace Sodium Treated and Vitamin E Supplemented Rat Testis

(A) Control: (1) Seminiferous tubules; (2) Spermatids; (3) Spermatogonia; (4) Leydig cells

(B) Diclofenac Sodium Treated: (5) Shrunken and damaged tubules; (6) Interstitial region

(C) Vitamin E Supplemented: (7) Partially recovering tubules; (8) Developing spermatogonia; (9) Forming spermatids.

key enzymes that regulate citrate flow to the cytoplasm. Increased production seen in the present study causes increased production of NADPH and increased utilization of citrate in the mitochondria thus increasing the supply of citrate in the cytoplasm. An increase in the levels of activity of A TP, citrate lyase, MDH and malic enzyme was observed on DS treatment. This indicates the slight initial idiosyncratic reduction in fatty acid synthesis. Vijaya (1996) has observed an increase in activity of ICDH–NADP$^+$ on caffeine treatment while acetaminophen decreased the enzyme activity.

Table 26.3: Effect of Diclofenac Sodium and Vitamin E Supplementation on Steroidogenic and Lipogenic Enzymes of Albino Rats

Enzyme/Group	*Control*	*Diclofenac Treated*	*Vitamin E Supplemented*
3ß-HSD (n moles/mg/min)	0.055 ± 0.003	0.031±0.003**	0.051±0.003
17ß-HSD (n moles/mg/min)	0.057±0.002	0.022±0.002**	0.036±0.004**
ATP-Citrate lyase (units/mg protein)	0.047±0.002	0.037±0.002**	0.039±0.002*
NADP-Isocitrate dehydrogenase (units/mg protein)	0.039±0.065	0.065±0.002**	0.063±0.005**
Malate dehydrogenase (units/mg protein)	0.347±0.025	0.244±0.036*	0.264±0.016
Malic enzyme (units/mg protein)	0.043±0.002	0.030±0.002**	0.036±0.002
Glucose-6-phosphate dehydrogenase (µ moles/min)	1.887±0.028	2.242±0.086**	2.116±0.093*
6-Phosphogluconate dehydrogenase (µ moles/min)	1.787±0.036	2.118±0.073**	2.032±0.051**

Descriptive statistics given are mean ± SE.

*: Significant at 5 per cent level by Dunnett's test.
**: Significant at 1 per cent level by Dunnett's test.

Malate Dehydrogenase (MDH) transfers reducing equivalents from NADH to oxaloacetate to form malate. This is then oxidatively decarboxylated to pyruvate by malic enzyme using $NADP^+$ thus recycling the reduced substrates for reductive synthesis. In general ICDH activity and MDH activity are higher than the other oxidation systems in testicular tissue (Free, 1970). Vitamin E supplementation plays an important role in maintenance of structure and function of several organ systems protecting the biological membranes against lipid peroxidation (Diplock, 1985). In the present study, vitamin E supplementation was not effective in restoring the normal levels of enzyme activity. Therefore, the action of DS through inhibition of PG synthesis seems to be responsible for the stimulation of enzymes of malate shuttle. Adverse effects of DS on lipogenic enzymes like NADPH isocitrate dehydrogenase, malic enzyme, malate dehydrogenase, and ATP citrate lyase on the epididymal segments of rats has been reported by Selvaraj (2002).

Glucose converted to glucose-6-phosphate enters the glycolytic and pentose phosphate pathway (Mayes, 1996). In the present study, as the germ cells were found to be markedly depleted, as assessed from histological studies, the glucose metabolism through the glycolytic pathway may be depressed. The decreased metabolism of glucose-6-phosphate through glycolytic pathway might have increased the availability of substrate for the pentose phosphate pathway, thus resulting in enhanced activities of G-6-PDH and 6-PGDH. In the present study, an increase in activities of both G-6-PD and 6-PGD on DS treatment, was observed in an effect similar to the effect of acetaminophen, observed by Vijaya (1996). This may be due to the entry of more of glucose entering the HMP shunt, due to demand of ribulose sugars required for nucleic acid synthesis. Itinose *et al.* (1994) have also reported about the inhibition of glycogen deposition in the liver and hence the adverse effect on energy metabolism. Supplementation of vitamin E brought about a nearly complete recovery of the activities of HMP shunt pathway enzymes.

The major lipid components of the testis are phospholipids, which constitute about 80 per cent of the total lipid (Oshima and Carpenter, 1968). Any alteration in this lipid fraction reflects on the total lipid concentration. Thus, the reduction in testicular phospholipid levels in addition to cholesterol and triglyceride levels observed in the present study, might be due to reduction in total lipid concentration. The total lipid concentration was also observed to be decreased (about 10–20 per cent) on DS administration. Partial recovery on vitamin E supplementation was observed. Cholesterol serves as the structural component of the secretory epithelium. Cholesterol is required by the testis for the synthesis of testosterone derived from HDL (Pignataro *et al.*, 1983). In the testis a small portion of cholesterol is esterified by ACAT, stored as lipid droplets serving as reservoirs of cholesterol (Mori and Christensen, 1980) (Table 26.4).

Table 26.4: Effect of Diclofenac Sodium and Vitamin E Supplementation on Lipid Classes of Albino Rats

Lipid/Groups	*Control*	*Diclofenac Treated*	*Vitamin E Supplemented*
Total lipids (mg/100 mg tissue)	22.743±0.704	18.330±051*	18.214±0.264*
Total cholesterol (mg/100 mg tissue)	0.095±0.704	0.167±0.504**	0.144± 0.204**
Total phospholipids (mg/100 mg tissue)	17.670±0.769	11.490±0.378**	11.926±0.504**
Total glyceride glycerol (mg/100 mg tissue)	6.027±0.062	3.500±0.202**	4.134±0.153**

Descriptive statistics given are mean ± SE.

*: Significant at 5 per cent level by Dunnett's test.
**: Significant at 1 per cent level by Dunnett's test.

In the present study, administration of DS caused a significant reduction (30–40 per cent) in cholesterol level. The low cholesterol concentration has been suggested to be due to the low transportation of triglycerides and phospholipids in the degenerating cells (Anfinsen, 1954). The histological profile showing the degeneration of tubules further corroborates this observation. Vijaya (1996) and Selvaraj (2002) have also reported about the reduction in cholesterol levels on administration of AP, caffeine and DS respectively. The consistent reduction in cholesterol may reflect the reduction in testosterone levels and loss of membrane cholesterol. Phospholipids are the integral components of the sperm membrane and play an important role in maintaining their structural and functional integrity (Ewing and Brown, 1977). In the present study, as there was a heavy loss of germinal elements, particularly spermatocytes, spermatids and spermatozoa as evidenced from the histological studies, it is therefore not surprising to get a decrease in phospholipid concentration.

Conclusion

From the above study on diclofenac sodium, it is well evident that this anti-inflammatory drug has adverse effects on bodyweight, testis as well as accessory sex organs. Changes in hormonal level of serum gonadotropins and testosterone concentration expose the drug's adverse effects on the hypothalamo-pituitary-gonadal axis. Severe adverse changes were also seen in the histoarchitecture of testis, with changes in the activities of steroidogenic enzymes, pentose phosphate cycle enzymes, pyruvate-malate shuttle enzymes and the energy providing as well as structural lipids. Vitamin E supplementation could only partially restore the histological, biochemical and hormonal response of the tissue, and so a greater dose can be advised for counteracting the drug's effect. Therefore, a judicious use of diclofenac sodium with Vitamin E supplementation has been recommended in needy conditions.

References

Achi Raman, S., 1997. Effect of Voveran on the testosterone level in the serum and histological profiles of the testis and accessory sex organs in rats. *M.Phil Thesis*, Bharathiar University Coimbatore India.

Agarwal, A., Saleh, R.A. and Bedaiwy, M.A., 2003. Role of reactive oxygen species in the pathophysiology of human reproduction. *Fertil. Steril.*, 79: 829–843.

Aitken, R.J. and Fisher, H., 1994. Reactive oxygen species generation and human spermatozoa: The balance of benefit and risk. *Bioassays*, 16: 259–267.

Anbalagan, C., 2004. Rat testicular and prostatic responses to oxidative stress induced by ciprofloxacin and the antioxidant activities of vitamin A, C and E. *Ph.D. Thesis*, Bharathiar University, Coimbatore, India.

Anfinsen, C.B. Jr., 1954. In: *Symposium on Artherosclerosis, Discussion.* National Research Council, Washington, D.C., pp. 167–179.

Badri Sriman Narayanan, 1995. Effects of methotrexate on testes of adult Albino rats. *Ph.D. Thesis*, Bharathiar University, Coimbatore, India.

Baud, L. and Ardaillou, R., 1986. Reactive oxygen species: Production and role in the kidney. *Am. J: Physiol.*, 251: F765–F776.

Bensoussan, K., Morales, C.R. and Hermo, L., 1998, Vitamin E deficiency causes incomplete spermatogenesis and affects structural differentiation of epithelial cells of the epididymis in the rat. *J. Androl.*, 19(3): 266–288.

Bergh, A., 1983. Paracrine regulation of Leydig cells by the seminiferous tubules. *Int. J. Androl.*, 6: 57–65.

Boelsterli, U.A., Zimmerman, H.J and Rommel, K.A., 1995. Idiosyncratic liver toxicity of non-steroidal anti-inflammatory drugs: Molecular mechanisms and pathology. *Crit. Rev. Toxicol.*, 25: 207–235.

Boguth, W., 1970. *Vitam. Horm.*, 27: 1.

Brezezinska-Slebodzinska, E., A.B., Pietras, B., Wieczorek G., 1995. Antioxidant effect of vitamin E and glutathione on lipid peroxidation in boar semen plasma. *Boil. Trace Elem. Res.*, 47(1–3): 69–74.

Brigelius-Flohe R., 1999. Tissue-specific functions of individual glutathione peroxidases. *Free Radic. Bioi. Med.*, 27: 951–965.

Bustos-Obregon, N.Z., Van rysselberghe, N., and Rodriguez, H., 1994. The role of FSH in adult rat spermatogonial proliferation. In: *Fertility and Sterility: The Proceeding of the 14th World Congress on Fertility and Sterility* (Ed.) O. Rodriguez-Armes. The Parthenon Publishing Group, pp. 51–60.

Cameron, D.F., Muffly, K.E. and Naziam, S.J., 1993. Testosterone stimulates spermatid binding to competent Sertoli cells *in vitro*. *Endocr. J.*, 1: 61–65.

Carlsen, E., Giwercman, A., Keiding, N. and Skakkebaek, N.E., 1992. Evidence for decreasing quality of semen during past 50 years. *EMF*, 305: 609–613.

Chan, L.Y., P.Y. Chiu, N.S.S. Siu and T.K. Lau, 2001. A study of diclofenac induced terratogenicity and increased lipid peroxidation during organogenesis using a whole rat embryo culture model. *Frontiers In fetal Health*, 3(11/12): 27–28.

Christensen, A.K. and Mason, N.R., 1965. Comparative ability of seminiferous tubules and interstitial tissue of rat testes to synthesize androgens from progesterone-4^{14}C *in vitro*. *Endocrinology*, 76: 646–656.

Cicero, T.J., Meyer, E.R., Bell, R.D., Koch, G.A., 1976. Effects of morphine and methadone on serum Testosterone and Luteinizing hormone levels and on the secondary sex organs of the male rat. *Endocrinology*, 98: 367.

Data Sheet, 2002. Information for Health Professionals. At www.medsafe.govt.nz.

Deleuran, H.W., 2001. NSAID-COX-2 inhibitors, where is the difference? Focus on the modes of action, Ugeskr. *Laeger.*, 163(32): 4185–4189.

Diplock, A.T., 1985. Vitamin F. In: *Fat-soluble Vitamin: Their Biochemistry and Applications.* Heinermann, London, pp. 154–224.

Doehmer, J., Goeptar, A.R. and Vermeulen, N.P.E., 1993. Cytochromes 450 and Drug Resistance. *Cytotechnology* 12(1–3): 357–366.

Dun, N.J., Dun, S.L., Wong, P.Y.D., Yang, J., and Chang, J.K., 2000. Cocaine and Amphetamine-regulated transcript peptide in the Rat epididymis: An immuno-histochemical and electrophysiological study. *Biology of Reproduction*, 63: 1518–1524.

Ewing, L.L., and Brown, B., 1977. Testicular steroidogenesis. In: *The Testis*, Vol. 4, (Eds.) Johnson A.D. and Gomes, W.R. Academic Press, New York, pp. 239.

Ezzat, A.R. and Gohary, Z.M., 1994. Hormonal and histological effects of chronic caffeine administration on the pituitary gonadal and pituitary-adreno-cortical axis in male rabbits. *Func. Dev. Morphol.*, 4: 45–50.

Fitch, K.L. and Johnson, P.B., 1977. Nature of foods fat soluble vitamins. In: *Human Life Sciences*, Chapter 12, Holt, Rinehart and Winston. New York, pp. 182–203.

Free, M.J., 1970. Carbohydrate metabolism in the testis. In: *The Testis*, Vol. 2, (Eds.) A.D. Johnson, W.R. Gomes and Van de Mark, N.L. Academic Press, New York, pp. 125–192.

Fritz, I.B., 1978. Sites of actions of androgens and follicle stimulating hormone on cells of the seminiferous tubule. In: *Biochemical Actions of Hormones, Vol. 5,* (Ed.) G. Litwack. Academic Press, New York, pp. 249–278.

Gokcimen, Alpaslan, Gulsen Aydin, Erdal Karaoz, M. Ali Malas and Meral Oncu, 2001. Effect of Diclofenac Sodium administration during pregnancy in the postnatal period. *Fetal Diag. Ther.*, 16(6): 417–422.

Goldstein, J.L., 2000. Significant upper gastrointestinal events associated with conventional NSAID versus celecoxib. *J. Rheumatol. Suppl.*, 60: 25–28.

Halliwell, B. Gutteridge, J.M.C. (Eds.) 1999. *Free Radicals in Biology and Medicine,* 3rd edn. Oxford University Press, New York, USA.

Hamilton, D.W., 1975. Structure and function of the epithelium lining the ductuli efferentes, ductus epididymidis and ductus deferens in the rat. In: *Handbook of Physiology, Endocrinology* Section 7, Vol. V, (Eds.) Hamilton, D.W. and Hreep, R.O. American Physiological Society, Washington, D.C., pp. 259–301.

Harrison, R., 2002. Structure and function of Xanthine oxidoreductase: Where are we now? *Free Radic. Biol. Med.*, 33: 774–797.

Hsu, P.C., Liu, M.Y., Hsu, C.C., Chen, L.Y. and Y.L., 1998. Effects of vitamin E on reactive oxygen species-related lead toxicity in the rats sperm. *Toxicol.*, 128: 169–179.

Huckins, C., Mills, N., Besch. P., Mans. A., 1973. Spermatogonial differentiation in immature testis: Regulation by FSH. *Endocrinology.*, 92(Abstr): A91.

IPCS, 2002. *Global Assessment of the State-of-the-Science of Endocrine Disruptors.* (Eds.) Damstra S., Barlow S., Bergrnan, A., Kavlock, R. and Van Der Kraak, G.

Itinose, A.M., Dolsakuno, M.I. and Bracht, A., 1994. N-acetyl cysteine stimulate hepatic glycogen deposition in the rate. *Res. Commn. Chem. Pathol.* (US), 83: 87–92.

Joris, J., Reuter, A.M., Gevaert, V.Y., Meuleman, G.R., Hargreaves, K. and Franchimont, P., 1998. Effect of diclofenac on plasma levels of immunoreactive prolactin, FSH, LH, thyrotrophin and β-endorphin in humans. *Horm. Res.*, 29: 143–146.

Kessopoulou, E., Powers, H.J., Sharma, K.K., Pearson, M.J, Russell, J.M., Cooke, I.D. and Barratt, C.L., 1995. A double-blind randomized placebo cross-over controlled trial using the antioxidant vitamin E to treat reactive oxygen species associated male infertility. *Fertil. Steril.*, 64(4): 825–831.

Kirchner, T., Aparicio, H., Argentieri, D.C., Lau, C.Y. and Ritchie, D.M., 1997. Effects of tepoxalin, a dual inhibitor of cyclooxygenase/5-lipoxygenase on events associated with NSAID induced gastrointestinal inflammation. *Prostaglandins, Leukot Essent. Fatty Acids*, 56(6): 417–423.

Kwan, T.K., Foon, S.L., Lim, Y.T. and Gower, D.B., 1993. Effect of some non-steroidal anti-inflammatory drugs (NSAIDS) on steroidogenesis in rat and porcine testis, pp. 305–330.

Lehninger, A.L., Nelson, D.L and Cox, M.M., 1993. Bioenergetics and metabolism. In: *Principles of Biochemistry*, 2nd ed. (Eds.) Lehninger, A.L., Nelson, D.L and M.M. Cox. Worth Publishers, New York, pp. 559–788.

Mann and Lutwak-Mann, 1981. *Male Reproductive Function and Semen.* Springer-Verlag, Berlin Heiderberg, New York, pp. 357.

Martin, G.G., Domiguez, A.R. and Geravasa, A., 1997. Pharmacokinetics and hepatotoxicity of diclofenac using an isolated perfused rat liver. *Biomed. Pharmacother.*, 51: 170–175.

Masubhuchi, Y., Saito, H. and Horie, T., 1998. Structural requirements for the hepatotoxicity of non-steroidal anti-inflammatory drugs in isolated rat hepatocytes. *J. Pharmacol. Exp. Ther.*, 287: 208–213.

Mayes, P.A., 1996. Structure and function of the lipid soluble vitamins. In: *Harper's Biochemistry*, (Eds.) Murray R.K., Granner, D.K., Mayer, P.A. and Rodwell, W.W., 24th edn. Appleton and Lange, Connecticut, pp. 619–622.

Minakata, K., Asano, M., Takahashi, Y. and Harada, N., 1987. Enhancement of testicular cysteine proteinase inhibitor level in vitamin E-deficient rats. *J. Nutr.*, 117(8): 1416–1421.

Mori, H. and Christensen, A.K., 1980. Morphometric analysis of Leydig cells in the normal rat testis. *J. Cell Biol.*, 84: 340–354.

O'Donnell, L., McLachlan, R.L., Wreford, H.S., Robertson, D.M., 1994. Testosterone promotes the conversion of round spermatids between stages III and VIII of the rat spermatogenic cycle. *Endocrinology,* 135: 2608–2614.

Oshima, M. and Carpenter, M.P., 1968. The lipid composition of the perpubertal and adult rat testis. *Biochem. Biophys. Acta,* 152: 479–497.

Pignataro, O.P., Radicella, J.P., Calvo J.C. and Charreau, E.H., 1983. Mitochondiral biosynthesis of cholestrol in Leydig cells from rat tesis. *Mol. Cell. Endo.,* 33: 53–67.

Plant, T.M., Hess, D.L., Hotchkiss, J. and Knobil, E., 1978. Testosterone and the control of gonadotropin secretion in the male rhesus monkey (*Macaca mulatta*). *Endocrinology,* 103: 535–541.

Rajalakshmi, M., 1985. Compounds affecting motility and fertilizing ability of spermatozoa in the male. In: *Advances in Fertility Regulation in the Male,* (Eds.) Qian, S.Z and Waitas, G.M.H. The People's Medical Publishing House, Beijing, China, pp. 135–138.

Rosen, C.H., 1995. The fat-soluble vitamins. In: *Normal and Therapeutic Nutrient,* 16th ed. Macmillan Publishing Co. Inc., New York, pp. 305–327.

Selvaraj, M., 2002. Studies on diclofenac sodium: Endocrine changes, Epididymal Biochemistry and Histoarchitecture in rats. *Ph.D. Thesis,* Bharathiar University, Coimbatore, India

Sivakumar, R., Poomarippan, A., Rengarajan, S., Govindarajulu, P. and Balasubramanian, K., 2004. Ethanol toxicity has adverse effects on the expression of LH receptor and its mRNA in rat Leydig cells. In: 22nd *Annual Symposium on Reproductive Biology and Comparative Endocrinology* (Abstr.), pp. 34.

Skakkebaek, N.E., Rajpert-De Meyts, E. and Main, K.M., 2001. Testicular dysgenesis syndrome: An increasingly common developmental disorder with environmental aspects. *Human Reproduction,* 16: 972–978.

Suleiman, S.A., Ali, M.E., Zaki, Z.M., el Malik, E.M. and Nasr, M.A., 1996. Lipid peroxidation and human sperm motility: Protective role of vitamin E. *J. Androl.,* 17: 530–537.

Tang, E., 2003. The metabolism of diclofenac-Enzymology and toxicology perspectives. *Curr. Drug Metab.,* 4(4): 319–329.

Tillbrook, A.J., de Kretser, D.M., Cummins, J.T. and Clarke, J.J., 1991. The negative feedback effect of testicular steroids on gonadotropin secretion in the ram is due to reduced GnRH secretion and not a direct effect on the pituitary gland. *Endocrinology*, 129: 3080–3092.

Umeda, F., Kato, K., Muta, K and Ibayashi, H., 1982. Effect of vitamin E on function of pituitary-gonadal axis in male rats and human subjects, *Endocrin. Jpn.*, 29(3): 287–292.

Van Thiel, D.H., Gavaler, J.S., Gobb C.F., Sherins. R.J., Lester, R., 1979a. Alcohol induced testicular atrophy in the adult male rats. *Endocrinology*, 105: 888.

Veena Somanatha Bhat, 1995. Effect of voveran on the haemotology and histological and biochemical profiles of liver, kidney and pancreas of adult albino rats. *M.Phil Thesis*, Bharathiar University, Coimbatore, Tamil Nadu, India.

Verma, A. and Kanwar, K.C., 1999. Effect of vitamin E on human sperm motility and lipid peroxidation *in vitro. Asian J. Androl.*, 1(3): 151–154.

Vijaya, D., 1996. Studies on caffeine and acetaminophen, effect on liver, kidney and tests of albino rats. *Ph.D. Thesis*, Bharathiar University, Coimbatore, India.

Yousef, M.I., Abdallahb, A.G.A. and K.I. Kamel, 2003. Effect of ascorbic acid and vitamin E supplementation on semen quality and biochemical parameters of male rabbits. *An. Reprod. Sci.*, 76(1–2): 99–111.

Chapter 27

Environmental and Occupational Exposure to Metals and Reproduction

Sunil Kumar* and Minal Mankad

Deputy Director (Sr. Grade), Division of Reproductive Toxicology and Histochemistry, National Institute of Occupational Health, Indian Council of Medical Research, Meghani Nagar, Ahmedabad – 380 016, India

ABSTRACT

A number of metals are essential for the normal growth, development, reproduction and also various other physiological functions. However, their excess or deficiency in the body might lead to adverse biological effects. A number of heavy metals even at a low level are having toxic potential not only for the exposed individual but also to the offspring.

Sufficient data is available with regards to the adverse effects of lead, cadmium, mercury and chromium, etc. on male and female reproduction of various experimental animal species. The studies on effects of heavy metals on human reproduction are scanty to reach any firm conclusion except some reasonable data on lead toxicity in male and female. It

* Corresponding author: Deputy Director (Sr. Grade), National Institute of Occupational Health (ICMR), Meghani Nagar, Ahmedabad – 380 016, Phone: (079) 22868351, 352; Fax: 22686110; E-mail: sunilnioh@yahoo.com

can be inferred from existing human studies coupled with experimental studies that in addition to lead, other metals like cadmium, mercury and chromium might also affect human reproduction. However, well-planned epidemiological studies are needed on these metals after controlling all the confounders. Further, detailed experimental studies on other metals, which are reported to have reproductive toxic potential in few animal studies, are also required to confirm the earlier findings. In addition, some of the metals, such as lead, cadmium and mercury have been reported to have endocrine disruptive potential affecting normal reproductive and endocrine function.

Keywords: *Environmental and occupational exposure, Metals, Lead, Cadmium, Mercury, Chromium, Reproduction, Spermatogenesis.*

Introduction

During the last 50–60 years a large number of synthetic chemicals have entered in the environment due to industrial, developmental and agricultural activities. Persistent nature of these chemicals in the environment and bioaccumulation in the body raises the possibility of long-term deterimental effects on aquatic, and wildlife fauna as well as on the health of other animals including humans. The effect of these persistent contaminants on health has become all the more important since scientists recognized a new kind of hazard known as "endocrine disruption". Now most of the researchers agreed on the fact that some widespread, man-made chemicals called 'endocrine disrupters', can interfere with the body's own hormones and affect health including reproductive system. Metals like lead, cadmium and mercury are reported to have endocrine disruptive potential.

Human beings as well as aquatic and wild life species are exposed to certain metals and their oxides, which are persistent in atmosphere. Some of the living beings are exposed to higher doses of these metals, through occupational or accidental exposure or through contaminated soil, water and food. Therefore, all the stages of gametogenesis would be the potent target of chronic exposure to very low doses of these metals. Heavy metals may enter the human body through inhalation, ingestion of contaminated food and water, and some by absorption through the skin when these metals come in contact with humans in agriculture and in manufacturing, pharmaceutical, industrial, or residential settings. Experimental studies indicated that both male and female reproductive system is vulnerable to the effects of some of the metals, which depend upon dose, duration, and timing of exposure. However, such data on human female reproductive system are less documented as compared to male. This may be due to the fact that human male reproductive endpoints can be studied easily. Thus, it is difficult to pinpoint which sex is more or less vulnerable to exposure of metals. Further, reproductive toxicity is an area of toxicity, which is different from other forms of toxicity. In reproductive toxicity both male and female or a reproductive unit is involved and impairment is confined not only to the exposed individual but it could also be maifested in the offsprings. In this article we have provided comprehensive current status of the effect of metals in human reproduction.

Material and Methods

The literature was assimilated by searching various databases such as Pub. Med, Toxline, Google and certain other websites as well as consulting various journals especially on environmental and occupational health and also incorporated some of our personal observations. This article is divided in to several headings based on the metals exposure and associated reproductive outcome in animals and human. Earlier overview or reviews published on metals and reproductive toxicity generally written either only on male or female reproduction or confined to animal studies. The present article provides recent trends of reproductive impairments for both sexes with regards to metal exposure.

Lead and Reproduction

All the living beings including human are exposed to lead to some extent due to contaminated air, water and food. A large number of workers are exposed to high doses of lead in various occupations such as lead based paints, construction, battery manufacturing or recycling, automobile repair, electronics, printing, welding and soldering, jewellery making and repair, stained glass window making, etc. Earlier, vehicular exhaust used to be a major source of lead exposure for the general population, but nowadays due to the use of unleaded gasoline, the problem has subsided to some extent. However, leaded petrol is still in use in many of the underdeveloped areas of the world. The general population may also be exposed to lead through leaded pipes that supply drinking water, lead based paints, lead glazed ceramics, plants grown in lead contaminated soil, ayurvedic medicines, toys, pencils, etc.

A number of experimental studies are available on the effects of lead on various animal species of both sexes. Studies carried out in mice indicated that high exposure to lead caused testicular atrophy and impairment of spermatogenesis (Eyden *et al.*, 1978). A study carried out at author's laboratory on toxic effects of lead acetate on testicular tissues and sperm morphology of Swiss mice after three different doses of lead acetate *i.e.* 180, 200 and 250 mg/kg b.wt. indicated conspicuous degenerative changes in testicular tissues and elevation in sperm head shape abnormalities (Gautam *et al.*, 2001). Several investigators have reported that oral administration of lead to animals even at doses in the µg/kg range can cause changes in spermatogenesis (Egorova *et al.*, 1966; Golubovic *et al.*, 1968). It has been reported very recently that an administered dose of 0.1 ppm via drinking water ingestion by neonatal male mice is sufficient to produce PbB of 20–60 µg/dL compromised reproductive function in these mice as adults. (Pace *et al.*, 2005). However, earlier Johansson and Wide (1986) mentioned that sexually mature male mice exposed to lead in the drinking water for 3 months attained a mean blood level of 32 µg/100 ml. There was no significant difference with respect to body weight, plasma level of testosterone, number of epididymal sperms, and weight of reproductive organs when compared to controls. Lead has been reported to cause oxidative cellular damage in reproductive tissues of adult male (Table 27.1), which may be closely associated with the reactive oxygen species (ROS) production. (Marchlewicz *et al.*, 2004). Earlier, Agarwal *et al.* (2003) mentioned that ROS plays an essential role in the pathogenesis of many reproductive processes. In male factor infertility oxidative stress attacks the fluidity of the sperm plasma membrane and the integrity of DNA in the sperm nucleus.

Lead is also reported to affect female reproductive system (Table 27.2) in various experimental animal species. A study carried out in monkeys indicated that lead acetate exposure causes adverse effects on ovarian activity and fertility (Franks *et al.*, 1989). Animal studies also support the contention that behavioral deficiencies can occur in infants and newborn as a result of intrauterine exposure to lead via their mothers (EHC, 1977).

Table 27.1: Human Male Reproductive Impairment Data Due to Metal Exposure

Metal	*Effect*	*Reference*
Lead	Significant decrease infertility	Gennart *et al.*, 1992
	Lower sperm count and motility	Roy Chowdhury *et al.*, 1986
	Reduction in fertility	Matthies *et al.*, 1989
	Impaired reproductive function by reducing sperm count, motility, morphology	Apostoll *et al.*, 1998
	Adverse effects on sperm	Bonde *et al.*, 2002
	Spermatogenesis impairment	Alexander *et al.*, 1998
	Reduced semen quality	Telisman *et al.*, 2000
Mercury	Reduced sperm counts, Terato and asthenozoospermia	Popschu, 1978
	Impairment of the fertility index	Matthies, 1989
Chromium	Moderate decrease in the sperm count per ejaculate, degree of sperm motility and the linear penetration rate of the sperm	Bonde, 1990
	Deterioration in sperm morphology	Kumar *et al.*, 2005
	Changes in semen quality and reproductive hormone levels	Li *et al.*, 1999
Nickel and Chromium	Negative correlation with semen concentration among welders	Danadevi, 2003
Cadmium	Inverse correlation between blood Cd level and sperm density among oligospermic men and between seminal plasma Cd and semen volume	Xu *et al.*, 1993
	Significant Association of high Cd level with varicocele related infertility	Benoff *et al.*, 1997
	No difference between cadmium-exposed men and unexposed controls in serum Testosterone, FSH and LH levels	Mason *et al.*, 1990
Lead, Cadmium and Aluminium	Lowering of sperm viability	Dawson *et al.*, 1998

Table 27.2: Human Female Reproductive Impairment Data Due to Metal Exposure

Metal	*Effect*	*Reference*
Lead	Increased risk of low birth weight and neural tube defects	Irgens *et al.*, 1998
	Preterm birth, stillbirth and spontaneous abortions	Saxena *et al.*, 1994
Mercury	Stored in pituitary gland and affect the production of gonadotropins	Dancher *et al.*, 1990
	Risk of spontaneous abortions for female dentists	Kaste, 1997
	Hazardous for reproductive processes	Sikorski *et al.*, 1987
Cadmium	Early delivery, lower birth weight	Nishijo *et al.*, 2002
	Preterm infants	Laudanski *et al.*, 1991
Elevated Heavy metals body load	Greater risk of miscarriages, foetal malformations premature births	McMichael *et al.*, 1986; Laudanski *et al.*, 1991

Epidemiological and case studies are available which indicated that occupational exposures to lead have adverse effects on human reproduction. A study on workers of a newspaper printing press carried out by Roychodhury *et al.* (1986) indicated that the average sperm counts were significantly lower and lesser proportion of them were found to be motile in the exposed subjects as compared to control. These changes were associated with the level of lead in blood. In a recent review on male reproductive toxicity of lead in animals and humans, the authors reported that human studies focused mainly on semen quality, endocrine function, and birth rates in occupationally exposed subjects. These human studies showed that exposure to inorganic lead greater than 40 μg/dl in blood impaired male reproductive function by reducing sperm count, volume, and density or changing sperm motility and morphology. However, no relevant effects were detected on endocrine profile (Apostoll *et al.*, 1998). Moderate exposure to Pb (Blood Pb < 40 μg/dl) and Cd (Blood Cd < 10 μg/l) could significantly reduce human semen quality without conclusive evidence of impairment of male reproductive endocrine "function also reported by Telisman *et al.* (2000). Very recently, Bonde *et al.* (2002) reported the adverse effects of lead on sperm concentration and explained that susceptibility to acid induced denaturation of sperm chromatin are unlikely at blood lead concentrations below 45 μg/dl. In addition, on the basis of a cross sectional study among men employed at a lead smelter unit, Alexander *et al.* (1998) concluded that blood lead concentrations below the currently accepted workers protection criteria seems to adversely affect spermatogenesis (Table 27.1).

Painters and construction workers supposed to be exposed to lead and solvents during occupation have been reported to have adverse effects on both male and female reproduction (Daniell and Vaughan, 1988). Battery workers exposed to lead indicated a significant decrease in fertility (Gennart *et al.*, 1992) and significant levels of asthenospermia and teratospermia (Lerda, 1992). Mehta and Anandkumar (1997) reported a decline in sperm count in a study of population from Bangalore, India and they have correlated this decline with the changes in various pollution indices such as suspended particulate matter, sulphur dioxide and lead. All the available studies clearly suggest that blood lead levels above 40 μg/dl might have adverse effects on male reproductive function. Lin *et al.* (1998) studied the paternal occupational lead exposure and low birth weight/pre-maturity, and reported that there were no significant differences in birth weight or gestational age between the exposed and control groups.

There are some reports that lead affects human female reproductive system. Higher concentrations of blood lead level in the Indian population have been reported by various investigators (Annapurna *et al.*, 1985; Parikh, 1990; Raghunath *et al.*, 1999). Lead absorbed by the pregnant mother is readily transferred to the developing foetus (Ong *et al.*, 1985). A mean blood lead level of 22.52 μg/dL was observed among pregnant women with various adverse. outcomes such as pre-term delivery, stillbirth and spontaneous abortion as compared to 19.4 μg/dL in normal delivery cases reported in a study carried out in Lucknow (Saxena *et al.*, 1994). Irgens *et al.* (1998) studied the associations between occupational lead exposure and reproductive outcome. They observed that offspring of lead-exposed mothers had an increased risk of low birth weight and neural tube defects. Reports from a period spanning more than a century, and covering occupationally exposed women in several countries, support an increase in pregnancy loss from high maternal lead exposures. However, further research is needed to confirm this association, to delineate the role of maternal vs. paternal exposures. (Hertz-Picciotto, 2000). On the basis of data available it can be inferred that lead is having reproductive toxic potential which depend upon dose, duration and time of exposure. Further, data available revealed that in human beings, a blood lead level of 40 µg/dL causes deterioration in semen quality. Thus use of lead in various walks of life should be restricted or if possible eliminated.

Mercury and Reproduction

There are three species of mercury: (1) Organic; (2) Inorganic and (3) Elemental mercury. Elemental mercury is used in the production of batteries, thermometers, barometers etc. It is also known that dental amalgam fillings contain mercury concentrations up to 50 per cent. Organic mercury is most toxic form, however all forms of mercury exhibit toxicity. Bacteria in the environment may methylate inorganic mercury in to organic mercury, and this methyl mercury is accumulated in fishes and finds its way to the food chain. People may be exposed to mercury through eating contaminated seafood, incinerator burning, contaminated soil, coal or oil based thermal power plant emissions, etc.

Experimental animal studies have revealed that high doses of mercury increases the risk of reproductive disorders, *e.g.* infertility, spontaneous abortion, stillbirth and congenital malformations (Schhurs, 1999). Methyl mercury has been reported to disturb spermatogenesis in rats and in several other experimental animal species (Roy Chowdhury and Arora, 1992). Mercuric chloride (inorganic mercury) at the dosage of 50 and 100 µg/kg (0.25 per cent and 0.5 per cent less than LD50 values respectively) i.p. administered to rats over a period of 90 days, caused a gradual decrease in testicular growth rate along with degeneration of spermatogenesis. Further, inhibition of 3 β Δ 5 hydroxysteroid dehydrogenase along with decrease in testicular ascorbic acid and high cholesterol values indicated the impairment in steroidogenesis (Roy Chowdhury *et al.*, 1989).

Davis *et al.*, (2001) studied the effect of mercury vapours on female reproduction. They reported that exposure to mercury vapour altered estrous cyclicity, but had no significant effect on ovulation, implantation, or maintenance of first pregnancy in female rats. In another study, it was reported that maternal methyl mercury (MeHg) exposure did not affect the length of the menstrual cycle or the conception rate or offspring size at birth. However MeHg exposure did significantly reduce the number of viable deliveries at blood Hg concentrations abuse 1.5 ppm. (Burbacher *et al.*, 1988). Recently, it has been reported that inhaled Hg vapour is distributed to all maternal and foetal tissues in a dose-dependent manner (Morgan *et al.*, 2002). Thus mercury may affect the normal development of the offspring.

Mercury is also reported to have deleterious effects on human male reproductive system but the evidence is limited. Occupational alkyl mercury exposure has been reported to cause reduced sperm counts and terato and asthenozoospermia (Popschu, 1978). In another study, twenty-two men occupationally exposed to mercury were reported to have impairment of the fertility index and concentration of the seminal mercury was ten times higher than serum concentration (Matthies 1989). These findings indicate that mercury may accumulate in the testicular or epididymal tissues, which may be responsible for its higher concentration in semen. Kaste (1997) carried out an extensive survey among dental professionals in United States. The data suggests little, if any alterations in the risk of spontaneous abortions for female dentists working with amalgam and nitrous oxide. However, Rowland *et al.* (1994) reported that women with high occupational exposure to mercury were less fertile than unexposed controls. Earlier Sikorski *et al.* (1987) also reported that dental work might represent an occupational hazard with reference to reproductive processes. Cordier *et al.* (1991) studied the potential reproductive toxicity of mercury vapour among the wives of 152 occupationally exposed to mercury vapour and 374 control subjects. The results indicated an increase in the rate of spontaneous abortions with an increasing concentration of mercury in the father's urine before pregnancy. Recently, Schuurs (1999) reviewed the work on occupational mercury exposure and reproductive toxicity. He mentioned that it seems warranted to conclude that the negative reproductive effects from exposure to mercury in the dental office are unproven and safe levels have not been established. He further suggested that more research is needed concerning the effects of occupational elemental mercury concentrations lower than the TLV on the menstrual cycles, conception, male fertility and children's behavior. Keeping in view of inconsistent findings in human there is a need to conduct well designed comprehensive studies after controlling all the confounders on occupationally exposed workers to find out the reproductive potential after exposure to low doses of mercury.

Chromium and Reproduction

Chromium is a naturally occurring element found in rocks, plants, animals, soil and in volcanic dust and gases. Chrome ore mining and chrome-based industries also release chromium into the environment. Human beings may be exposed to high doses of chromium during certain occupational and environmental situations. Occupational exposure has been shown to give rise to elevated levels of chromium in blood, urine and some body tissues, inhalation being the main route of chromium exposure (IARC, 1990).

Chromium is also implicated as a reproductive toxicant in experimental studies. A number of reports are available on alteration in reproductive function in experimental system due to chromium and few data are also available on human reproduction. Hexavalent chromium administered by intraperitoneal injection to male rats 5 days/week for 8 weeks resulted in reduced testicular atrophy, reduced sperm count and motility, decreased testosterone levels, and increased levels of follicle-stimulating hormone (Bonde and Ernst, 1992a). Experimental study of Saxena *et al.* (1990) suggested risk to growing testis if rats are exposed to hexavalent chromium during prepubertal stages of development, which in turn, may disturb normal testicular physiology at adulthood. A recent report of Li *et al.* (1999) suggested that occupational exposure to chromium might lead to changes in semen quality and reproductive hormone level (Table 27.1). Recent study carried out at our laboratory on chromium-exposed workers in a chemical industry revealed non-significant alteration in sperm concentration, motility, and viability as compared to control. However, abnormality in sperm morphology was higher in chromium-exposed workers with respect to control and these changes were associated with the level of chromium in the blood (Kumar *et al.*, 2005).

Welders comprise a major occupational group with known exposure to toxic fume containing certain metals and their oxides, gases, intense heat etc. Danadevi *et al.* (2003) measured blood level of nickel and chromium among workers from a welding plant in

South India and reported that there was a negative correlation between blood chromium and semen concentration. They observed that semen abnormalities correlated with the number of years of exposure to welding fumes containing nickel and chromium. Recently the author's study carried out among welders indicated that welding had some adverse effects on sperm motility, morphology and physiologic function, although sperm concentration was within the normal range (Kumar *et al.*, 2003). However, earlier, Bonde and Ernst (1992b) studied the level of sex hormones and semen quality in welders exposed to hexavalent chromium and reported that low level exposure to hexavalent chromium associated with tungsten inert gas stainless steel and mild steel welding do not appear to be a major hazard for human spermatogenesis.

Cadmium and Reproduction

Cadmium exposure can occur through contaminated air, water and food. Cadmium can reach in to the body by ingestion with food, such as fish and rice, from areas with contaminated groundwater. Occupational exposure to cadmium occurs in industries such as metal plating, semiconductor manufacture, wire, plastic, or battery manufacture, welding, soldering, ceramics etc. Another major source of cadmium exposure in the environment is cigarette smoke. Cadmium (Cd) is an ubquitous element and a significant inorganic pollutant that has been found to bioaccumulate in reproductive organs of fish and disrupts important endocrine processes, especially those involved in synthesis, release and metabolism of hormones (Tilton *et al.*, 2003).

Effects of cadmium on the testis have been extensively reviewed by Gunn and Gould (1970). Laskey *et al.* (1984) documented decreased testicular weights and decline in testosterone levels in rats at very low doses after oral administration. Semen quality and testicular characteristics were measured in male rabbits exposed to cadmium chloride at 12 or 27 weeks of age. The results indicated that Cd with all the three mode of exposure (0.08 μmol/kg s.c:, 0.20 μmol/kg orally, and 0.02 μmol/kg i.v. of Cd) resulted to depress sperm output of these males when adults. Necropsies confirmed previous findings of hyperemia, hemorrhaging, necrosis, and destruction of all spermatogenic elements in severely affected males (Foote, 1999). The semen volume, density and the total number of spermatozoa were significantly reduced in 3-month-old ram lambs by the administration of cadmium chloride. No differences were observed in the viability grade motility of spermatozoa, or the percentage of dead and morphologically abnormal spermatozoa between exposed and control. Histopathological examination in the cadmium-treated animals revealed the presence of lesions in the Sertoli cells, the seminiferous tubules, primary and the secondary spermatocytes and spermatids, whereas in the Leydig cells no significant lesions were evident (Lymberopoulos *et al.*, 2000). The marked effects on the testis after cadmium injection are probably the result of endothelial damage. In the small vessels this damage gives rise to increased capillary permeability. This leads to vascular escape of fluids and blood plasma substances into the interstitium, which results in oedema, decreased capillary blood flow, ischaemia, and testicular cell necrosis (Aoki and Hoffer, 1978; Francavilla *et al.*, 1981).

A single injection of cadmium salts at a dose that induces testicular haemorrhagic necrosis has been shown to induce haemorrhages and necrosis in the ovaries of prepubertal rats (Kar *et al.*, 1959), and in the ovaries of adult rats in persistent oestrus (Parizek *et al.*, 1968). In a recent study on the developmental and reproductive effects of low cadmium concentration on female shrimps, the authors reported a significant decrease in the pre-reproductive period and a reduction in the percentage offer the females as a consequence of the metal exposure (Sarabia *et al.*, 2003). In another study conducted in adult female rats treated subcutaneously (0.05 mg/kg body wt./ day) with lead acetate and cadmium acetate separately and in combination during the gestational and lactational periods with a pre-exposure before mating. No change in the reproductive cyclicity was observed in any of the treated groups. The observed biochemical alterations in the liver of rats co-exposed to lead and cadmium may result from an independent effect of lead and/or cadmium and also from their interaction. However, these results suggest that when lead and cadmium are administered together in similar concentrations, the major effects are mediated by cadmium (Pillai and Gupta, 2005).

The available experimental studies points that cadmium causes testicular necrosis in several experimental animal species, although there is scanty data on its possible effects in humans. Xu *et al.* (1993) reported a significantly inverse correlation between blood cadmium level and sperm density among oligospermic men and between seminal plasma cadmium and semen volume among men without known occupational exposure to cadmium. Mason (1990) found no difference between cadmium-exposed men in a manufacturing plant and unexposed controls in serum Testosterone, FSH and LH levels. However, higher levels of cadmium have been reported in infertile men compared to fertile men by Saaranen *et al.* (1997). A significant, inverse correlation was found between Cd and sperm density, sperm number per ejaculum in a study carried out in China among non-smokers, which indicated that Cd in seminal plasma could affect semen quality and oxidative DNA damage in human spermatozoa (Xu *et al.* 2003). In females also maternal exposure to Cd seems to increase early delivery, which leads to a lower birth weight. Also, the Cd is transferred in part to the next generation through breast milk after birth (Nishijo *et al.*, 2002). Laudanski *et al.* (1991) found that the mean blood concentration of Cd in mothers delivered of pre-term infants was higher than that of women who went to full term in an area with high amounts of lead and cadmium in the soil. These data point towards the reproductive toxic potential of cadmium.

Other Metals and Reproduction

There are several other metals that may also affect reproductive and related endocrine functions. The data on reproductive potential of other metals are scanty. Copper is an essential element for several metabolic pathways and adverse health effects are related to deficiency as well as excess. It is present as a prosthetic group in several enzymes. Catalase, peroxidase, cytochrome, C oxidase, Superoxide dismutase, Alcohol dehydrogenase and Alkaline phosphatase all require copper for normal activity.

There is limited evidence that exposure to copper compounds can affect reproduction in animals (EHC, 1998). In some studies of rats exposed by the oral route, the weights and/or histology of the testes, seminal vesicles, uterus or ovaries have been affected by chronic intakes of 27–120 mg Cu/kg body weight per day as copper (II) sulfate, acetate, or gluconate, although the results are inconsistent between studies and the reporting of some studies is deficient. In mice, there were no effects on male or female reproductive organs at 398–537 mg Cu/kg body weight per day as copper (II) sulfate in the diet. In a single study of rats inhaling copper (II) chloride aerosol, there were effects on sperm, testis weight and circulating levels of reproductive hormones (EHC, 1998). High levels of copper have been reported to affect reproduction in *Daphnia magna* spp. (Atienzar *et al.*, 2001). Copper fumes are among a number of exposures encountered by welders, who may experience decline in semen parameters (Bonde, 1990). However, welding fumes contain a large number of reproductive toxic metals and their oxides. Thus it is not possible to pin point the role of copper exposure in declining seminal parameters. However, there is an *in vitro* study where incubation with the metal caused a fall in the percentage of human motile sperm (Battersby *et al.*, 1982).

Manganese is an essential nutrient and is used in steel production, in battery manufacture, in fungicides, as chemical compounds, in tanning agents, in fertilizers and also in welding rods. Very limited data is available on the reproductive toxic potential of manganese. Laskey *et al.* (1982) studied the effects of manganese in rats by exposing the dams a diet containing high level (3,500 ppm) of Mn_3O_4. Male offsprings showed delayed reproductive development and reduced fertility. In another study, female rats whose diets contained 1,000 ppm manganese developed atretic follicles and persistence of the corpus luteum (Correla Vargas, 1984). In these experimental studies very high doses have been used by the investigators and significance of these studies to human is unclear. Studies on human with reference to the manganese exposure and reproductive dysfunction is scanty. No conclusion can be drawn as on today on the basis of the isolated study.

No data is available on the reproductive and developmental effects of nickel in human (EHC, 1991). However, some experimental data is available on male and female reproductive system following a single intracutaneous injection of 0.04 µmol nickel sulfate/kg body weight, histological examination, 18 h after exposure, revealed shrinkage of the tubules and complete degeneration of the spermatozoa. Infertility was observed in rats after 120 days of daily ingestion of 25 mg nickel sulfate/kg (Waltschewa *et al.*, 1972). Later Mathur *et al.* (1978) studied the dermal exposure of male rats to nickel sulfate at daily levels of 40, 60, or 100 mg nickel/kg body weight for 15 and 30 days. Tubular damage and spermatozoal degeneration were observed in the testis following exposure for 30 days. These changes were more severe with exposure at 100 mg/kg. There were no effects on the testis following exposure to 40 mg/kg for 30 days or at any dose level after 15 days exposure. No effects on fertility were observed in 3 generations, when breeding rats were exposed to 5 mg nickel/litre as a soluble salt in the drinking water (Schroeder and Mitchener, 1971). Nickel has been reported to localize within the pituitary and the hypothalamus of rats and to inhibit prolactin secretion (La Bella *et al.*, 1973a,b). It may, therefore, modify interactions between the hypothalamus and the pituitary, needed to maintain pregnancy.

Zinc is necessary for growth, sexual maturation and reproduction. A number of animal studies are available on various zinc compounds. Low (12 mg/kg b.w) or medium (120mg/kg b.w.) dose of zinc appeared to enhance reproductive function whereas lowering of reproductive function would be from high dose (240 mg/kg b.w.) of zinc (Wei *et al.* 2003). Zinc has a protective role against lead and cadmium induced toxicity in rats reported by Batra *et al.* (2004). However, very high levels of zinc have been reported to have a detrimental effect on reproduction. Dietary zinc specifically targets reproduction on *D. magna* through accumulation in particular target sites (De Schamphelacre *et al.*, 2004). Following an initial exposure of rats, mice, and guinea pigs to concentrations as high as 119.3 or 121.7 mg zinc/m^3 as zinc chloride smoke (which also contained other compounds) for 1 hour/day, 5 days/week, for 20 weeks, no adverse effects on the mammary glands, ovaries, fallopian tubes, or uteri were observed at 18 months (Marrs *et al.* 1988).

Zinc in human semen seems to play an important role in the physiology of spermatozoa. It is secreted by the prostate in small vesicles called protasomes. There are conflicting reports regarding the levels of zinc in seminal plasma and different semen quality parameters. It has been reported that seminal zinc level has a significant positive correlation with sperm count and sperm motility (Madding *et al.*, 1986; Chia *et al.*, 2000). However, Carreras and Mendoza (1999) reported a negative correlation of zinc with sperm motility. A recent study was conducted at author's laboratory to examine the relationship between concentration of zinc in seminal plasma and semen quality. A total of 50 subjects was analyzed for semen quality. The results showed that mean zinc levels in seminal plasma were lower among azoospermics as compared to oligozoospermic and normospermic groups. A significant correlation was observed between zinc levels and sperm count ($r = 0.22, p < 0.05$) (Figure 27.1). Earlier, a significant relationship between

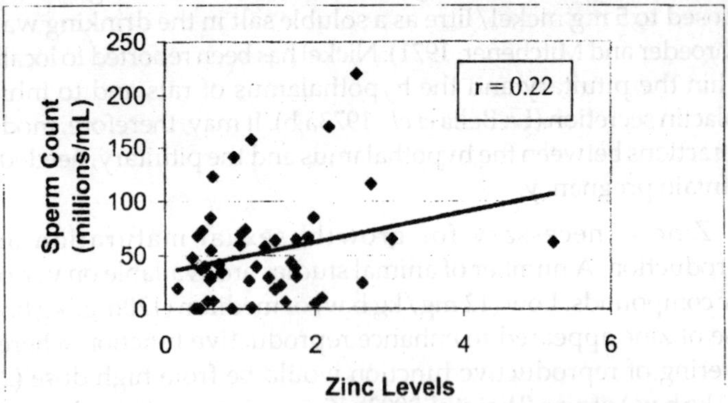

Figure 27.1: Relationship of Zinc in Seminal Plasma with Sperm Count

zinc level in seminal plasma and alpha-glucosidase which is an epididymal marker was also observed at our laboratory (Mankad and Kumar, 2004). These studies indicate that zinc has an important role in maintaining the normal male reproductive function. On the basis of data available and coupled with earlier studies it is suggested that zinc might have significant role in male reproduction, however the level is important for this aspect.

There is scanty information regarding reproductive toxicity in humans following exposure to aluminum (EHC, 1997). However, recently Dawson *et al.*, (1998) compared the sperm viability with metal levels in seminal plasma among the sixty-four apparently healthy men. Significant differences were observed between the high and low live sperm groups for Pb ($p < 0.01$) and Al ($p < 0.05$) but not for Cd. Further, linear regression between the live sperm counts and level of the three metals in semen showed inverse correlation with the percentage of live sperm and metal level. This study suggests that the presence of these metals in the environment of workers may affect the normal spermatogenesis. Only few studies are also available on effect of aluminum on animal reproduction. Domingo *et al.* (1987a) administered aluminium nitrate at 0, 13, 26 or 52 mg Al/kg body weight per day by incubation to male rats for 60 days prior to mating and to virgin females treated for 14 days prior to mating. They reported no reproductive effects on fertility, litter size, or intrauterine or postnatal offspring mortality. Numbers of corpus lutea on day 13 of gestation were significantly lower at 52 mg Al/kg body weight per day. However a dose-dependant delay in the growth of the pups was observed in all treatment groups; female offspring were affected at 13 mgAl/kg body weight per day and males at 26 and 52 mg Al/kg body weight per day. Dosing of pregnant animals from 14 days gestation to 21 days lactation with 13, 26 or 52 mg Al/kg body weight per day did not produce overt fetotoxicity, but growth of offspring was significantly delayed from birth to weaning (Domingo *et al.* 1987b). The data available on aluminium suggests that no or very little effect on reproduction as it is difficult to conclude reproductive toxic potential of aluminium on the basis of an isolated positive study.

Some other metals such as calcium, cobalt, iron, magnesium, manganese, molybdenum, rubidium, selenium, vanadium are also reported to be present in seminal plasma (Abou Shakra *et al.*, 1989). However, there are very few reports relating these metals to reproductive toxicity. Moreover, there is insufficient data on the levels at which these metals exert their effect in the human reproductive system.

The available data indicate that lead is having adverse effects on reproduction. Other metals like cadmium, mercury and chromium have also been reported to adversely affect the reproductive system of various animal species but the data available on human studies is not adequate to reach firm conclusion. Nevertheless, experimental data points their reproductive toxic potential for human. Further, their potential to affect endocrine and immune system might also play a role in infertility and subfertility. Numerous variable findings in human may be due to varying sample populations, confounding factors such as age, smoking habits, consumption of herbal medicines, susceptibility, etc and in most of the cases no exposure assessment. Therefore, there is a need to conduct well-defined epidemiological studies taking care of all possible confounding factors and adequate sample size to reach firm conclusion. Further, experimental data is also needed on other metals to report the effect on reproductive system even in isolated animals. Prevention of infertility from reproductive toxicants can only be feasible through education, information and by investigating reproductive toxicants through well-planned *in vivo* and *in vitro* studies before we know about the toxic potential of any chemical in human reproduction.

Acknowledgement

The authors are thankful to Dr. H.N. Saiyed, Director, NIOH for encouragement.

References

Abou-Shakra, F.R., Ward, N.I. and Everard, D.M., 1989. The role of trace elements in male infertility. *Fertil. Steril.*, 52(2): 307–310.

Agarwal, A., Saleh Ramadan, A. and Bedaing Mohamed, A., 2003. Role of reactive oxygen species in the pathophysiology of human reproduction. *Fertil. Steril.*, 79: 829–843.

Alexander, B.H., Checkoway, H., Faustman, E.M., van Netten, C., Muller, C.H. and Ewers, T.G., 1998. Contrasting associations of blood and semen lead concentrations with semen quality among lead smelter workers. *Am. J. Ind. Med.*, 34(5): 464–469.

Annapurna, V., Mahajan, B.A. and Murti, M.V., 1985. Studies of blood lead levels in Bombay's citizens. *Ind. J. Environ. Studies*, 24: 239–243.

Aoki, A. and Hoffer, A.P., 1978. Re-examination of the lesions in rat testis caused by cadmium. *Biol. Reprod.*, 18: 579–591.

Apostoll, P., Kiss, P., Porru, S., Bonde, J.P., and Vanhoone, M., 1998. Male reproductive toxicity of lead in animals and humans. Occup. Environ. Med., 55: 364–374.

Atienzar, F.A., Cheung, V.V., Jha, A.N. and Michael, H., 2001. Depledg fitness parameters and DNA effects are sensitive indicators of copper-induced toxicity in *Daphnia magna*. *Toxicological Sciences*, 59: 241–250.

Batra, N., Nehru, B. and Bansal, M.P., 2004. Reproductive potential of male Portan rats exposed to various levels of lead with regard to zinc status. *Br. J. Nutr.*, 91(3): 387–391.

Battersby, S., Chandler, J.A. and Morton, M.S., 1982. Toxicity and uptake of heavy metals by human spermatozoa. *Fertil. Steril.*, 37(2): 230–235.

Benoff, S., Hurley, I.R., Barcia, M., Mandel, F.S., Cooper, G.W. and Hershlag, A., 1997. A potential role for cadmium in the etiology of varicocele-associated infertility. Fertil. Steril., 67(2): 336–347.

Bonde, J.P., 1990. Semen quality and sex hormones among mild steel and stainless steel welders: A cross sectional study. *Br. J. Ind. Med.*, 47: 508–514.

Bonde, J.P. and Ernst, E., 1992a. Sex hormones and epididymal sperm parameters in rats following subchronic treatment with hexavalent chromium. *Hum Exp Toxicol.*, 11: 255–258.

Bonde, J.P. and Ernst, E., 1992b. Sex hormones and semen quality in welders exposed to hexavalent chromium. *Hum. Exp. Toxicol.*, 11: 259–263.

Bonde, J.P., Joffe, M., Apostoll, P., Dale, A., Kiss, P., Spano, M. and Caruso, F. *et al.*, 2002. Sperm count and chromatin structure in men exposed to inorganic lead: Lowest adverse effect levels. *Occup. Environ. Med.*, 59: 234–242.

Burbacher, T.M., Mohammed, M.K. and Karle, M.N., 1988. Methylmercury effects on reproduction and offspring size at birth. *Reproductive Toxicology*, 1(4): 267–278.

Carreras, A. and Mendoza, C., 1990. Zinc levels in seminal plasma of fertile and infertile men. *Andrologia*, 22(3): 279–283.

Chia, S.E., Ong, C.N., Chua, L.H., Ho, L.M. and Tay, S.K., 2000. Comparison of zinc concentrations in blood and seminal plasma and the various sperm parameters between fertile and infertile men. *J. Androl.*, 21(1): 53–57.

Cordier, S., Deplan F., Mandereau, L. and Hemon, D., 1991. Paternal exposure to mercury and spontaneous abortions. *Br. J. Ind. Med.*, 48(6): 375–381.

Corella-Vargas, R., 1984. High levels of manganese in the diet of rats (*Ratlus norvegicus albinicus*) I. Effect on reproduction. *Arch. Latinoan Nutr.*, 34: 457–465.

Danadevi, K., Rozati, R., Reddy, P.P. and Grover, P., 2003. Semen quality of Indian welders occupationally exposed to nickel and chromium. *Reprod Toxicol.*, 17(4): 451–456.

Daniell, W.E. and Vaughan, T.L., 1988. Paternal employment in solvent related occupations and adverse pregnancy outcome. *Br. J. Ind Med.*, 45: 193–197.

Dancher, G.P., Horstedt, Bindslev, P. and Rungby, J., 1990. Traces of mercury in organs from primates with amalgam fillings. *Exp. Mol. Pathol.*, 52: 291–299.

Davis, B.J., Pricetic, O'Conner, R.W., Fernando, R., Rowland, A.C. and Morgan, D.L., 2001. Mercury vapour and female reproductive toxicity. *Toxicol. Sci.*, 59: 291–296.

Dawson, E.B., Ritter, S., Harris, W.A., Evans, D.R. and Powell, L.C., 1998. Comparison of sperm viability with seminal plasma metal levels. *Biol. Trace Element Res.*, 64: 215–219.

De Schamphelaere, K.A., Canli, M., Van Lierde, V., Forrez, I., Vanhaecke, F. and Janssen, C.R., 2004. Reproductive toxicity of dietary zinc to Daphnia magna. *Aquat. Toxicol.*, 70(3): 233–244.

Domingo, J.L, Paternain, J.L., Llobet, J.M., and Corbella, J., 1987a. Effects of oral aluminum administration on perinatal and postnatal development in rats. *Res. Commun. Chem. Pathol. Pharmacol.*, 57: 129–132.

Domingo, J.L., Paternain, J.L., Llobet, J.M., and Corbella, J., 1987b. The effect of aluminium ingestion on reproduction and postnatal survival in rats. *Life Sci.*, 41: 1127–1131.

Environmental Health Criteria, 1997. IPCS Geneva; *Aluminium*, pp. 194.

Environmental Health Criteria, 1977. IPCS Geneva; *Lead*, pp. 3.

Environmental Health Criteria, 1998. IPCS Geneva; *Copper*, pp. 200.

Environmental Health Criteria, 1992. IPCS Geneva; *Nickel*, pp. 134.

Egorova, G.M., Ivanov, N.G. and Sanocskij, J.V., 1966. Specificity of the effect of lead on spermatogenesis. *Toksikol., Nov. Prom. Him., Vescestv.*, 8: 33–41.

Foote, R.H., 1999. Cadmium affects testes and semen of rabbits exposed before and after puberty. *Reprod. Toxicol.*, 13(4): 269–277.

Eyden, B.P., Maisin, J.R. and Mattelin, G., 1978. Long term effects of dietary lead acetate on survival, body weight and seminal cytology in mice. *Bull. Environ. Contam. Toxicol.*, 19: 266–272.

Francavilla, S., Moscardelli, S., Francavilla, F., Casasanta, N., Properzi, G., Martini, M., and Santiemma, V., 1981. Acute cadmium intoxication: Influence of cyproterone acetate on the testis and epididymis of the rat. *Arch. Androl.*, 6: 1–11.

Franks, P.A., Laughlin, N.K., Dierschke, D.J., Bowman, R.E. and Meller, P.A., 1989. Effects of lead on luteal function in rhesus monkeys. *Biol. Reprod.*, 41(6): 1055–1062.

Gautam, A.K., Agarwal, K., Shah, B.A., Kumar, S. and Saiyed, H.N., 2001. Lead induced spermatoxicity in mouse and MPG treatment. *J. Environ. Biol.*, 22(4): 287–291.

Gennart, J.P., Buchet, J.P., Roels, H., Ghyselen, P., Ceulemans, E. and Lauwerys, R., 1992. Fertility of male workers exposed to cadmium, lead, or manganese. *American Journal of Epidemiology*, 135(11): 1208–1219.

Golubovic, E.Y., Avhimehko, M.M. and Cirkova, E.M., 1968. Biochemical and morphological changes in testicles of rats under the effect of low doses of lead. *Toksikol. Nov. Prom. Him. Vescestv*, 10: 64–72 (in Russian).

Gunn, S.A. and Gould, T.C., 1970. Cadmium and other mineral elements. In: *The Testis*, (Eds.) Johnson, A.D., Gomes, W.R., and VanDeMark, N.L. Academic Press, New York, London, 3: 377–381.

Hertz-Picciotto, I., 2000. The evidence that lead increases the risk for spontaneous abortion. *Am. J. Ind. Med.*, 38(3): 300–309.

IARC, 1990. Monographs on the evaluation of carcinogenic risks to in humans. In: *Chromium, Nickel and Welding*, 49: 20.

International Chromium Development Association (ICDA), 1997. Criteria document for hexavalent chromium, Paris.

Irgens, A., Kruger, K., Skorve, A.H. and Irgens, L.M., 1998. Reproductive outcome in offspring of parents occupationally exposed to lead in Norway. *Am. J. Ind. Med.*, 34: 431–437.

Johansson, L. and Wide, M., 1986. Long-term exposure of the male mouse to lead: Effects on fertility. *Environ. Res.*, 41(2): 481–487.

Kaste, L.M., 1997. Occupation and reproductive health of female dentists: The relationship of nitrous oxide and amalgam (mercury) with spontaneous abortion. *Diss. Abstr. Int. Sci.*, 57(12): 7489.

Kar, A.B., Das, R.P. and Karkun, J.N., 1959. Ovarian changes in prepubertal rats after treatment with cadmium chloride. *Acta. Biol. Med. Germ.*, 3: 372–399.

Kumar, S., Zaidi, S.S.A., Gautam, A.K., Dave, L.M. and Saiyed, H.N., 2003. Semen quality and reproductive hormones among welders: A preliminary study. *Environmental Health and Preventive Medicine*, 8: 64–67.

Kumar, S., Sathwara, N.G., Gautam, A.K., Agarwal, K., Shah, B., Kulkarni, P.K., Patel, K., Patel, A., Dave, L.M., Parikh, D.J., Saiyed, H.N., 2005. Semen quality of industrial workers occupationally exposed to chromium. *Journal of Occupational Health* (In press).

Labella, F., Dular, R., Lemon, P., Vivian, S. and Queen, G., 1973a. Prolactin secretion is specifically inhibited by nickel. *Nature* (Lond.), 245: 331–332.

Labella, F., Dular, R., Vivian, S. and Queen, G., 1973b. Pituitary hormone releasing or inhibiting activity of metal ions present in hypothalamic extracts. *Biochem. Biophys. Res. Commun.*, 52: 786–791.

Laskey, J.W., Rehnberg, G.L., Hein, J.F. and Carter, S.D., 1982. Effects of chronic manganese (Mn_3O_4) exposure on selected reproductive parameters in rats. *J. Toxicol Environ Health*, 9: 677–687.

Laskey, J.W., Rehnberg, G.L., Laws, S.C. and Hein, J.F., 1984. Reproductive effects of low acute doses of cadmium chloride in adult male rats. Toxicol. Appl. Pharmacol., 73(2): 250–255.

Laudanski, T., Sipowich, P., Modzelewski, J. *et al.*, 1991. Influence of high lead and cadmium soil content on human reproductive outcome. *Int J. Gynecol Obstet.*, 366: 309–315.

Lerda, D., 1992. Study of sperm characteristics in persons occupationally exposed to lead. *Am. J. Ind. Med.*, 22(4): 567–571.

Li, H., Chen, Q. and Li, S., *et al.*, 1999. Studies on male reproductive toxicity caused by hexavalent Chromium. *Zhonghua Yufang Xue Za Zhi*, 33: 351–353.

Lin, S., Hwang, S.A., Marshall, E.G. and Marion, D., 1998. Does paternal occupational lead exposure increase the risks of low birth weight or prematurity? *Am. J. Epidemiol.*, 148: 173–181.

Lymberopoulos, A.G., Kotsaki-Kovatsi, V.P., Taylor, A., Papaioannou, N., Brikas, P., 2000. Effects of cadmium chloride administration on the macroscopic and microscopic characteristics of ejaculates from Chios ram-lambs. *Theriogenology*, 54(7): 1145–1157.

Madding, C.I., Jacob, M., Ramsay, V.P. and Sokol, R.Z., 1986. Serum and semen zinc levels in normozoospermic and oligozoospermic men. *Ann. Nutr. Metab.*, 30(4): 213–218.

Mankad, M. and Kumar, S., 2004. Demonstration of relationship between zinc levels in seminal plasma with semen quality. In: *Proc. 24th Annual Session of Academy of Environ. Biol.*, pp. 34.

Marchlewicz, M., Michalska, T. and Wiszniewska, B., 2004. Detection of lead-induced oxidative stress in the rat epididymis by chemiluminescence. *Chemosphere*, 57(10): 1553–1562.

Marrs, T.C., Colgrave, H.F. and Edginton, J.A.G., *et al.*, 1988. The repeated dose toxicity of a zinc oxide/hexachloroethane smoke. *Arch Toxicol.*, 62: 123–132.

Mason, H.J., 1990. Occupational cadmium exposure and testicular endocrine function. *Hum. Exp. Toxicol.*, 9: 91–94.

Mathur, A.K., Dikshith, T.S.S, Lal, M.M. and Tandon, S.K., 1978. Distribution of nickel and cytogenetic changes in poisoned rats. *Toxicology*, 10: 105–113.

Matthies, J. Van, Donat, H. and Schwarz, I., 1989. Einfluss Von Schwarmetallionen auf die mannliche fertilitat. *Zentralbl Gynakol.*, 111: 155–166 (in German).

McMichael, A.J., Timpani, G.V., Robertson, E.F., Baghurst, P.A. and Clark, P.D., 1986. The port Pirie cohort study: Maternal blood lead and pregnancy outcomes. *J. Epidemiol. Commun. Health*, 40: 18–25.

Mehta, R.H. and Anandkumar, T.C., 1997. Declining semen quality in Bangaloreans: A preliminary report. *Curr. Sci.*, 72: 621–623.

Morgan, D.L., Chanda, S.M., Price, H.C., Fernando, R., Liu, J., Brambila, E., O'Connor, R.W., Beliles, R.P. and Barone, S., Jr., 2002. Disposition of inhaled mercury vapor in pregnant rats: Maternal toxicity and effects on developmental outcome. *Toxicological Sciences*, 66: 261–273.

Nishijo, M., Nakagawa, H., Honda, R., Tanabe, K., Saito, S. and Teranishi, K. *et al.*, 2002. Effects of maternal exposure to cadmium on pregnancy and breast milk. *Occup. Environ. Med.*, 59: 394–397.

Ong, C.N., Phoon, W.O., Law, H.Y., Tye, C.Y. and Lim, H.H., 1985. Concentration of lead in maternal blood, cord blood and breast milk. *Arch. Dis. Child.*, 60: 756–759.

Parikh, D.J., 1990. Human exposure assessment to lead, an overview in India. *Ind. J. Indust. Med.*, 36: 63–75.

Parizek, J., Ostadalova, I., Belles, I., and Pitha, J., 1968. The effect of a subcutaneous injection of cadmium salts on the ovaries of adult rats in persistent oestrus. *J. Reprod. Fertil.*, 17: 559–562.

Pace, B.M., Lawrence, D.A., Behr, M.J., Parsons, P.J. and Dias, J.A., 2005. Neonatal lead exposure changes quality of sperm and number of macrophages in testes of BALB/c mice. *Toxicology*, 210(2–3): 247–256.

Pillai, A. and Gupta, S., 2005. Effect of gestational and lactational exposure to lead and/or cadmium on reproductive performance and hepatic oestradiol metabolising enzymes. *Toxicol. Lett.*, 155: 179–186.

Popschu, H., 1978. Poisoning with alkyl mercury compounds. *Br. Med. J.*, 1: 1347.

Raghunath, R., Tripathi, R.M., Kumar, A.V., Sathe, A.P., Khandekar, R.N., Nambi, K.S., 1999. Assessment of Pb, Cd, Cu and Zn in exposures of 6 to 10 year old children in Mumbai. *Environ. Res.*, 80: 215–221.

Rowland, A.S., Baird, D.D., Weinberg, C.R., Shore, D.L., Shy, C.M. and Wilcox, A.J., 1994. The effect of occupational exposure to mercury vapour on the fertility of female dental assistants. *Occupational and Environmental Medicine*, 51: 28–34.

Roychowdhury, A., Chinoy, N.J., Gautam, A.K., Rao, R.V., Parikh, D.J., Shah, G.M., Highland, H.M., Patel, K.G. and Chatterjee, B.B., 1986. Effect of lead on human sperm. *Adv Contra Deliv Syst*, 11: 208–210.

Roychowdhury, A., Makhija, S. and Vachhrajani, K.D., 1989. Methyl mercury induced biochemical and histochemical alterations in rat testis. *Indian J. Physiol. Pharmacol.*, 33(4): 219–222.

Roychowdhury, A. and Arora, U., 1982. Toxic effect of mercury on testes in different animal species. *Ind. J. Physiol. Pharmacol.*, 26: 246–249.

Saaranen, M., Suistomaa, U. and Kantola, M., 1987. Lead, magnesium, selenium and zinc in human seminal fluid: Comparison with semen parameters and fertility. *Human Reprod.*, 2: 475–479.

Sarabia, R., Del Ramo, J., Diaz-Mayans, J. and Torreblanca, A., 2003. Developmental and reproductive effects of low cadmium concentration on Artemia parthenogenetica. *J. Environ. Sci. Health, A Tox. Hazard Subst. Environ.*, Eng., 38(6): 1065–1071.

Saxena, D.K., Singh, C., Murthy, R.C., Mathur, N. and Chandra, S.V., 1994. Blood and placental lead levels in an Indian city: a preliminary report. *Arch. Environ. Health*, 49: 106–110.

Saxena, D.K., Murthy, R.C., Lal, B., Srivastava, R.S. and Chandra, S.V., 1990. Effect of hexavalent chromium on testicular maturation in the rat. *Reprod. Toxicol.*, 4: 223–228.

Schuurs, A.H., 1999. Reproductive toxicity of occupational mercury: A review of the literature. *J. Dent.*, 27(4): 249–256.

Schroeder, H.A. and Mitchener, M., 1971. Toxic effects of trace elements on the reproduction of mice and rats. *Arch. Environ. Health*, 23: 102–106.

Sikorski, R., Juszkiewicz, T., Paszkowski, T. and Szprengier-Juszkiewicz, T., 1987. Women in dental surgeries: Reproductive hazards in occupational exposure to metallic mercury. *International Archives of Occupational and Environmental Health*, 59(6): 551–527.

Telisman, S., Cvitkovic, P., Jurasovic, J., Pizent, A., Gavella, M. and Rocic, B., 2000. Semen quality and reproductive endocrine function in relation to biomarkers of lead, cadmium, zinc and copper in men. Environ. Health Perspect 108: 45–53.2000

Tilton, S.C., Foran, C.M. and Benson, W.H., 2003. Effects of cadmium on the reproductive axis of Japanese medaka (*Qryzias latipes*). *Comp. Biochem. Physiol. C. Toxicol. Pharmacol.*, 136(3): 265–276.

Waltschewa, W., Slatewa, M. and Michailow, I., 1972. Testicular changes due to long-term administration of nickel sulphate in rats. *Exp. Pathol.*, 6(3): 116–120 (in German).

Wei, Q., Fan, R., Yang, X. and Chan, T., 2003. Effect of zinc on reproductive toxicity in rats. *Wei Sheng Yan Jiu*, 32(6): 618–619.

Xu, B., Chia, S.E., Tsakok, M. and Ong, C.N., Trace elements in blood and seminal plasma and their relationship to sperm quality. *Reprod. Toxicol.*, 7(6): 613–618.

Xu, D.X., Shen, H.M., Zhu, Q.X., Chua, L., Wanmg, Q.N., Chia, S.E. and Ong, C.N., 2003. The associations among semen quality, oxidative DNA damage in human spermatozoa and concentrations of cadmium, lead and selenium in seminal plasma. *Mut. Res.*, 534(1–2): 155–163.

Index

B

C

Previous Volumes

— Volume 1 —

2002, xiii+308p., figs., tabls., ind., 23 cm Rs. 650

ISBN 81-7035-272-X

Section I: Animal Ecology

Section II: Animal Reproduction

— Volume 2 —

2004, xi+387p., figs., tabls., ind., 23 cm **Rs. 850**

ISBN 81-7035-322-X

Section I: Animal Ecology

Section II: Animal Reproduction

— Volume 3 —

2006, xvii+443p., figs., tabls., ind., 23 cm **Rs. 1150**

ISBN 81-7035-424-2

Section I: Animal Ecology

Section II: Animal Reproduction